PERFORMANCE MODELING AND ANALYSIS OF BLUETOOTH NETWORKS

OTHER AUERBACH PUBLICATIONS

Agent-Based Manufacturing and Control Systems: New Agile Manufacturing Solutions for Achieving Peak Performance
Massimo Paolucci and Roberto Sacile
ISBN: 1574443364

Curing the Patch Management Headache
Felicia M. Nicastro
ISBN: 0849328543

Cyber Crime Investigator's Field Guide, Second Edition
Bruce Middleton
ISBN: 0849327687

Disassembly Modeling for Assembly, Maintenance, Reuse and Recycling
A. J. D. Lambert and Surendra M. Gupta
ISBN: 1574443348

The Ethical Hack: A Framework for Business Value Penetration Testing
James S. Tiller
ISBN: 084931609X

Fundamentals of DSL Technology
Philip Golden, Herve Dedieu, and Krista Jacobsen
ISBN: 0849319137

The HIPAA Program Reference Handbook
Ross Leo
ISBN: 0849322111

Implementing the IT Balanced Scorecard: Aligning IT with Corporate Strategy
Jessica Keyes
ISBN: 0849326214

Information Security Fundamentals
Thomas R. Peltier, Justin Peltier, and John A. Blackley
ISBN: 0849319579

Information Security Management Handbook, Fifth Edition, Volume 2
Harold F. Tipton and Micki Krause
ISBN: 0849332109

Introduction to Management of Reverse Logistics and Closed Loop Supply Chain Processes
Donald F. Blumberg
ISBN: 1574443607

Maximizing ROI on Software Development
Vijay Sikka
ISBN: 0849323126

Mobile Computing Handbook
Imad Mahgoub and Mohammad Ilyas
ISBN: 0849319714

MPLS for Metropolitan Area Networks
Nam-Kee Tan
ISBN: 084932212X

Multimedia Security Handbook
Borko Furht and Darko Kirovski
ISBN: 0849327733

Network Design: Management and Technical Perspectives, Second Edition
Teresa C. Piliouras
ISBN: 0849316081

Network Security Technologies, Second Edition
Kwok T. Fung
ISBN: 0849330270

Outsourcing Software Development Offshore: Making It Work
Tandy Gold
ISBN: 0849319439

Quality Management Systems: A Handbook for Product Development Organizations
Vivek Nanda
ISBN: 1574443526

A Practical Guide to Security Assessments
Sudhanshu Kairab
ISBN: 0849317061

The Real-Time Enterprise
Dimitris N. Chorafas
ISBN: 0849327776

Software Testing and Continuous Quality Improvement, Second Edition
William E. Lewis
ISBN: 0849325242

Supply Chain Architecture: A Blueprint for Networking the Flow of Material, Information, and Cash
William T. Walker
ISBN: 1574443577

The Windows Serial Port Programming Handbook
Ying Bai
ISBN: 0849322138

PERFORMANCE MODELING AND ANALYSIS OF BLUETOOTH NETWORKS

POLLING, SCHEDULING, AND TRAFFIC CONTROL

JELENA MIŠIĆ
VOJISLAV B. MIŠIĆ

Auerbach Publications
Taylor & Francis Group
Boca Raton London New York Singapore

Published in 2006 by
Auerbach Publications
Taylor & Francis Group
6000 Broken Sound Parkway NW, Suite 300
Boca Raton, FL 33487-2742

Printed in the United States of America on acid-free paper
10 9 8 7 6 5 4 3 2 1

International Standard Book Number-10: 0-8493-3157-9 (Hardcover)
International Standard Book Number-13: 978-0-8493-3157-2 (Hardcover)
Library of Congress Card Number 2005045358

Library of Congress Cataloging-in-Publication Data

Misic, Jelena.
Performance modeling and analysis of Bluetooth networks : polling, scheduling, and traffic control / Jelena Misic, Vojislav B. Misic.
p. cm.
Includes bibliographical references and index.
ISBN 0-8493-3157-9 (alk. paper)
1. Bluetooth technology. 2. Network performance (Telecommunication) I. Misic, Vojislav B. II. Title

TJ5103.3.M57 2005
004.6'2—dc22 2005045358

T&F informa
Taylor & Francis Group
is the Academic Division of T&F Informa plc.

Visit the Taylor & Francis Web site at
http://www.taylorandfrancis.com

and the Auerbach Publications Web site at
http://www.auerbach-publications.com

to Bratislav and Velibor

Contents

List of Figures

List of Tables

Preface

Bluetooth is a recent wireless communication technology specifically intended for short range ad hoc networking. It was initially envisaged as a simple cable replacement technology to connect various mobile, portable, and fixed devices such as PDAs, mobile phones, laptops, and printers, but its use has soon spread to include various general purpose networking tasks, and a number of Bluetooth-enabled devices have appeared on the market. In order to facilitate the development and acceptance of Bluetooth devices, systems, and applications, the development and promotion of Bluetooth technology has been coordinated through the Bluetooth Special Interest Group (SIG). The Bluetooth SIG was founded in 1999 by Agere Systems, IBM, Intel, Microsoft, Motorola, Nokia, and Toshiba, and subsequently joined by other companies. The SIG has issued a series of specifications that describe the operation of Bluetooth devices and networks in detail. Version 1.1 of the specification has been adopted in 2001 [Bluetooth SIG, 2001*b*], version 1.2 in 2003 [Bluetooth SIG, 2003*a*], and the most recent one, version 2.0, has just been adopted [Bluetooth SIG, 2004] as we write this book. Furthermore, the IEEE Wireless Personal Area Network (WPAN) Working Group has adopted the Bluetooth specification, with slight modifications, as the IEEE standard for medium rate WPANs under the designation 802.15.1.

Thanks to this wealth of information, the characteristics of Bluetooth communications and the operation of Bluetooth devices may appear to be well defined and well understood. Well defined they certainly are, but understood they are to a much lesser degree. First, detailed performance analysis of Bluetooth networks, as is the case with other networks, is certainly beyond the scope of relevant specifications. Second, while the specification goes into great detail to explain some details of Bluetooth operation, some of the important issues are mentioned only casually, or not at all. Most notable among such issues are the intra-piconet polling scheme—the manner in which the piconet master should poll its slaves—and details of the operation of Bluetooth scatternets, in particular the scheduling of shared devices – bridges. All of these are vital factors that affect both the design of Bluetooth devices-hardware, software, and firmware – but also the design, operation, and performance of Bluetooth networks. As a result, the developers as well as users are left with a number of open issues but little guidance about the possible answers and their relative advantages and disadvantages.

It should come as no surprise, then, that a number of researchers have focused their attention on the performance analysis of Bluetooth networks and on algorithms that complement the official Bluetooth specification, striving toward better understanding of the operation of Bluetooth networks and, consequently, toward even wider acceptance for Bluetooth devices and applications. Performance of Bluetooth networks has also been the focus of our research since late 2001, and this book presents the summary of the work that we have done in this area in the last three years.

What the book is about, and what it's not about

Before we say what this book contains, let us state what it does not contain. The book is not intended to describe all the details of Bluetooth technology; the specifications should be used to that effect. The book is also not intended to describe possible usage scenarios of Bluetooth-enabled devices in various networking environments; there are several books that deal with those topics. Finally, the book does not deal with issues related to Bluetooth communications at the physical layer; quite a few research papers describe the interaction of Bluetooth networks with one another, as well as with other wireless networks operating in the same frequency band.

In most succinct terms, the goal of this book is twofold: first, to provide insights into the performance of Bluetooth networks using a dual foundation of rigorous analytical approach based on queueing theory and discrete event simulation. Second, to propose and validate solutions for a number of important issues that are not prescribed by the official Bluetooth specifications. In this manner, the readers—developers and researchers alike—can enrich their knowledge about performance related issues in Bluetooth, and thus be better equipped to solve the problems they might encounter in the design, deployment, and operation of Bluetooth networks.

The book consists of fourteen chapters. It begins with an introductory treatment of Bluetooth data communications; while far from being exhaustive—version 2.0 of the Bluetooth specification has over 1,200 pages!—the information presented in Chapter 1 should equip the readers with basic tenets of Bluetooth data communications and allow them to easily follow subsequent discussions.

The remainder of the book consists of two parts. The first part is devoted to performance analysis of simple piconets, and it starts with an overview of intra-piconet polling techniques in Chapter 2. The queueing theoretic analysis of exhaustive, 1-limited, and E-limited polling is presented in Chapter 3. This analysis assumes that the device buffers are of infinite size; the impact of finite device buffers on performance is dealt with in Chapter 4. The next two Chapters discuss the issue of admission control and the performance of TCP traffic in Bluetooth piconets. Finally, Chapters 7 and 8 discuss the performance of two polling schemes specifically designed to provide tight delay bounds required by voice and multimedia traffic.

The second part of the book is devoted to Bluetooth scatternets. Chapter 9 presents an overview of the many scatternet formation algorithms available, while Chapter 10 discusses issues related to scatternet operation and introduces the problem of bridge scheduling. The next three Chapters present in more detail and analyze the three different approaches to bridge scheduling: rendezvous-based, adaptive, and walk-in, respectively. As in the first part, these analyses are based on the assumption that device buffers are of infinite size. Chapter 14 analyzes the impact of finite device buffers on the performance of scatternets operating under walk-in bridge scheduling.

A short Appendix presents the definitions and notation related to probability generating functions and Laplace-Stieltjes transforms thereof. An extensive bibliography is also provided. While we have done our best to make sure that none of the important contributions are left out, we certainly make no claim as to its exhaustiveness.

Acknowledgments

Books like this cannot be written without the help, assistance, and encouragement of others, and these contributions should be duly acknowledged. But before doing so, we want to express our indebtedness to each other. Neither of us could, or would, have written this book alone, and the book is indeed a result of our collaborative work and complementary expertise. Our collaboration was not as smooth a process as it may appear, and few decisions have been made without a thorough discussion or, on occasions, a heated debate. Ideas were constantly proposed, questioned, refined, and validated; some of them were sound enough to make it through to the final version, many much less so and hence left out. Yet we have learned through the process, and the experience thus gained has been invaluable for both of us.

We are deeply indebted to Professor Ivan Stojmenovic and Professor Nejib Zaguia from University of Ottawa, who have kindly agreed to share their expertise on scatternet formation algorithms in the form of Chapter 9.

We are also indebted to Mr. Ka Lok Chan for the numerous simulation experiments on piconets and scatternets, done both before and after his MSc thesis work at the Hong Kong University of Science and Technology. Throughout our collaboration, Mr. Chan has always been s source of healthy criticism, and his suggestions and comments have helped improve many facets of the work presented in this book.

We also thank Mr. Eric W. S. Ko, who has implemented the ACLS scheme and related simulations as part of his MSc thesis at the Hong Kong University of Science and Technology, and Mr. G. R. Reddy, who has conducted the simulations of the scatternet with finite buffers while working on his MSc thesis at the University of Manitoba.

We wish to express our heartfelt gratitude to Professor Attahiru S. Alfa and Professor Terrence Todd, who have found time to read the book and make suggestions that led to improvements to both the material itself and the presentation thereof.

Those contributions notwithstanding, this book has been devised and written by us only, and we remain responsible for any errors that you may find in the final version.

Last but not least, we would like to thank Bratislav and Velibor who have had the patience to endure our absence, logical and (quite often) physical, through many hours devoted to the work presented in this book.

In Winnipeg, November 2004 Authors

1

Introduction to Bluetooth

In this Chapter we will describe the basic operation of the Bluetooth core system and outline the most important characteristics of Bluetooth communications, in particular those that are relevant to our main goal – performance analysis of Bluetooth networks. It may be interesting to note that the Bluetooth specification has undergone a few major updates, the most recent official one being, at the time of this writing, version 1.2 [Bluetooth SIG, 2003*b*; Bluetooth SIG, 2003*c*; Bluetooth SIG, 2003*d*]. As the material in this Chapter is mostly based on this specification, explicit references to it will be omitted, except where necessary to explain the differences from the earlier version 1.1 [Bluetooth SIG, 2001*b*]. Still earlier versions of the specification are not referred to in this book. An updated version of the specification is currently in the process of being adopted by the Bluetooth SIG [Bluetooth SIG, 2004].

The Bluetooth core system on a host device consists of the blocks shown in Fig. 1.1; together, they provide services to support connection of different devices and exchange of application data over short range wireless links. The Host-Controller Interface, or HCI, is a command interface to the baseband controller and link manager that provides a uniform method of accessing the Bluetooth baseband capabilities.

1.1 Lower layers of the architecture: RF and baseband

The RF unit performs the functions corresponding to the PHY (physical layer) of the IEEE 802.11 protocol stack [IEEE, 2001]. It operates in the unlicensed ISM (Industrial, Scientific and Medical) band at 2.4 GHz. A total of 79 RF frequencies are used in the ISM band. The RF transceiver hops through the available channels in a pseudo-random fashion, which is known as Frequency Hopping Spread Spectrum, or FHSS. This technique reduces the interference from and to other systems operating in the ISM band, possibly including other Bluetooth systems as well.

The physical radio channel is shared by a group of devices that are synchronized to a common clock and hopping sequence. The device that provides the synchronization reference – the common clock and the hopping sequence – is known as the master. All other devices are known as slaves. This group of devices, master and its slaves, is referred to as the piconet, and it is the fundamental communication arrangement in Bluetooth. Due to clock drift, periodic communication between the master

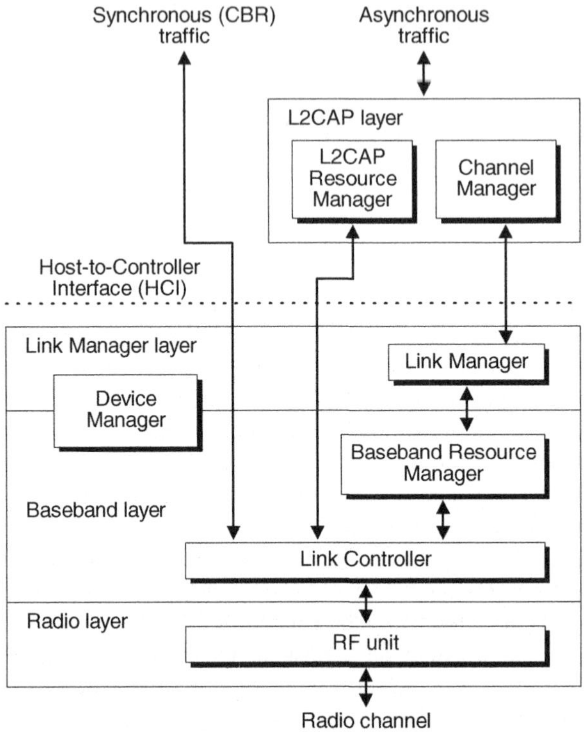

FIGURE 1.1
Basic blocks of the Bluetooth core system architecture.

and the slaves may be necessary to keep the slaves synchronized to the master, even in the absence of actual data to exchange.

The pseudo-random hopping sequence, which is sometimes referred to as the physical channel, is algorithmically derived from the device address of the piconet master. Thanks to the frequency hopping technique, a number of piconets may exist and operate in the same physical space with minimum interference, as shown in Fig. 1.2. Note that the black and white circles depict piconet masters and their slaves, respectively, while lines represent master-slave links.

We note that version 1.2 of the Bluetooth specification relaxes the requirements on the hopping sequence. In particular, the slave can respond at the same frequency at which the master has addressed it, and some frequencies may be explicitly excluded from the sequence in order to reduce interference to and/or from other devices operating in the vicinity. The modified sequence is known as an *adapted channel* [Bluetooth SIG, 2003*b*].

All communications in the piconet take place under the control of the piconet master. Downlink transmissions are those sent from the master to a slave, while

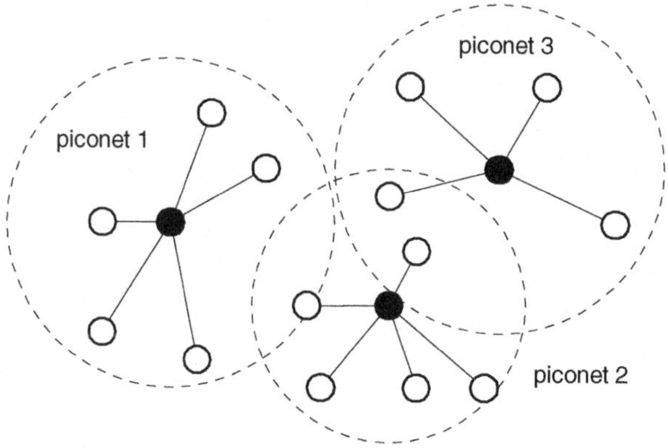

FIGURE 1.2
Bluetooth piconet is a group of devices within the radio range that share the physical radio channel.

uplink transmissions are those sent in the opposite direction, i.e., from a slave to the master. The action of the master sending a packet to the slave will be referred to as polling, since it is analogous to the concept of polling used in queueing theory [Kleinrock, 1972; Takagi, 1991].

The basic time unit is known as a slot, the duration of which is $T = 625\mu s$; consequently, there are exactly 1600 slots per second. Data is transmitted in packets that take one, three, or five slots; packets that contain management information from higher layers, such as Link Manager (LM) layer and Logical Link Control and Adaptation Protocol (L2CAP) layer, also take one slot each.

If the master needs to poll a slave, but has no actual data or management information to send, it will send an empty packet. Such packets are labeled as POLL packets [Bluetooth SIG, 2003*b*]. If the slave has been polled by the master and has to respond, but has no data or administrative information to send back to the master, it will send an empty packet. Such packets are referred to as NULL packets [Bluetooth SIG, 2003*b*].

Since the same hopping sequence is used for both transmission and reception of packets, simultaneous transmission and reception (i.e., duplex operation) is not possible. Instead, the master and the slaves access the channel in alternate slots. A downlink packet and the subsequent uplink packet are commonly referred to as a frame. By default, all master transmissions start in even-numbered slots, whilst all slave transmissions start in odd-numbered slots. Therefore, the master and the addressed slave use the same communication channel, albeit not at the same time. This communication mechanism is known as Time Division Duplex, or TDD for short; it is schematically shown in Fig. 1.3.

It should be noted that the RF frequency does not change during the transmission of a single packet, even for three- or five-slot packets. The transmission in the next

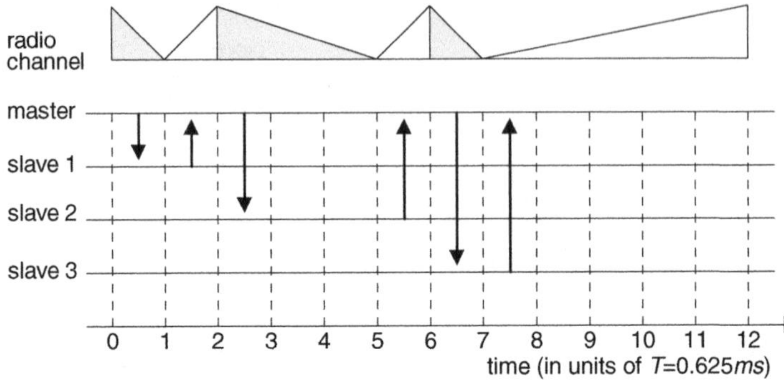

FIGURE 1.3
TDD master-slave communication in Bluetooth. Gray triangles denote data packets, white triangles denote empty (POLL and NULL) packets.

time slot uses the next frequency from the original hopping sequence (i.e., the two or four frequencies from the original sequence are simply skipped).

The master should poll each slave at least once in every T_{poll} time slots. The duration of this interval, known as the poll interval, is set by the Link Manager. This constraint aims to provide a rudimentary Quality of Service capability.

1.2 Higher layers of the architecture: LMP and L2CAP

The Logical Link Control and Adaptation Layer (L2CAP) contains two main architectural blocks. The Channel Manager is responsible for creating, managing, and destroying L2CAP channels for the transport of service protocols and application data streams [Bluetooth SIG, 2003*a*]. Channel Managers at different communicating devices create the L2CAP channels, through which the applications executing on those devices communicate. Furthermore, it collaborates with the local Link Manager to create new logical links if necessary and configure them accordingly.

The L2CAP Resource Manager deals with the ordering of submission of protocol data units (PDUs) to the baseband, and to monitor the inter-channel scheduling so as to ensure that QoS commitments are respected. Optionally, it may perform traffic shaping or policing [Bertsekas and Gallager, 1991] to make sure that the applications submitting L2CAP service data units (SDUs) conform to the allocated QoS settings.

The Device manager controls the behavior of the Bluetooth device, including operations that are not directly related to data transport. In particular, it is responsible for participation in the inquiry and paging procedures which are described in Section 1.5 below. It may request access to the transport medium when it is necessary to

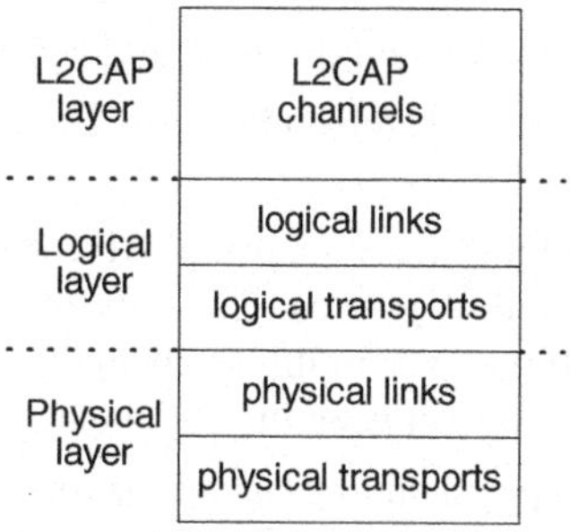

FIGURE 1.4
Generic data transport architecture, adapted from [Bluetooth SIG, 2003*c*].

carry out the desired function.

The Link Manager is responsible for creation, modification, and release of logical links and the associated logical transports, when necessary. Similar to the L2CAP Channel Manager, it collaborates with the Link Manager at other device (or devices) in order to create LMP links through which the applications executing on those devices communicate. It is also responsible for monitoring and control of logical link and transport attributes such as the enabling of encryption, the adapting of transmit power, or the adjusting of QoS settings.

The Resource Manager in the Baseband layer manages access to the radio medium, including a scheduler that allocates time on the physical channels to all the entities that need to access it.

The Link Controller is responsible for the encoding and decoding of Bluetooth packets according to the parameters of the physical channel, logical transport, and logical link. It also carries out signalling tasks for the link control protocol, which includes flow control as well as acknowledgment and retransmission requests.

Finally, the RF block is responsible for transforming data streams to and from the physical channel, under the control of appropriate entities in the baseband block.

1.3 Data transport and link types

Overall, the Bluetooth data transport architecture follows a layered approach schematically depicted in Fig. 1.4. In this architecture, L2CAP channels provide logical connections at the L2CAP level between two devices serving a single application or a higher-layer protocol. Both logical and physical layers are subdivided into the transport and link sublayers, for reasons of efficiency and legacy compatibility.

Logical links identify independent data transport services to clients of the Bluetooth system, while logical transports represent common features shared among those links, such as acknowledgment protocol and link identifiers.

Physical link is a baseband level connection between two devices established through the paging.

Finally, a physical channel is the sequence of RF carrier frequencies shared by two or more devices. A Bluetooth device can use only one of the channels at any given time; if concurrent operations over multiple channels is necessary, the device must use time division multiplexing between the channels.

A number of logical and physical links and transports with different purposes exist, as shown in the simplified hierarchy depicted in Fig. 1.5. Logical transports can be used to carry packet-structured user data, control data from the Link Manager layer, and stream-oriented synchronous data such as voice.

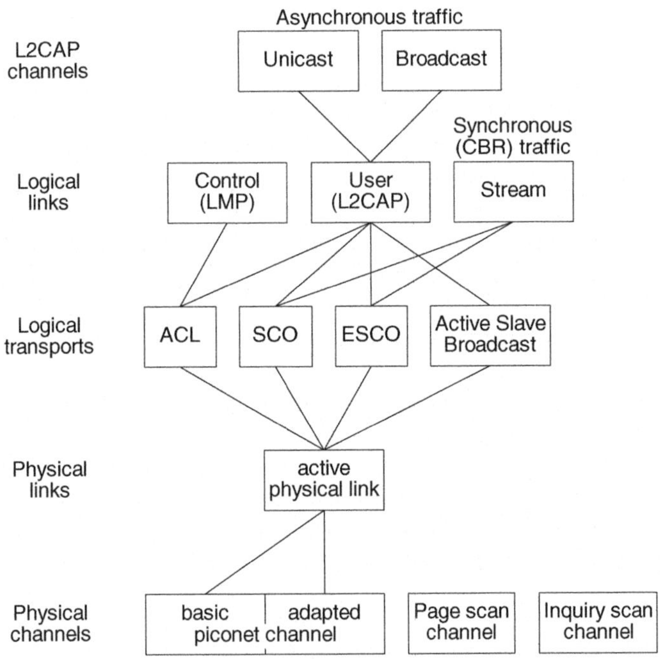

FIGURE 1.5
Overview of transport architecture entities and hierarchy, adapted from [Bluetooth SIG, 2003*c*].

A more detailed description of these concepts may be found in the Bluetooth specifications; where there are differences between versions, our presentation has followed the more recent one.

In the following, we will describe the most important logical transport types, ACL,

SCO, and eSCO. For brevity, these will be referred to as 'links' despite the subtle difference between the two concepts, as explained above.

ACL logical link

Once the devices enter the master-slave relationship through the inquiry and paging procedures (described in more detail in Section 1.5 below), a so-called Asynchronous Connectionless (ACL) link is established by default between the two. As part of the paging procedure, the master assigns one of the seven available active member addresses (AM_ADDR) to the slave, which is then referred to as an active slave.

Being an active slave with the ACL transport essentially means being synchronized to the physical channel defined by the piconet master – the slave keeps its hopping sequence and the phase of its clock synchronized to those of the piconet master. In this manner, the slave is able to follow the hopping sequence of that particular piconet, to listen to and receive all transmissions from the master, and to talk back to the master when polled. Of course, the slave decodes only the transmissions explicitly addressed to it (i.e., those in which AM_ADDR matches the value assigned to that slave), as well as broadcasts targeting active slaves. All other transmissions from the master are ignored.

The master, on the other hand, is free to poll the slave at will, or perhaps not poll it for a prolonged period of time (but shorter than T_{poll}). All active slaves listen to downlink transmissions from the master. The slave may reply with an uplink transmission of its own if and only if it has been explicitly polled by the master, and only immediately after being polled by the master.

TABLE 1.1
ACL packet types.

Type	Slot(s)	Payload (bytes)	FEC	Asymmetric data rate (kbps total)
DM1	1	17	2/3	217.6
DH1	1	27	none	341.6
DM3	3	121	2/3	516.2
DH3	3	183	none	692.0
DM5	5	224	2/3	514.1
DH5	5	339	none	780.8

The ACL link may use different types of packets, with different lengths (1, 3, or 5 slots) and different information-carrying capacity. CRC protection allows the receiver to detect packet damage due to interference and/or noise, and request retransmission if necessary. Some types of packets offer forward error correction (FEC) as well. The available packet types for ACL traffic are listed in Table 1.1.

The maximum packet length, expressed in slots of the Bluetooth clock, is a negotiable parameter. Each ACL connection created after page, page scan, role switch, or unpark operations, initially has the maximum packet length set to one slot by default. The value of maximum packet length can then be changed through negotiation. This provision is due to the fact that, according to the specification, Bluetooth devices are not required to support packets of three- and five-slots [Bluetooth SIG, 2001*b*], and this restriction is not likely to be removed for reasons of backward compatibility.

SCO link

Once an ACL link to a slave is in place, the master can establish a Synchronous Connection-Oriented (SCO) link to that slave. This is accomplished by sending an appropriate setup message through its Link Manager. SCO links are specifically designed to support synchronous, constant bit rate (CBR) traffic such as voice. They use different types of packets, known as HV-type packets, the characteristics of which are listed in Table 1.2; the packet type to be used is determined at the time the SCO link is established. HV-type packets are not protected with a CRC, and there can be no retransmission in case a packet is damaged. (Note that, in voice communications, packet loss is generally preferred to packet delays.)

TABLE 1.2
SCO packet types.

Type	Payload (bytes)	Speech duration (ms)	FEC	SCO interval (slots)
HV1	10	1.25	1/3 (repetition)	2
HV2	20	2.5	2/3 (polynomial)	4
HV3	30	3.75	none	6

In order to satisfy the bandwidth constraints for the transmission of one 64kbps voice channel, a strict timing scheme has to be observed. This means that consecutive time slots are reserved for the SCO link, with even-numbered slots assigned to the master, and the odd-numbered slots to the slave. However, either participant may choose to skip the transmission in its assigned slot if there is no actual data to send; and the slave can send data even if the master has skipped its slot, which differs from the strict TDD protocol that must be followed in ACL links. However, the master may send single-slot ACL packets of the DM1 type in its reserved slot, but this packet must be addressed to the slave with the SCO link for which the slot is reserved. In this case, the slave is allowed to respond with a single-slot ACL packet, rather than the HV-type packet normally used in this slot.

More than one SCO link may be established between the master and any given slave. At any given time, at most three SCO links with packets of appropriate type can be active in the piconet, for reasons that should be obvious.

eSCO link

From version 1.2 of the Bluetooth specification on, another type of logical transport suitable for the transmission of stream-oriented, isochronous (CBR) traffic has been introduced: the extended SCO (eSCO) logical transport. It uses special packet types listed in Table 1.3. Those packets carry larger payloads than their SCO counterparts, which allows them to support higher data rates with longer polling intervals.

TABLE 1.3
Packet types for communication over an eSCO link.

Type	Slot(s)	Payload (bytes)	CRC	FEC	Data rate (kbps)
EV1	1	1-30	yes	none	96
EV3	3	1-120	yes	2/3	192
EV4	3	1-180	yes	none	288

Polling of slaves with eSCO links differs somewhat from that of the slaves with ordinary SCO links, since limited retransmission is allowed within a specified retransmission window. The performance of eSCO links is beyond the scope of this book; more details may be found in the official specification [Bluetooth SIG, 2003*a*].

1.4 Connection state and related modes

While in the connection state, the slave is active in the piconet and retains its active mode address within the piconet, known as AM_ADDR [Bluetooth SIG, 2003*a*]. However, the slave need not listen to master transmissions all the time; it may choose to switch to one of the so-called modes in which it may detach itself from the piconet for prolonged periods without having to surrender its piconet address; these modes are known as HOLD and SNIFF. The slave may also switch to a parked state, in which it releases its active mode address; this switch may be initiated by either the master or the slave itself. Broadcast as well as unicast messages can target active or parked slaves only, as appropriate. The connection states and modes are schematically shown in Fig. 1.6.

An active slave may temporarily detach itself from the piconet through entering the so-called HOLD mode, the operation of which is shown in Fig. 1.7. In this mode, the master will not poll the slave for a specified time interval, referred to as the *holdTO*, or hold timeout. The inactivity period due to the HOLD mode affects only ACL links established between the master and the slave; SCO and/or eSCO links, if any, remain operational even when the slave is in the HOLD mode. During

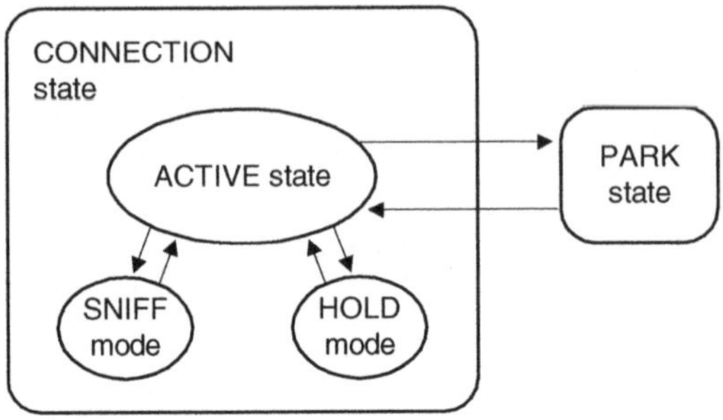

FIGURE 1.6
Connection states and modes.

the HOLD mode, the slave can engage in other activities such as scanning, paging, or joining another piconet. The slave can also enter a low power mode in order to conserve energy.

The actual duration of the HOLD mode is negotiated between the master and the slave; the negotiation process can be initiated by either the master or the slave itself. The initiating party proposes the switch to the HOLD mode as well as the hold timeout; the responding party may accept it, or respond with a counterproposal of its own. The messages exchanged during the negotiation are schematically depicted in Fig. 1.8.

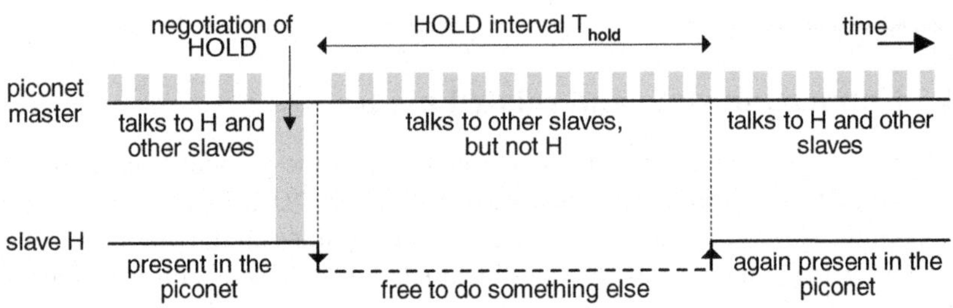

FIGURE 1.7
Pertaining to the operation of the HOLD mode.

Another mode which can be entered from the active connection state is the SNIFF mode, the operation of which is shown in Fig. 1.9. In this mode, the slave is absent from the piconet for a specified time, during which the master will not poll it. The slave periodically joins the piconet in order to listen to master transmissions. If no transmission is initiated, or even detected, during the predefined window, the slave again detaches itself from the piconet for another interval of absence. During this

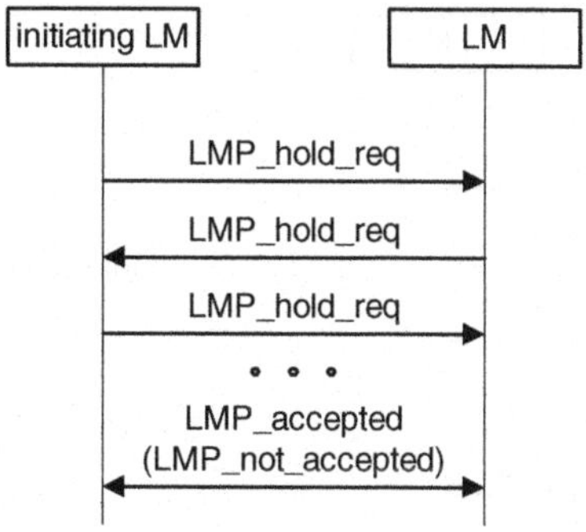

FIGURE 1.8
Message exchange during the negotiation of the switch to HOLD mode.

time interval, the slave can engage in other activities, similar to the HOLD mode described above. Again, the inactivity due to SNIFF mode affects only ACL link or links that may be set between the master and the slave in question, but not the SCO or eSCO ones, if any.

As is the case with the HOLD mode, the SNIFF mode and its parameters are negotiated between the master and the slave. The negotiation process can be initiated by either party; the initiating Link Manager (LM) proposes the SNIFF mode and its parameters to the corresponding LM of the other participant, as shown in Fig. 1.10(a). Once the switch and the parameters are accepted, the slave can start alternating between active and SNIFF mode.

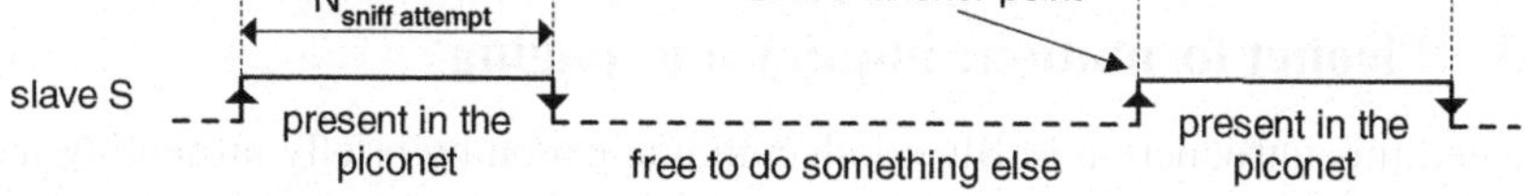

FIGURE 1.9
Pertaining to the operation of the SNIFF mode.

Unlike the HOLD mode, which is a one-off event, the SNIFF mode lasts until one of the participants explicitly requests its termination, as shown in Fig. 1.10(b).

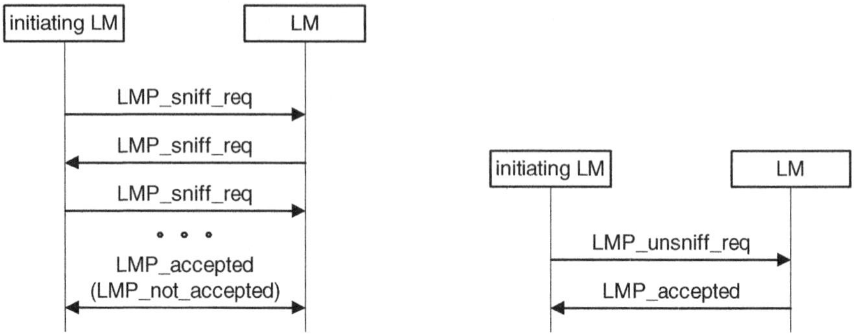

(a) Negotiating the SNIFF mode. (b) Terminating the SNIFF mode.

FIGURE 1.10
Message exchange to negotiate and terminate SNIFF mode.

Connection loss and supervision timer

A connection may break down for different reasons, including power failure, user movement, or severe interference. In order to detect the loss of connection, the traffic on each link must be monitored on both the master and the slave side. This is accomplished through the so-called supervision timer, $T_{supervision}$, which is reset to zero every time a valid packet is received on the associated physical link. If the timer reaches the *supervisionTO* timeout, the value of which is negotiated by the Link Manager, the link is considered to be lost, and the associated active piconet member address may be reassigned to another device. The value of the supervision timeout should be longer than negotiated HOLD and SNIFF periods. The same link supervision timer is used for all logical transports carried over the same physical link.

1.5 Piconet formation: inquiry and paging

We end this introduction to Bluetooth communication by briefly presenting the two procedures, inquiry and paging, through which piconets and scatternets are formed. In the first step, some Bluetooth devices enter the *inquiry* mode with the goal of making their presence known to other devices that have chosen to enter the *inquiry scan* mode. The choice of mode is performed randomly, in order to maximize the possibility of discovery and subsequent connection; but it may also be deterministic, if the device has specific reasons to become the master (or the slave) in the piconet formed afterward.

The inquiry procedure follows a special protocol: the device performing inquiry repeatedly sends the inquiry message (also known as the ID packet with a specific *inquiry access code* (IAC), using a slower frequency hopping pattern through a subset of available RF frequencies. This code may be the general code (GIAC), in which case any device may respond to it, or a dedicated one (DIAC), in which case only the devices that possess the specific capability that corresponds to the used access code may respond. There is a number of possible IAC codes but their meaning is currently unassigned, except for the GIAC which all devices must recognize. Between individual inquiry messages, the inquiring device listens for inquiry response messages.

The device performing inquiry scan listens at the frequencies from the same set, but at an even slower hopping rate. This increases the probability that the transmission from an inquiring device will be detected. Once an inquiring device has been detected, the scanning device responds with the FHS packet containing its device address and other parameters. The FHS packets are sent after a random delay, so as to reduce the probability of collisions of two or more inquiry response messages from different devices in the vicinity.

The inquiring device continues to send inquiry messages (and record responses) until it decides it has received a sufficient number of responses, when a time-out has been reached, or when instructed to cancel the inquiry procedure by the host.

In the second step, the devices formerly performing inquiry may decide to enter the *page* mode, while the devices formerly performing inquiry scan may decide to enter the *page scan* mode. In the page mode, the paging device attempts to connect to a specific device by repeatedly transmitting the page message (ID packet) containing the slave device access code. This is done using a number of different RF frequencies at the hopping rate of 3200 $hops/s$, as the Bluetooth clocks of the two devices are not synchronized yet.

The device that has responded to an inquiry may decide to enter the *page scan* mode, in which it listens to paging massages from the device that has performed the inquiry. Once a page message with the correct device access code is received, a page response (another ID packet) may be sent. The master then may send a FHS message containing its device address and other parameters, to which the scanning device responds with another ID packet. This exchange is needed in order to allow the page scanning device to adjust its clock and hopping sequence to those of the paging device. Following the successful paging, the devices enter the connection state described above. Thereafter the paging device acts as the master of a piconet while the scanning device acts as the slave in that piconet, and regular communication between the two may take place.

Due to the increased workload during the inquiry and paging procedures, devices are allowed to temporarily suspend other activities through the use of HOLD, SNIFF, or low power mode. This does not apply to SCO transmissions, which must be sent or listened to even during the inquiry and paging procedures.

A device may connect to several piconets in turn, thus enabling a scatternet to be formed. Unfortunately, Bluetooth specifications provide little guidance in this respect, and many algorithms for scatternet formation have been proposed. A detailed overview of most important among those algorithms is given in Chapter 9.

The device may disconnect from the piconet explicitly, or by simply failing to respond to the transmissions of the other device (master or slave) within the timeout defined by the supervision timer, $T_{supervision}$. If the device happened to be the master of the piconet, the piconet effectively ceases to exist, and the remaining slaves will have to connect to another piconet. (This may not be necessary in cases where the slave has already been connected to another piconet as a bridge.)

Under certain circumstances, the piconet master and one of its slaves may negotiate the so-called role switch, in which the slave takes up the role of the master. As the result, a new piconet is created using the physical channel defined by the new master's clock. However, other slaves and their existing links, if any, are not automatically transferred to the new piconet, and other commitments or modes are not preserved either. Renegotiation of those commitments is a time-consuming process, the duration of which is comparable to that of the process through which those commitments were initially established. On account of this, the role switch has found virtually no use in practice.

Beginning with version 2.0 of the specification [Bluetooth SIG, 2004], the data rate of Bluetooth networks can be increased through the Enhanced Data Rate (EDR) mode. This mode may be enabled independently on each logical transport, and for certain types of logical transports only. Once enabled, the EDR mode uses slightly different packet headers and different modulation in the packet payload, allowing data rates of up to 2.6Mbps to be achieved. However, all other aspects of network operation, including piconet formation, polling, TDD, and packet length, are not affected. Our analyses thus apply to both the standard and the EDR modes, the only difference being the maximum achievable data rate; consequently, we do not consider the operation of the EDR mode in this book.

This concludes our brief introduction to Bluetooth communication technology; more detailed descriptions of the concepts mentioned here, as well as many others that are not of immediate concern to our analysis, may be found in the Bluetooth specification [Bluetooth SIG, 2003*a*].

In subsequent discussions we will assume that the piconet or scatternet has already been formed through inquiry and paging, and focus our attention on the performance of data communications within that take place therein.

2

Intra-piconet polling schemes

We begin by reviewing the basic tenets of data communication in Bluetooth piconets. As will be seen, the Bluetooth networks operate in a rather different manner from other wireless networks. Therefore, the known performance analysis results from other wireless networks cannot be directly applied in the Bluetooth environment, and in-depth analyses of the performance are necessary. To that end, we will also review and roughly classify existing algorithms for intra-piconet polling.

2.1 Bluetooth communications and intra-piconet polling

Bluetooth devices must form networks before the actual communication can start [Bluetooth SIG, 2001*b*]. The simplest form is a piconet: a small, centralized network with up to eight active nodes or devices. One of the nodes is designated as the master, while the others are slaves. The TDD communication mechanism, outlined in the previous Chapter, requires all communications in the piconet to be routed through the master. The operation of the piconet, therefore, resembles the traditional polling system: the queueing system in which the server cyclically services a number of input queues. The piconet master would, then, correspond to the server, while the slaves would correspond to input devices, as shown in Fig. 2.1.

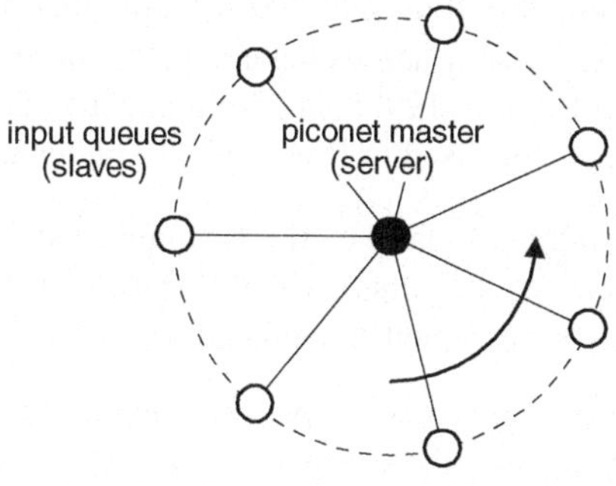

FIGURE 2.1
Bluetooth piconet as a single-server, multiple-input polling system.

The master polls the slave by sending a data packet or an empty packet when no data packets are available. The slave responds by sending a data packet or an empty packet. In Bluetooth parlance, empty packets are designated as POLL packets in the downlink, and NULL packets in the uplink direction [Bluetooth SIG, 2003*a*]. As the process of polling the slaves is actually embedded in the data transmission mechanism, we will use the term 'polling' for every downlink transmission from the master to a slave.

Since packets must wait at the slave and/or at the master before they can be delivered to their destinations, they will experience queueing delays. Therefore, the performance of the data traffic will be mostly dependent on the choice of the polling scheme used by the master to poll the active slaves in the piconet. As usual, the main performance indicator we shall use is the end-to-end packet delay, with lower delays corresponding to better performance.

The polling scheme in a multiple-input, single-server system [Kleinrock, 1972; Takagi, 1991] defines the answers to the following questions: how long to stay with the current queue (i.e., when to move on to some other queue), what to do when the current queue is empty (i.e., stay with the current queue or move on to another one), and which queue is to be serviced next when the server decides to move on from the current one. At first, the correspondence may seem easy to establish, with the server corresponding to the piconet master and the input queues corresponding to the uplink queues at individual slaves. But one must keep in mind that the input queue is actually two queues, the downlink *and* the uplink queue (which are serviced in immediate succession) and that the action of polling the slave and obtaining its response takes time, even when there are no data packets to send in either direction.

The polling scheme is obviously the main determinant of performance of Bluetooth piconets. (It is also one of the main determinants of performance of Bluetooth scatternets, as will be seen later.) Note that the terms "polling" and "scheduling" are used as synonyms by many authors. However, in this book we will use the term "polling" exclusively for intra-piconet communication, while the term "scheduling" will be reserved for inter-piconet communication.

The current Bluetooth specification does not require or prescribe, or indeed even propose any specific polling scheme [Bluetooth SIG, 2001*b*]. This may not seem to be too big a problem, since optimal polling schemes for a number of similar single-server, multiple-input queueing system are well known for multiple-input, single server systems [Levy, Sidi and Boxma, 1990; Liu, Nain and Towsley, 1992]. However, the communication mechanisms used in Bluetooth are rather specific, because

- All communications are bidirectional (i.e., there cannot exist a downlink packet without an uplink packet, or vice versa), which means that there are actually two multiple-input single-server queues to consider.

- The master polls the slaves using regular packets, possibly without data payload (i.e., all polls and responses thereto take at least one slot each).

- All slave-slave communications have to be routed through the master (i.e., there can be no direct slave-to-slave communication).

- The master does not know the status of queues at the slaves, because there are no provisions for exchange of such information in the Bluetooth packet structure. In fact, the master can request such information from a slave, but the request and response both take time, and the status of other slaves' queues can change. Therefore, it's impossible for the master to know the status of all the uplink queues at any given time.

As a consequence, the existing results cannot be applied, and the performance of different polling schemes has to be re-assessed, taking the aforementioned characteristics of the Bluetooth communication mechanisms. It should come as no surprise, then, that a number of polling schemes have been proposed and analyzed [Capone, Kapoor and Gerla, 2001; Das, Ghose, Razdan, Saran and Shorey, 2001; Johansson, Körner and Johansson, 1999; Kalia, Bansal and Shorey, 1999]. Many of the proposed schemes are simply variations of the well-known 1-limited and exhaustive service polling, but several improved adaptive schemes have been described as well.

The polling scheme has, however, at least two other requirements to satisfy. One of those is fairly general and holds in all networks, while the other is specific to Bluetooth and other wireless technologies. The first is fairness: the piconet master should try to maintain fairness among the slaves, so that all slaves in the piconet receive equal attention in some shorter or longer time frame. Of course, their traffic load should be taken into account, if known. The second one is computational simplicity: Bluetooth devices are, by default, low power devices, and the polling scheme should be sufficiently simple in terms of computational and memory requirements. As will be seen, some of the schemes require considerable computational overhead and are thus unsuitable for power-limited devices.

2.2 Classification of polling schemes

In the discussion that follows, we present a rough classification of piconet polling schemes based on three important characteristics. First, the polling scheme determines the number of frames exchanged during a single visit to the slave. This number may be set beforehand to a fixed value, or it may be dynamically adjusted on the basis of current and historical traffic information.

Second, different slaves may receive different portions of the bandwidth; again, the allocation may be done beforehand, or it may be dynamically adapted to varying traffic conditions. The latter approach is probably preferable in Bluetooth piconets, which are ad hoc networks formed by mobile users, and the traffic may exhibit considerable variability. In fact, due to users' mobility, even the topology of the piconet may change on short notice. However, the fairness of polling may be more difficult to maintain under dynamic bandwidth allocation.

Finally, the sequence in which slaves are visited may be set beforehand, or it may change from one piconet cycle to another according to a specified algorithm that takes current and/or historical traffic information into account. In either case, slaves that had no traffic in the previous cycle(s) may be skipped for one or more cycles, but the polling scheme should ensure that the fairness is maintained.

Traditional polling schemes

The simplest polling schemes use a fixed ordering of the slaves and fixed bandwidth allocation per slave. The only variable parameter left, then, is the duration of master's visit to each slave.

Under *1-limited service* polling, the master visits each slave for exactly one frame, and then moves on to the next slave. Data packets are sent if there are any, otherwise empty packets (POLL or NULL) are sent. The piconet cycle lasts for exactly $2(m - 1)$ packets. The scheme is sometimes referred to as (Pure) Round Robin [Capone et al., 2001] or simply limited service.

Under *exhaustive service* polling, the master stays with the slave as long as there are packets to exchange in either downlink or uplink direction. The absence of packets is detected by a POLL-NULL frame, which is a signal for the master to move on to the next slave. Note that the duration of the piconet cycle is not limited; a slave will be polled by the master as long as any of them has packets in the corresponding queue, or queues. Therefore, this scheme cannot guarantee fairness.

Under the *E-limited service* polling, the master stays with a slave until there are no more packets to exchange, or for a fixed number M of frames ($M > 1$), whichever comes first. (We will refer to M as the *polling parameter*.) Packets that arrive during the master's visit are allowed to enter the uplink queue at the slave, and may even be serviced – provided the total limit of M frames per master's visit is not exceeded [Takagi, 1991]. This scheme is also referred to as Limited Round Robin [Capone et al., 2001; Lee, Kapoor and Gerla, 2002]. In this case, the piconet cycle can last at most $2M(m - 1)$ packets.

In fact, 1-limited and exhaustive service polling may be considered as special cases of E-limited service, where the limit M equals 1 and ∞, respectively. In all three cases, the sequence of slaves is fixed and does not change.

In traditional polling systems, exhaustive service has been shown to perform better than either 1-limited or E-limited service [Levy et al., 1990]. As Bluetooth piconet is not a traditional polling system, this result does not hold. Several authors have found that 1-limited performs better than exhaustive service under high load [Capone et al., 2001; Mišić and Mišić, 2003*b*]. Furthermore, E-limited service has been found to offer better performance than either limited or exhaustive service, and the value of M may be chosen to achieve minimum delays for given traffic burstiness [Mišić, Chan and Mišić, 2004]. A detailed analysis of all three polling schemes is presented in Chapter 3.

Dynamic reordering of slaves

In fact, traditional polling systems have been shown to achieve best results – i.e., the shortest delays – when the server follows the so-called Stochastically Largest Queue

(SLQ) policy [Liu et al., 1992]. The server should always service the queue (i.e., master-slave channel) which has the highest sum of master and slave queues. For obvious reasons, the lengths of all queues must be known to the server at all times.

The application of this policy in Bluetooth piconet would require the piconet master to know the current status of the uplink queues of *all* of its slaves at all times. However, the Bluetooth packet structure has no provisions for simple exchange of relevant information: packet headers do not have any fields to carry this information, nor are there any spare bits to be used to that effect [Bluetooth SIG, 2001*b*; Bluetooth SIG, 2003*a*; Bluetooth SIG, 2004]. Therefore, if the information on queue status is to be exchanged between the master of a piconet and its slaves, it will have to be done at the expense of actual packet payload. Such exchange will take at least two cycles per slave, during which the state of *other* slave queues may change. Consequently, the piconet master cannot know the status of all slave uplink queues at any given time.

The information which is readily accessible to the master are the lengths of the downlink queues, and the next slave to be polled may be chosen as the slave for which the corresponding downlink queue is the longest. This policy will be referred to as LDQF (Longest Downlink Queue First). Note that slave selection (and, effectively, reordering) may be done for each poll or once in each piconet cycle. In the latter case, polling of individual slaves may follow the 1-limited, exhaustive, or E-limited scheme.

One problem with dynamic reordering is that fairness among the slaves cannot be guaranteed if reordering is done for every poll. When reordering is done on a per cycle basis, fairness may still be difficult to maintain under the exhaustive scheme, as two slaves that talk to each other can virtually monopolize the piconet and starve all the others.

The Exhaustive Pseudo-cyclic Master (EPM) queue length scheme, proposed in

TABLE 2.1
A comparison of non-adaptive piconet polling schemes.

	Polling scheme			
Property	1-limited	Exhaustive	E-limited	EPM
Slave reordering	no	no	no	yes
Slave skipping	no	no	no	yes
Burst-preserving	no	yes	depends on M	no
Efficiency at low loads	low to medium	high	medium to high	medium to high
Efficiency at high loads	high	high	high	high
Short-term fairness	yes	no	yes	no
Long-term fairness	yes	no	yes	no

[Capone et al., 2001], uses dynamic reordering coupled with 1-limited polling. At the beginning of each cycle, the slaves are reordered according to the decreasing length of downlink queues. Each slave is then polled once per piconet cycle.

Main properties of the non-adaptive polling schemes described above are summarized in Table 2.1.

Adaptive polling

The third parameter that characterizes the polling scheme is the duration of the visit to each slave. This duration may be fixed or adjustable according to the current or historical traffic information. The adjustment can be done by rewarding slaves that have more traffic, or by penalizing slaves that have less traffic. In the former case, the master is allowed to stay longer, and thus exchange more frames, with the slave that has some data to send and/or receive. In the latter case, the slave that had less traffic or no traffic at all, or the slave which is expected to have no traffic, will simply be ignored for a certain number of piconet cycles.

The former approach is exploited in the Limited and Weighted Round Robin (LWRR) [Capone et al., 2001], which tries to increase efficiency by reducing the rate of visits to inactive slaves. Initially, each slave is assigned a weight equal to the so-called maximum priority, MP. Each slave is polled in E-limited fashion with up to M frames. Whenever there is a data exchange between the slave and the master, the weight of the slave is increased to the value of MP. On the other hand, when a POLL-NULL sequence occurs, the weight for that particular slave is reduced by one. If the slave weight drops to one (which is the lowest value), the slave has to wait a maximum of $MP-1$ cycles to be polled again.

A variation of this scheme, labeled Pseudo-Random Cyclic Limited slot-Weighted Round Robin [Lee et al., 2002], uses both slave reordering and poll rate reduction. The sequence in which slaves will be polled is determined in a pseudo-random fashion at the beginning of every cycle, and less active slaves are not polled for a certain number of slots (not cycles). In addition, the maximum number of frames that may be exchanged during a single visit to any slave is limited.

Poll rate reduction is also utilized in the Fair Exhaustive Polling (FEP) scheme [Johansson et al., 1999], where a pool of 'active' slaves is maintained by the master. Slaves are polled with one frame per visit, as in 1-limited service. When an empty (POLL-NULL) frame occurs, that slave will be dropped from the pool and will not be polled for some time. The 'inactive' slave may be restored to the pool when the master downlink queue that corresponds to that slave contains a packet, or when the entire pool is reset to its original state. The pool is reset when the last slave in the pool is to be dropped, or after a predefined time-out. In this manner, the slaves that have more traffic will receive a proportionally larger share of the bandwidth as long as they have traffic to send or receive.

The Adaptive Cycle-Limited Service scheme [Mišić, Mišić and Ko, 2004] strives to keep the duration of the piconet cycle as close to a predefined value as possible. Bandwidth is allocated dynamically, partly on the basis of historical data (i.e., the amount of traffic in the previous piconet cycle), and partly on the basis of current traffic (i.e., whether they have some data to exchange or not). However, each of the m slaves is guaranteed a fair share of the available bandwidth over the period of $m - 1$ piconet cycles, plus a certain minimum bandwidth in each piconet cycle. This scheme appears well-suited for piconets in which some of the slaves have tight bandwidth and latency constraints, as is often the case with multimedia traffic.

Recently, Lapeyrie and Turletti [2003] have proposed the Fair Predictive poller with QoS support, or FPQ. FPQ strives to minimize the number of POLL/NULL packets in the presence of upstream and downstream traffic. It also tries to maintain fairness while respecting the predefined delay bounds, labeled Maximum Delay (MD) request in the original paper. Since it is not possible to achieve maximum efficiency while maintaining fairness, the algorithm uses a variable parameter α to adjust the tradeoff between the two. Also, it may be difficult to achieve the maximum efficiency while respecting the required delay bounds. In the extreme case where every slave has a maximum delay request, the FPQ turns into a simple exhaustive service scheme.

The authors of FPQ observe that low delay values can be achieved if all the baseband packets resulting from an IP packet are transmitted without interleaving with packets from other slaves' packet streams. To that end, they assign priorities to uplink and downlink baseband packets, and thus prevent interleaving of IP packets as much as possible. However, the FPQ approach is unable to prevent interleaving of IP packets (or other PDUs generated in higher layers of the protocol stack) in case of their simultaneous arrivals at multiple slaves.

Finally, the computational complexity of the FPQ scheme is much higher than for the other schemes, due to the need for the Analyzer module to calculate the

TABLE 2.2
A comparison of adaptive piconet polling schemes.

	Polling scheme			
Property	LWRR	FEP	ACLS	FPQ
Slave reordering	no	no	no	yes
Slave skipping	no	no	no	yes
Burst-preserving	no	yes	depends on M	no
Short-term fairness	yes	no	yes	no
Long-term fairness	yes	no	yes	adjustable
Efficiency at low loads	low to medium	high	medium to high	medium to high
Efficiency at high loads	high	high	high	high
Predefined delay bounds	none	none	optional cycle control	adjustable
Computational complexity	low	low	low	high

probability of packet arrivals in slave queues, as well as for the Selector module which calculates priorities of baseband packets.

Table 2.2 summarizes the features of adaptive polling schemes described above.

2.3 On segmentation and reassembly policies

Most, if not all, of the polling schemes described above focus on the optimization of performance of Bluetooth baseband traffic. However, other considerations, in particular, the segmentation and reassembly algorithm, may also affect the performance of the chosen polling scheme.

Namely, the traffic to be transported over Bluetooth links is generated and consumed by the applications running on Bluetooth (and other) devices. This traffic will be formatted as a stream of packets (or protocol data units, PDUs) conforming to the rules of the protocol used by the application. In most practical cases, that protocol will belong to the ubiquitous TCP/IP family [Kurose and Ross, 2005].

As is customary in such cases, packets sent from the higher layers will have to be repackaged or *segmented* into Bluetooth packets, with appropriate administrative information added to the Bluetooth packet headers. At the receiving device, the complementary *reassembly* operation will take place.

As the packets used in the TCP/IP family of protocols are generally longer than the Bluetooth ones, each TCP packet will be segmented and transmitted as the payload of a number of Bluetooth packets. An example of such a segmentation approach is the Bluetooth Network Encapsulation Protocol, or BNEP [Bluetooth SIG, 2001*a*], which we will briefly present in the following.

The BNEP protocol is designed to transmit Ethernet traffic through Bluetooth networks; Fig. 2.2 shows the corresponding protocol stack. Ethernet packets are segmented into a number of Bluetooth baseband packets for transmission, and reassembled at the destination, as shown in Fig. 2.3. Note that each TCP message generated will require a total of 59 bytes in appropriate headers throughout the protocol stack.

In BNEP, multi-hop traffic, including traffic between the slaves in the same piconet, is handled by Bluetooth masters and/or bridges acting as store-and-forward switches – i.e., entire Ethernet packets have to be stored in the device before being repackaged (if necessary) and forwarded to the next stop along the route. Routing in this case is done in the upper layers, possibly in the IP layer, transparently to Bluetooth. (It should be noted that this is a forced, rather than deliberate choice, as the Bluetooth does not define any routing capabilities [Bluetooth SIG, 2003*a*], and all routing functions must be provided by higher layers of the protocol stack.)

Neither BNEP nor indeed the Bluetooth specification itself do prescribe the actual segmentation algorithm, i.e., which exact Bluetooth baseband packets should be used. As single slot ACL packets have substantially lower data carrying capacity than their longer counterparts, as shown in Table 1.1, the throughput should be

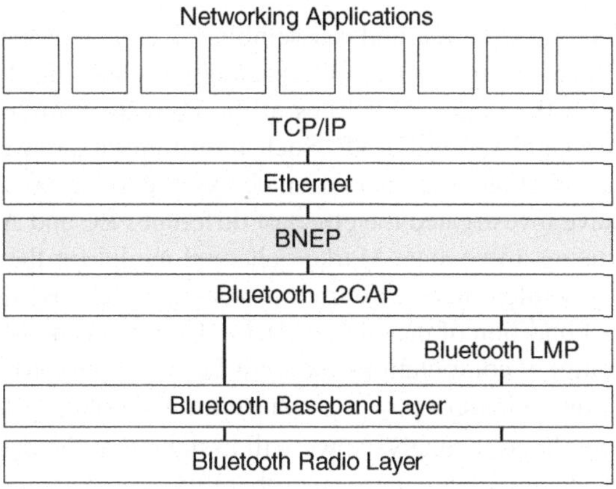

FIGURE 2.2
BNEP protocol stack, adapted from [Bluetooth SIG, 2001*a*].

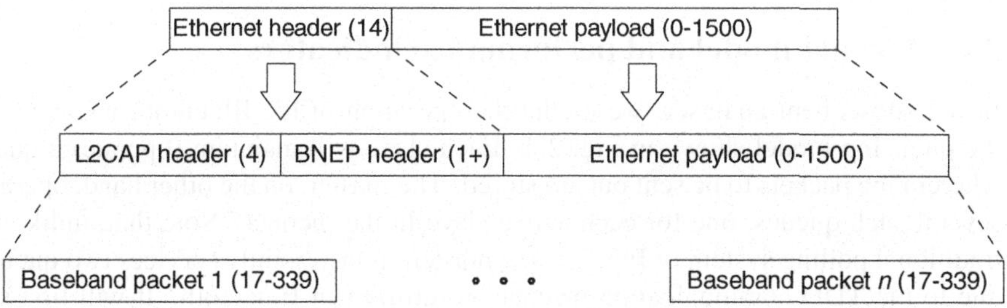

FIGURE 2.3
BNEP Ethernet packet segmentation (field sizes expressed in bytes), adapted from [Bluetooth SIG, 2001*a*].

higher if more of the longer, three- and five-slot packets are used.

However, noise and interference levels should be taken into account as well. The Bluetooth frequency hopping sequence has been shown to be fairly efficient, and a rather large number of Bluetooth piconets may coexist within the same radio range before the effects of interference become noticeable [Zürbes, 2000]. Still, packets can be damaged by noise and interference, and longer packets are more susceptible to damage, in which case frequent retransmissions will occur. (As Bluetooth devices operate as store-and-forward switches, packet loss and the need for retransmission might not be identified until the actual reassembly is attempted.) In noisy environments, shorter three-slot packets might be preferred over the longer, five-slot ones, despite their smaller payload, and DM-type packets might be preferred over DH-type packets, due to their higher resilience to noise provided by the FEC protection (see Table 1.1).

Algorithms for segmentation and reassembly have been proposed by several authors. Some of them try to optimize the packaging, choosing the baseband packet sizes depending on the data arrival rates at the corresponding queue [Kalia et al., 1999; Kalia, Bansal and Shorey, 2000]. Such algorithms are, however, likely to result in lower efficiency and longer transmission delays due to the excessive use of shorter packets. Some have investigated the effect of different FEC and ARQ schemes at the baseband level, using a two-state Markov channel model for the Bluetooth RF link [Das et al., 2001], while others have tried to adaptively choose the best packet type depending on the condition of the channel [Kim, Lim, Kim and Ma, 2001]. Note that schemes with complex computations may not be well suited for Bluetooth devices which are operating on battery power and have limited computational capabilities.

In our queueing theoretic analysis, we will assume that the packets of one, three, and five slots are generated with uniform probability, as explained in the next Section.

2.4 Piconet model and performance indicators

In our subsequent analyses, we model the operation of the Bluetooth piconet with the queueing model shown in Fig. 2.4. Each slave will maintain (operate) a queue wherein the packets to be sent out are stored. The master, on the other hand, operates several such queues, one for each active slave in the piconet. Note that, unlike the traditional polling system of Fig. 2.1, the master, in fact, jointly services two queues, due to the TDD communication mechanism utilized in Bluetooth. It will first service its own downlink queue that corresponds to the slave being polled, immediately followed by the uplink queue at that slave; this approach to polling is commonly referred to as a *bidirectional polling system.*

Note that, depending on the hardware architecture, the downlink queues may be implemented as separate physical queues, as shown in the diagram, or as a single queue from which packets can be serviced in the order of their destinations. However, the queueing model with separate downlink queues provides a convenient modeling framework that facilitates performance analysis of Bluetooth networks.

Our analysis will follow the theory of $M^{[x]}/G/1$ queueing systems with vacations. Where necessary, the descriptors for uplink and downlink queues will be distinguished through subscripts u and d, respectively. For example, the burst arrival rate for the uplink channel queue of slave i will be denoted with λ_i.

In order to obtain closed form solutions for the queue length probability distributions, we assume that device buffers have infinite size. This means that all packets received from the upper layers of the protocol stack will eventually be transmitted. This assumption is, of course, unrealistic, but it does simplify the analysis and allows different polling schemes to be compared. (The case where device buffers are finite will be addressed in Chapter 4.)

The main performance indicators for the piconet are two delay variables: access

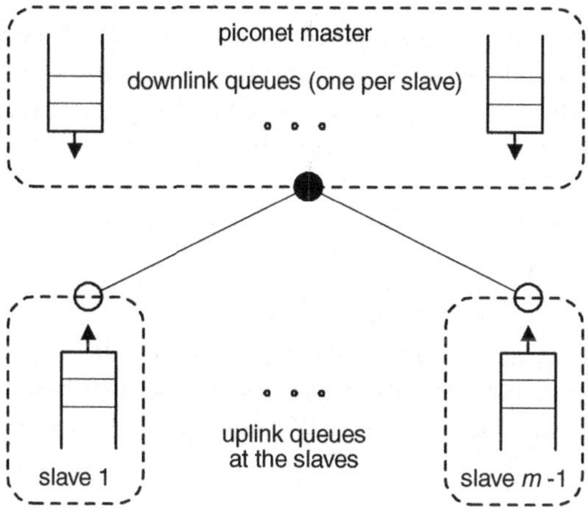

FIGURE 2.4
The queueing model of the Bluetooth piconet.

delay W_{ai} for slave i, the time a data packet has to wait in the uplink queue of the source device before it is serviced, and end-to-end delay W_{ije}, the time from the moment a packet enters the uplink queue at the source device i, to the time it arrives at its destination device j.

Each slave's application generates packets which are further segmented into a number of baseband packets [Bluetooth SIG, 2001*a*]. Application packet arrivals to the uplink queue of slave i follow a Poisson distribution with the arrival rate λ_{iu}, which has been shown to be a satisfactory approximation for the traffic of many Internet applications [Paxson and Floyd, 1995]. If the application packets themselves arrive in bursts, the probability generating function (PGF) of the burst size of application packets should be integrated into the corresponding PGF for the baseband burst size.

The length of the burst of baseband packets is geometrically distributed with the mean value of $\overline{B}$; other distributions can be easily accommodated in our model, provided that the corresponding PGF is known. Assuming that all slaves use the same segmentation/reassembly mechanism (which is reasonable, as this mechanism is commonly implemented in firmware), the burst length distribution will be the same for all slaves. We assume that the traffic goes from slaves to other slaves only, which simplifies our calculations without undue loss of generality. (Any traffic generated or received by the master can be easily modeled by increasing the packet arrival rates in the downlink queues without any other changes to the analytical model.) All packets within the given burst will have the same destination node, and the distribution of destinations is assumed to be uniform. In that case, the packet burst arrival rate to the downlink queue would be $\lambda_{jd} = \sum_{i \neq j} \frac{\lambda_{iu}}{m-2}$. In more complex cases, the resulting downlink arrival rates may be calculated from the matrix of probabilities for slave-to-slave communication, without changing the model itself. Under 1-limited and E-limited polling, the downlink burst size will be changed and therefore packet burst arrival rate will be changed. This will be discussed in sections which model these techniques.

The probability distribution of length of the packet burst may be described with the PGF of $G_b(z) = \sum_{k=0}^{\infty} b_k z^k$, where b_k is the probability that the burst will contain exactly k packets; the mean value of the burst length is $\overline{B} = G'_b(1)$. The equivalent Laplace-Stieltjes transform (LST) of the probability distribution may be obtained by substituting the variable z with e^{-s} [Takagi, 1991]; for example, the LST of the packet burst length PDF is $G^*_b(s) = \sum_{k=0}^{\infty} b_k e^{-ks}$.

The probabilities of packets being one, three, and five slots long are p_1, p_3, and $p_5 = 1 - p_1 - p_3$, respectively. The corresponding probability generating function (PGF) is $G_p(z) = p_1 z + p_3 z^3 + p_5 z^5$. First and second moments of the packet length distribution are equal to $\overline{L} = G'_p(1)$ and $\overline{L^2} = G''_p(1) + G'_p(1)$, respectively, and its LST is $G^*_p(s) = p_1 e^{-s} + p_3 e^{-3s} + p_5 e^{-5s}$.

We assume that piconet members have index numbers. Piconet master has number 1 while other slaves have numbers from 2 to m. Let S_i denote the time to service a single channel i which depends on the polling scheme used. The Probability Generating Function (PGF) for the channel service time will be denoted as $S_i(z)$ and its Laplace-Stieltjes Transform (LST) will be denoted as $S^*_i(s)$. Also, let us denote the piconet cycle time as C, its probability density function (pdf) as $c(x)$, and its PGF and LST as $C(z)$ and $C^*(s)$, respectively.

With the queueing model thus set up, we are ready to undertake detailed performance analysis of Bluetooth piconets.

3

Analysis of polling schemes

This chapter describes analytical modeling of the Bluetooth bidirectional polling system in a single piconet with one master device and $m - 1$ slave devices, as shown in Fig. 2.4. We first consider the simpler cases of exhaustive and 1-limited polling, and then move on to analyze the E-limited polling, which may be considered to be a generalization of the first two schemes. Neither of these requires that the piconet master knows the status of the slaves' queues, and therefore does not incur additional signaling overhead between the master and the slaves.

3.1 Performance of exhaustive service

Under exhaustive polling, a number of frames (packet sent in downlink, followed by another in uplink) may be exchanged between the master and a single slave during a single visit to that slave. The timing diagram of packet exchanges in this case is shown in Fig. 3.1 (note that packet lengths are not taken into account). The exchange ends when both queues are empty, i.e., when the slave responds with a NULL packet to a POLL packet sent by the master [Bluetooth SIG, 2001*b*]. The actual number of frames is equal to the number of packets in the downlink or the corresponding uplink queue, whichever is larger, plus one for the POLL-NULL sequence. The channel service time for the exhaustive polling is the time required to empty both channel queues for one particular slave.

The PGF $S_i(z)$ of the channel time can be found if we first find the mass probabilities of k packet arrivals in the i-th slave's and master's queue during the piconet cycle time.

The probability of l packet burst arrivals to the uplink queue of the i-th slave during the piconet cycle time is equal to

$$b_{l,iu} = \int_0^\infty \frac{(\lambda_{iu} x)^l}{l!} e^{-\lambda_{iu} x} c(x) dx \tag{3.1}$$

The probability that k data packets are sent from the slave i to the master may be obtained as

$$a_{k,iu} = \sum_{l=0}^{\infty} \frac{1}{k!} \frac{d^k}{dz^k} (G_b(z))^l \Big|_{z=0} \int_0^\infty \frac{(\lambda_{iu} x)^l}{l!} e^{-\lambda_{iu} x} c(x) dx \tag{3.2}$$

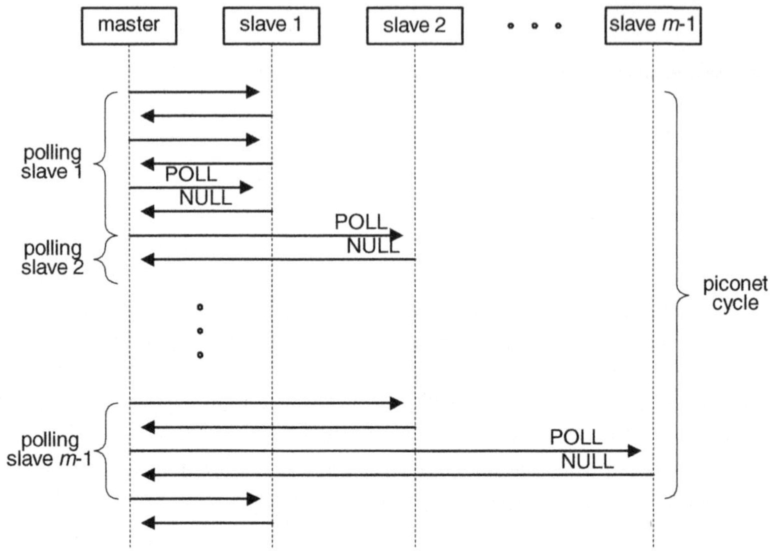

FIGURE 3.1
Timing diagram of exhaustive polling.

By definition, the LST of piconet cycle time is equal to

$$C^*(s) = \int_0^\infty e^{-sx} c(x) dx \tag{3.3}$$

By exchanging the order of summation and integration, we obtain the mass probability for k arrivals to the i-th slave's queue during the piconet cycle time as

$$a_{k,iu} = \frac{1}{k!} \frac{d^k}{dz^k} C^* \left(\lambda_{iu} - \lambda_{iu} G_b(z)\right)_{|z=0} \tag{3.4}$$

We also note that $C^*(\lambda_{iu} - \lambda_{iu} G_b(z))$ denotes the PGF for the number of data packet arrivals in the slave uplink queue during the piconet cycle time. By the same token, the equivalent conditional probability that k data packets are sent from the master to the slave is

$$a_{k,id} = \frac{1}{k!} \frac{d^k}{dz^k} C^*(\lambda_{id} - \lambda_{id} G_b(z))_{|z=0} \tag{3.5}$$

By combining these elements together, the PGF for the i-th channel service time becomes

$$\begin{aligned} S_i(z) &= \sum_{k=0}^{\infty} a_{k,iu} a_{k,id} G_p(z)^{2k} z^2 \\ &+ \sum_{j=1}^{\infty} \sum_{k=0}^{\infty} \left(a_{k,iu} a_{k+j,id} + a_{k+j,iu} a_{k,id}\right) G_p(z)^{2k+j} z^{(j+2)} \end{aligned} \tag{3.6}$$

Finally, the PGF for the piconet cycle time becomes

$$C(z) = \prod_{i=2}^{m} S_i(z) \tag{3.7}$$

Let us view the PGF for the piconet cycle time as the polynomial $C(z) = \sum_{i=0}^{\infty} r_i z^i$ and its LST as $C^*(s) \sum_{i=0}^{\infty} r_i e^{-si}$, where r_i represents the mass probability that piconet cycle has the duration of i Bluetooth time slots. In order to solve (3.7), we have to truncate the $C(z)$ to the polynomial of degree i_{max}. This is approximate solution but it should be sufficient for large i_{max}. Then, we have to set i_{max} equations in the form $r_i = \frac{1}{i!} \frac{d^i}{dz^i} C(z)|_{z=0}$. Each of the equations will have the term r_i on the left side and the function of all r_i mass probabilities on the right side. The resulting system of equations can be solved using numerical solvers (e.g., Waterloo Maple).

The PGF for the frame time is

$$\begin{aligned} F_{is}(z) = & \sum_{k=0}^{k_{max}} a_{k,iu} a_{k,id} (G_p(z)^{2k} z^2)^{1/(k+1)} \\ & + \sum_{j=1}^{j_{max}} \sum_{k=0}^{k_{max}} \left(a_{k,iu} a_{k+j,id} + a_{k+n,iu} a_{k,id} \right) \cdot (G_p(z)^{2k+j} z^{j+2})^{1/(k+j+1)} \end{aligned} \tag{3.8}$$

where k_{max} and j_{max} are determined according to the accuracy required. Then, the mean frame length will be $\overline{F_{is}} = F'_{is}(1)$.

In order to find the duration of the piconet cycle time, we will make use of the concept of *vacation*. In single server-multiple client systems, when the server finds an empty client queue, it goes on to service other clients – i.e., it takes a vacation, which lasts until the next visit to this client [Takagi, 1991]. In a Bluetooth piconet, a server vacation starts when the master, after having polled a slave, moves on to next one. From the viewpoint of a particular slave, the master is not available—i.e., takes a vacation—during the time it services other slaves. The vacation lasts until the next visit to the slave; if the slave queue is empty at that time, the master will immediately start a new vacation. The vacation time is, then, the time while the master is busy servicing other slave queues, as shown in Fig. 3.2. The vacation time for slave i is denoted as V_i, its PGF is $V_i(z)$ and its LST is denoted as $V_i^*(s)$.

The duration of the vacation period V_i may be described with the following PGF

$$V_i(z) = \prod_{\substack{j=2 \\ j \neq i}}^{m} S_j(z) \tag{3.9}$$

and its first and second moments are $\overline{V_i} = V_i'(1)$ and $\overline{V_i^2} = V_i''(1) + V_i'(1)$, respectively.

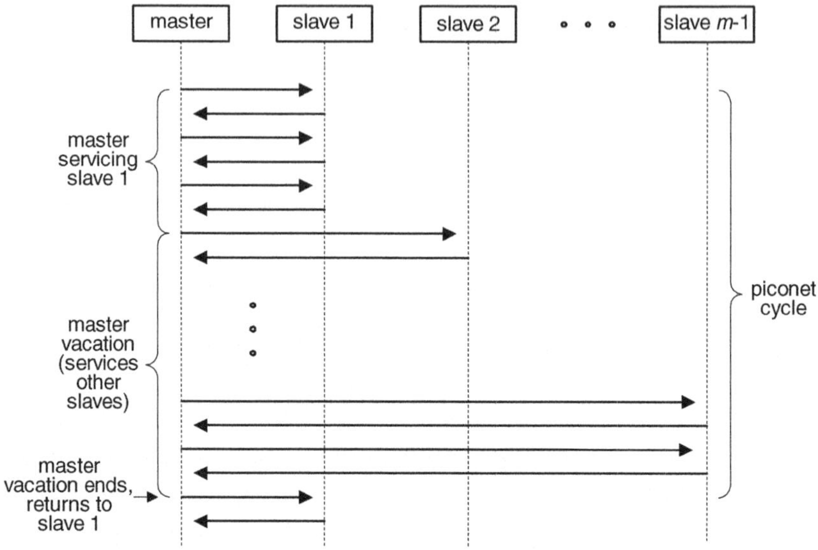

FIGURE 3.2
Pertaining to the concept of server vacation.

The LST for the access delay probability distribution at slave i is:

$$W_{ai}^*(s) = \frac{1 - V_i^*(s)}{s\overline{V_i}} \cdot \frac{1 - G_b(F_{is}^*(s))}{\overline{B}\,(1 - F_{is}^*(s))} \cdot \frac{s(1 - \lambda_{iu}\overline{B}\;\overline{F_{is}})}{s - \lambda_{iu} + \lambda_{iu}G_b(F_{is}^*(s))} \tag{3.10}$$

which translates into the mean access delay of the form

$$\overline{W_{ai}} = \frac{\lambda_{iu}\overline{B}^2\;\overline{F_{is}^2} + \overline{B^{(2)}}\;\overline{F_{is}}}{2\overline{B}(1 - \lambda_{iu}\overline{B}\;\overline{F_{is}})} + \frac{\overline{V_i^2}}{2\overline{V_i}} \tag{3.11}$$

Under exhaustive polling, the entire burst from the uplink slave will be transferred to the corresponding downlink queue. Therefore, the LST of delay at the master queue for slave j has the same form as that of access delay – with λ_{iu} replaced by λ_{jd}, of course. Consequently, the mean value of the end-to-end delay between slave i and slave j is $\overline{W_{ije}} = \overline{W_{ai}} + \overline{W_{dj}}$.

3.2 Performance of 1-limited service

Under 1-limited service, the master exchanges exactly one packet with each slave, and then moves on to the next one. The corresponding timing diagram is shown

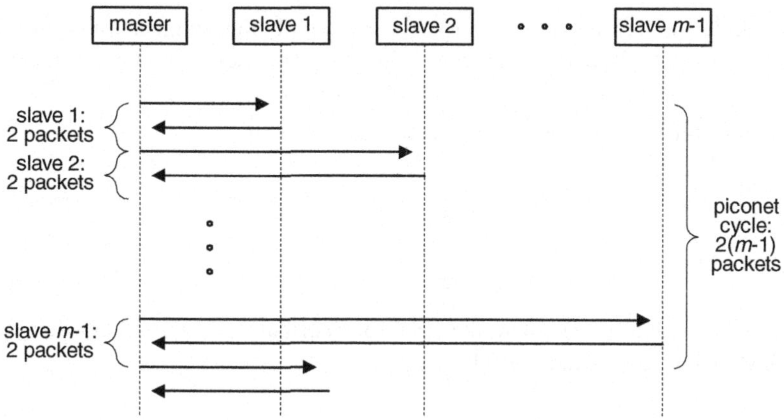

FIGURE 3.3
Timing diagram of 1-limited polling.

in Fig. 3.3. Using the notation introduced in the previous Section, if C denotes the length of the piconet cycle time, the probabilities that the channel queues are not empty will be $P_{id} = \lambda_{id}\overline{B}\,\overline{C}$ and $P_{iu} = \lambda_{iu}\overline{B}\,\overline{C}$ for master and slave queues, respectively. The probability that a data packet is sent from the slave to the master in a polling cycle is

$$a_{1,iu} = P_{iu} \tag{3.12}$$

while the complementary probability that there are no data packets at the slave queue (which means that a NULL packet will be sent) is

$$a_{0,iu} = 1 - P_{iu} \tag{3.13}$$

By the same token, the equivalent conditional probabilities that one empty (POLL) packet or one data packet is sent from the master to the slave are

$$\begin{aligned} a_{0,id} &= 1 - P_{id} \\ a_{1,id} &= P_{id} \end{aligned} \tag{3.14}$$

By combining these elements together, the PGF for the channel service time for one slave becomes

$$S_i(z) = a_{0,iu}a_{0,id}z^2 + (a_{1,iu}a_{0,id} + a_{0,iu}a_{1,id})G_p(z)z + a_{1,iu}a_{1,id}G_p(z)^2 \tag{3.15}$$

Then the PGF for the piconet cycle time becomes

$$C(z) = \prod_{i=2}^{m} S_i(z) \tag{3.16}$$

By taking the first derivative of (3.16), we can obtain the mean value of piconet cycle time $\overline{C} = C'(1)$ and, subsequently, its PGF. The second moment of the piconet cycle time may be obtained as $\overline{C^2} = C''(1) + C'(1)$.

The PGF for the duration of the vacation period V_i is

$$V_i(z) = \prod_{\substack{j=2 \\ j \neq i}}^{m} S_j(z) \tag{3.17}$$

We are now able to calculate the LST for distribution of the access delay at the i-th slave queue as [Takagi, 1991]:

$$W_{ai}^*(s) = \frac{1 - V_i^*(s)}{s\overline{V_i}} \cdot \frac{s(1 - \lambda_{iu}\overline{B}\,\overline{C})}{s - \lambda_{iu} + \lambda_{iu}G_b(C^*(s))} \cdot \frac{1 - G_b(C^*(s))}{\overline{B}\,(1 - C^*(s))} \tag{3.18}$$

where $V^*(s)$ and $C^*(s)$ denote the LST of the probability distributions of vacation time and piconet cycle time, respectively. The average access delay at the slave is then calculated as $\overline{W_{ai}} = -W_{ai}^{*'}(0)$, or:

$$\overline{W_{ai}} = \frac{\lambda_{iu}\overline{B}\,\overline{C^2}}{2(1 - \lambda_{iu}\,\overline{B}\,\overline{C})} + \frac{\overline{B^{(2)}}\,\overline{C}}{2\overline{B}(1 - \lambda_{iu}\overline{B}\,\overline{C})} + \frac{\overline{V_i^2}}{2\overline{V_i}} \tag{3.19}$$

where $\overline{B^{(2)}} = E[B(B-1)] = G_b''(1)$ is the second factorial moment of the burst length distribution.

Furthermore, the burstiness of the traffic in the downlink queue will differ from the one in the slave queue because the bursts from different source slaves might become interleaved in the same downlink queue, which will in turn lead to an equivalent decrease in burst length. Exact analysis of this phenomenon is fairly involved, so we will only present an approximate model for the decrease of the burst length. The probability that two packet bursts from different slaves will not have the same destination and, hence, will not be interleaved in the same downlink queue, will be $1 - 1/(m-1)^2$. Therefore, the equivalent mean burst length at the master (downlink) queue will be

$$\frac{1}{\overline{B_m}} = \frac{1}{\overline{B}}\left(1 - \frac{1}{(m-1)^2}\right) + 1 \cdot \frac{1}{(m-1)^2} \tag{3.20}$$

In order to maintain the same server utilization under decreased burst length, the burst arrival rate has to be scaled so that $\lambda_{idm}\overline{B_m} = \lambda_{id}\overline{B}$.

The queueing delay at the piconet master may be described with an expression similar to (3.18), except that λ_{id} is replaced with λ_{idm}, and mean burst length is replaced with $\overline{B_m}$. The LST of the end-to-end delay from slave i to slave j is $W_{ije}^*(s) = W_{ai}^*(s)W_{dj}^*(s)$ and, consequently, the mean value of the end-to-end delay is $\overline{W_{ije}} = \overline{W_{ai}} + \overline{W_{dj}}$.

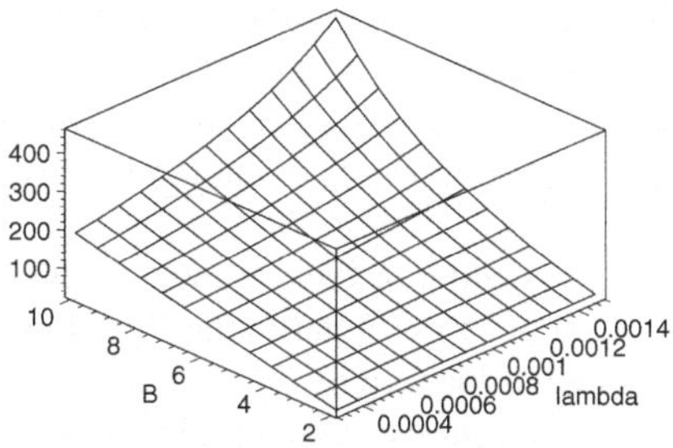

(a) Limited polling.

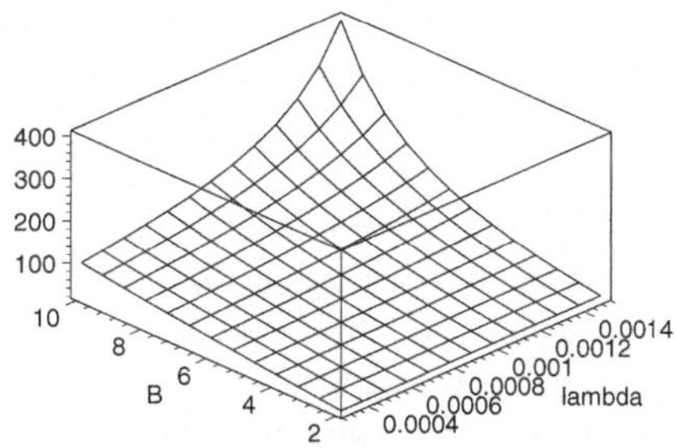

(b) Exhaustive polling.

FIGURE 3.4
End-to-end delay (in units of $T = 625\mu s$) vs. packet burst arrival rate λ and mean burst size $\overline{B}$. (From J. Mišić and V. B. Mišić, "Modeling Bluetooth piconet performance," *IEEE Comm. Lett.* **7**(1):18–20, © 2003 IEEE. Reprinted with permission.)

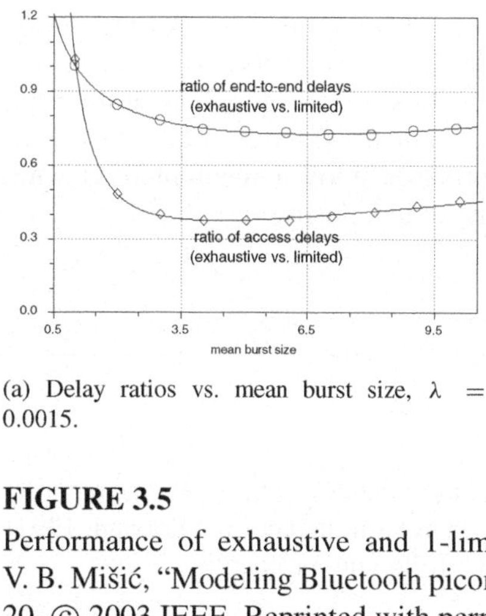

(a) Delay ratios vs. mean burst size, $\lambda = 0.0015$.

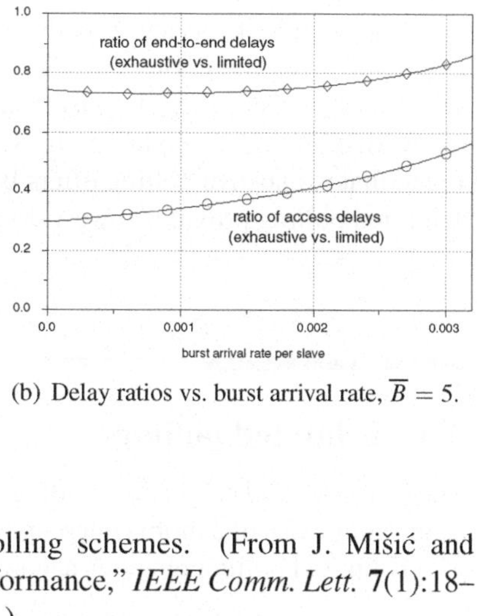

(b) Delay ratios vs. burst arrival rate, $\overline{B} = 5$.

FIGURE 3.5
Performance of exhaustive and 1-limited polling schemes. (From J. Mišić and V. B. Mišić, "Modeling Bluetooth piconet performance," *IEEE Comm. Lett.* **7**(1):18–20, © 2003 IEEE. Reprinted with permission.)

Performance comparison

We have performed both theoretical analysis as presented in the previous Section, and simulation using the simulator built using the Artifex object-oriented Petri Net simulation engine by RSoft-Design, Inc. [RSoft Design, Inc., 2003]. We analyzed the piconet with a maximum size of $m = 8$, with geometrically distributed packet burst size. Mean packet length was $\overline{L} = 3$, and $p_1 = p_2 = p_3 = 1/3$.

Analytical results for end-to-end delays as functions of burst arrival rate (which was the same for all slaves, $\lambda_{iu} = \lambda$) and mean burst size $\overline{B}$ are shown in Figs. 3.4(a) and 3.4(b), for 1-limited and exhaustive polling, respectively. Delays are expressed in Bluetooth time slots $T = 625\mu s$. Burst arrival rates are scaled so as not to exceed the piconet capacity: for example, the maximum values of λ in Fig. 3.4 correspond to offered load of $\rho = m(\lambda_{iu} + \lambda_{id})\overline{B}\,\overline{L} = 0.72$. Ratios of delays are shown in Figs. 3.5(a) and 3.5(b), where lines denote analytical solutions and diamonds correspond to simulation results.

As can be seen, exhaustive service clearly performs better in a wide range of traffic parameters, the difference being in the range of 20 to 25%. In particular, the shape of Fig. 3.4(b) suggests that exhaustive service is less influenced by bursty traffic than its 1-limited counterpart. At very high packet arrival rates and/or long burst sizes, this difference tends to diminish or even disappear. It should be noted that the case with very high traffic load has little practical value, as the end-to-end delays will be unacceptably long anyway.

Similar results have been reported by others. For example, Johansson et al. [1999] have concluded that limited polling offers lower waiting time than the exhaustive service one, especially under medium to heavy traffic load. On the other hand, Kalia et al. [1999] and Kalia, Bansal and Shorey [2000] have shown that more sophisticated polling schemes such as exhaustive service can improve performance compared to the simple limited service, round robin polling. However, their work relies on the assumption that the master is aware of the state of slaves' queues, so that optimum scheduling decisions can be made; such assumption is not appropriate at the Bluetooth MAC level, as explained in Section 2.2. Finally, Capone et al. [2001] have found that exhaustive polling offers best performance at low to medium loads, while 1-limited polling provides lowest overall delays at high loads.

3.3 E-limited polling

The 1-limited and exhaustive polling schemes, as explained before, may be regarded as limiting cases of a more general policy known as E-limited polling [Takagi, 1991]. The E-limited polling operates according to the following protocol:

1. The master polls the slave by sending data packets from the corresponding downlink queue, or empty (POLL) packets if the downlink queue is empty.

2. The slave responds to each downlink packet by sending the data packet from its uplink queue, or an empty (NULL) packet if the uplink queue is empty.

3. The exchange ends after exactly M frames have been exchanged. The exchange may end earlier if both downlink and uplink queues are empty, i.e., when the master sends a POLL packet and the slave responds with a NULL packet.

4. After the end of the exchange, the master moves on to poll the next slave; when all the slaves have been visited, the sequence is cyclically repeated.

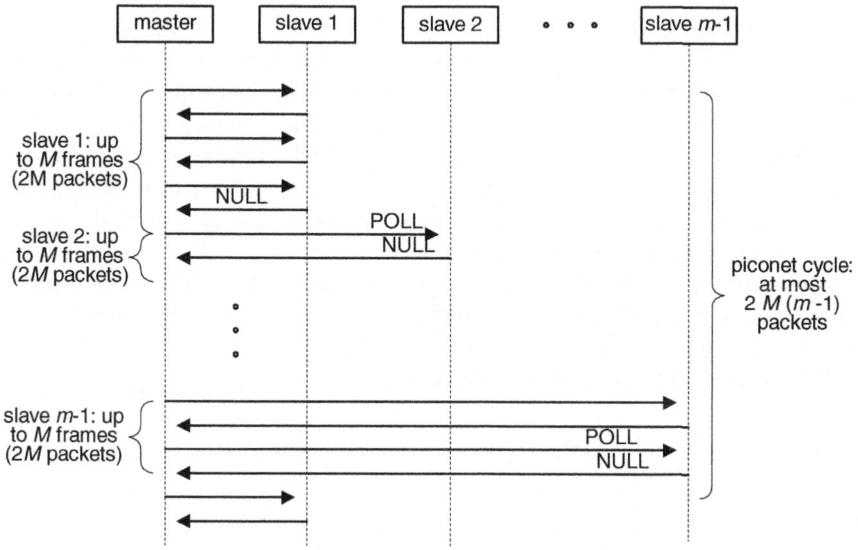

FIGURE 3.6
Timing diagram of E-limited polling.

The timing diagram of E-limited polling is shown in Fig. 3.6; again, the packet length is not taken into account. We will now proceed to analyze the performance of the piconet operating under E-limited polling using the queueing model from Section 2.4.

Queue length distributions

Thanks to the symmetry of the piconet with respect to the slaves, it suffices to consider the packet exchange between a single slave, say i, and the master. As before, we use the concept of master vacation. The number of packets at the uplink queue of a slave and the corresponding downlink queue at the master can be modeled with a set of imbedded Markov points [Takagi, 1991]. The Markov points correspond to vacation termination times – the times immediately before the master sends a first downlink packet to the slave – and frame service completion times – the times when the polled slave finishes one uplink packet transmission.

Let $q^i_{k_u,k_d}$ denote the joint probability that a Markov point in the uplink queue of slave i is a vacation termination time. Let us assume that, at the vacation termination time, there are $k_u = 0, 1, 2 \ldots$ packets at the uplink queue of the i-th slave and $k_d = 0, 1, 2 \ldots$ packets at the downlink queue corresponding to slave i. Also, let $\pi^{i,(\mu)}_{k_u,k_d}$ denote the joint probability that a Markov point is the μ-th frame transmission completion time and that there are k_u packets in the slave's i queue at that time, and k_d packets at the master's queue toward the slave i, where $\mu = 1 \,..\, M$ and

$k_d, k_u = 0, 1, 2, \ldots$.

Let $f_{is}(x)$ and $v_i(x)$ stand for the pdf-s (probability density functions) of the frame transmission time and vacation time, respectively, at the uplink queue of slave i; their LST transforms will be $F_{is}^*(s)$ and $V_i^*(s)$. We will also make use of the following probabilities: the probability of k_u packet arrivals at the slave i's uplink queue during the frame time, which will be denoted with a_{k_u}; the probability of k_d packet arrivals at the master's downlink queue during the frame time, which will be denoted with a_{k_d}; the probability of k_u packet arrivals at the slave i's uplink queue during the vacation time (i.e., while the master is servicing other slaves), which will be denoted with f_{k_u}; and the probability of k_d packet arrivals in the master's downlink during the vacation time, which will be denoted with f_{k_d}.

Those probabilities may be calculated as

$$\begin{aligned}
a_{k_u} &= \sum_{l=0}^{\infty} \frac{1}{k_u!} \frac{d^{k_u}}{dz^{k_u}} (G_b(z))^l \Big|_{z=0} \int_0^{\infty} \frac{(\lambda_{iu} x)^l}{l!} e^{-\lambda_{iu} x} f_{is}(x) dx \\
&= \frac{1}{k_u!} \frac{d^{k_u}}{dz^{k_u}} (F_{is}^*(\lambda_{iu} - \lambda_{iu} G_b(z)))\Big|_{z=0} \\
a_{k_d} &= \sum_{l=0}^{\infty} \frac{1}{k_d!} \frac{d^{k_d}}{dz^{k_d}} (G_b(z))^l \Big|_{z=0} \int_0^{\infty} \frac{(\lambda_{id} x)^l}{l!} e^{-\lambda_{id} x} f_{is}(x) dx \\
&= \frac{1}{k_d!} \frac{d^{k_d}}{dz^{k_d}} (F_{is}^*(\lambda_{id} - \lambda_{id} G_b(z)))\Big|_{z=0} \\
f_{k_u} &= \sum_{l=0}^{\infty} \frac{1}{k_u!} \frac{d^{k_u}}{dz^{k_u}} (G_b(z))^l \Big|_{z=0} \int_0^{\infty} \frac{(\lambda_{iu} x)^l}{l!} e^{-\lambda_{iu} x} v_i(x) dx \\
&= \frac{1}{k_u!} \frac{d^{k_u}}{dz^{k_u}} V_i^*(\lambda_{iu} - \lambda_{iu} G_b(z))\Big|_{z=0} \\
f_{k_d} &= \sum_{l=0}^{\infty} \frac{1}{k_d!} \frac{d^{k_d}}{dz^{k_d}} (G_b(z))^l \Big|_{z=0} \int_0^{\infty} \frac{(\lambda_{iu} x)^l}{l!} e^{-\lambda_{iu} x} v_i(x) dx \\
&= \frac{1}{k_d!} \frac{d^{k_d}}{dz^{k_d}} V_i^*(\lambda_{iu} - \lambda_{iu} G_b(z))\Big|_{z=0}
\end{aligned} \tag{3.21}$$

Note that $F_{is}^*(\lambda_{iu} - \lambda_{iu} G_b(z))$ and $V_i^*(\lambda_{iu} - \lambda_{iu} G_b(z))$ denote the PGFs for the number of packet arrivals in the uplink queue during the frame time and vacation time, respectively:

$$\begin{aligned}
A_{F,u}(z) &= \sum_{k_u=0}^{\infty} a_{k_u} z^{k_u} = \sum_{k_u=0}^{\infty} \frac{z_u^k}{k_u!} \frac{d^{k_u}}{dz^{k_u}} F_{is}^*(\lambda_{iu} - \lambda_{iu} G_b(z))\Big|_{z=0} \\
&= F_{is}^*(\lambda_{iu} - \lambda_{iu} G_b(z)) \\
A_{V,u}(z) &= \sum_{k_u=0}^{\infty} f_{k_u} z^{k_u} = \sum_{k_u=0}^{\infty} \frac{z^{k_u}}{k_u!} \frac{d^{k_u}}{dz^{k_u}} V_i^*(\lambda_{iu} - \lambda_{iu} G_b(z))\Big|_{z=0} \\
&= V_i^*(\lambda_{iu} - \lambda_{iu} G_b(z))
\end{aligned} \tag{3.22}$$

Similar expressions hold for the number of arrivals in the downlink queue. The probabilities that the uplink queue of slave i contains k_u packets and that the downlink queue toward slave i contains k_d packets in imbedded Markov points, satisfy the following equations:

$$
\begin{aligned}
\pi^{i,(1)}_{k_u,k_d} &= \sum_{j_u=1}^{k_u+1}\sum_{j_d=1}^{k_d+1} q^i_{j_u,j_d} a_{k_u-j_u+1} a_{k_d-j_d+1} + \sum_{j_d=1}^{k_d+1} q^i_{0,j_d} a_{k_d-j_d+1} a_{k_u} \\
&+ \sum_{j_u=1}^{k_u+1} q^i_{j_u,0} a_{k_u-j_u+1} a_{k_d} \\
\pi^{i,(\mu)}_{k_u,k_d} &= \sum_{j_u=1}^{k_u+1}\sum_{j_d=1}^{k_d+1} \pi^{i,(\mu-1)}_{j_u,j_d} a_{k_u-j_u+1} a_{k_d-j_d+1} + \sum_{j_d=1}^{k_d+1} \pi^{i,(\mu-1)}_{0,j_d} a_{k_d-j_d+1} a_{k_u} \\
&+ \sum_{j_u=1}^{k_u+1} \pi^{i,(\mu-1)}_{j_u,0} a_{k_u-j_u+1} a_{k_d}, \qquad \mu = 2\,..\,M \\
q^i_{k_u,k_d} &= \left(\sum_{\mu=1}^{M-1} \pi^{i,(\mu)}_{0,0} + q^i_{0,0}\right) f_{k_u} f_{k_d} + \sum_{j_u=0}^{k_u}\sum_{j_d=0}^{k_d} \pi^{i,(M)}_{j_u,j_d} f_{k_u-j_u} f_{k_d-j_d}
\end{aligned} \tag{3.23}
$$

The probability generating functions (PGFs) for the number of packets in the uplink and the corresponding downlink queue, in the imbedded Markov points, are defined by

$$
\begin{aligned}
\Pi_{i,\mu}(z,w) &= \sum_{k_u=0}^{\infty}\sum_{k_d=0}^{\infty} \pi^{i,(\mu)}_{k_u,k_d} z^{k_u} w^{k_d}, \qquad \mu = 1\,..\,M \\
Q_i(z,w) &= \sum_{k_u=0}^{\infty}\sum_{k_d=0}^{\infty} q^i_{k_u,k_d} z^{k_u} w^{k_d}
\end{aligned} \tag{3.24}
$$

which may be written as

$$
\begin{aligned}
\Pi_{i,1}(z,w) &= \sum_{k_u=0}^{\infty}\sum_{k_d=0}^{\infty} z^{k_u} w^{k_d} \sum_{j_u=1}^{k_u+1}\sum_{j_u=1}^{k_u+1} q^i_{j_u,j_d} a_{k_u-j_u+1} a_{k_d-j_d+1} \\
&+ \sum_{k_u=0}^{\infty}\sum_{k_d=0}^{\infty} z^{k_u} w^{k_d} \sum_{j_d=1}^{k_d+1} q^i_{0,j_d} a_{k_d-j_d+1} a_{k_u} \\
&+ \sum_{k_u=0}^{\infty}\sum_{k_d=0}^{\infty} z^{k_u} w^{k_d} \sum_{j_u=1}^{k_u+1} q^i_{j_u,0} a_{k_u-j_u+1} a_{k_d}
\end{aligned} \tag{3.25}
$$

continued on next page ...

... continued from previous page

$$\begin{aligned}
\Pi_{i,\mu}(z,w) &= \sum_{k_u=0}^{\infty}\sum_{k_d=0}^{\infty} z^{k_u} w^{k_d} \sum_{j_u=1}^{k_u+1}\sum_{j_u=1}^{k_u+1} \pi_{j_u,j_d}^{i,(\mu-1)} a_{k_u-j_u+1} a_{k_d-j_d+1} \\
&+ \sum_{k_u=0}^{\infty}\sum_{k_d=0}^{\infty} z^{k_u} w^{k_d} \sum_{j_d=1}^{k_d+1} \pi_{0,j_d}^{i,(\mu-1)} a_{k_d-j_d+1} a_{k_u} \\
&+ \sum_{k_u=0}^{\infty}\sum_{k_d=0}^{\infty} z^{k_u} w^{k_d} \sum_{j_u=1}^{k_u+1} \pi_{j_u,0}^{i,(\mu-1)} a_{k_u-j_u+1} a_{k_d},\ \mu = 2\,..\,M \qquad (3.25) \\
Q_i(z,w) &= \sum_{k_u=0}^{\infty}\sum_{k_d=0}^{\infty} z^{k_u} w^{k_d} \left(\sum_{\mu=1}^{M-1} \pi_{0,0}^{i,(\mu)} + q_{0,0}^{i}\right) f_{k_u} f_{k_d} \\
&= \sum_{k_u=0}^{\infty}\sum_{k_d=0}^{\infty} z^{k_u} w^{k_d} \sum_{j_u=0}^{k_u}\sum_{j_d=0}^{k_d} \pi_{j_u,j_d}^{i,(M)} f_{k_u-j_u} f_{k_d-j_d}
\end{aligned}$$

After exchanging the order of summation between the pairs (k_u, j_u) and (k_d, j_d), equations (3.24) may be simplified to

$$\begin{aligned}
\Pi_{i,1}(z,w) &= \frac{F_{is}^*(\lambda_{iu} - \lambda_{iu}G_b(z))F_{is}^*(\lambda_{id} - \lambda_{id}G_b(w))}{zw} \\
&\cdot \Big(Q_i(z,w) - (1-w)Q_i(z,0) - (1-z)Q_i(0,w) + q_{0,0}^{i}(1-z-w)\Big) \\
\Pi_{i,\mu}(z,w) &= \frac{F_{is}^*(\lambda_{iu} - \lambda_{iu}G_b(z))F_{is}^*(\lambda_{id} - \lambda_{id}G_b(w))}{zw} \\
&\cdot \Big(\Pi_{i,\mu-1}(z,w) - (1-w)\Pi_{i,\mu-1}(z,0) \\
&-(1-z)\Pi_{i,\mu-1}(0,w) + \pi_{0,0}^{i,(\mu-1)}(1-z-w)\Big), \qquad \mu = 2\,..\,M \\
Q_i(z,w) &= V_i^*(\lambda_{iu} - \lambda_{iu}G_b(z))V_i^*(\lambda_{id} - \lambda_{id}G_b(w)) \\
&\cdot \left(\sum_{\mu=1}^{M-1} \pi_{0,0}^{i,(\mu)} + q_{0,0}^{i} + \Pi_{i,M}(z,w)\right)
\end{aligned} \qquad (3.26)$$

When we substitute $z = 0$ in (3.26), we can find $\Pi_{i,1}(0,w) \ldots \Pi_{i,M}(0,w)$ and $Q_i(0,w)$ as functions of $\pi_{0,0}^{i,(\mu)}$ $\mu = 1 \ldots M$ and $q_{0,0}^{i}$. For example, for $z = 0$ the system (3.26) becomes

$$\begin{aligned}
\Pi_{i,1}(0,w) &= \frac{F_{is}^*(\lambda_{iu})F_{is}^*(\lambda_{id} - \lambda_{id}G_b(w))}{w} \cdot \Big(Q_i(0,w) - q_{0,0}^{i}\Big) \\
\Pi_{i,\mu}(0,w) &= \frac{F_{is}^*(\lambda_{iu})F_{is}^*(\lambda_{id} - \lambda_{id}G_b(w))}{w} \cdot \Big(\Pi_{i,\mu-1}(0,w) - \pi_{0,0}^{i,(\mu-1)}\Big), \\
&\qquad \mu = 2\,..\,M
\end{aligned} \qquad (3.27)$$

continued on next page ...

... continued from previous page

$$Q_i(0,w) = V_i^*(\lambda_{iu})V_i^*(\lambda_{id} - \lambda_{id}G_b(w))\left(\sum_{\mu=1}^{M-1}\pi_{0,0}^{i,(\mu)} + q_{0,0}^i + \Pi_{i,M}(0,w)\right) \tag{3.27}$$

From the system (3.27), we find $\Pi_{i,1}(0,w)\ldots\Pi_{i,M}(0,w)$ and $Q_i(0,w)$. In an analogous fashion, we can find $\Pi_{i,1}(z,0)\ldots\Pi_{i,M}(z,0)$ and $Q_i(z,0)$ as functions of $\pi_{0,0}^{i,(\mu)}$, $\mu = 1\ldots M$, and $q_{0,0}^i$. For clarity, we introduce additional substitutions:

$$\begin{aligned} Q_0(z,w) &= (1-w)Q_i(z,0) + (1-z)Q_i(0,w) - q_{0,0}^i(1-z-w) \\ \Pi_{\mu,0}(z,w) &= (1-w)\Pi_{i,\mu}(z,0) \\ &\quad + (1-z)\Pi_{i,\mu}(0,w)\pi_{0,0}^{i,(\mu)}(1-z-w), \qquad \mu = 1\,..\,M \end{aligned} \tag{3.28}$$

and the system (3.26) becomes

$$\begin{aligned} \Pi_{i,1}(z,w) &= \frac{F_{is}^*(\lambda_{iu} - \lambda_{iu}G_b(z))F_{is}^*(\lambda_{id} - \lambda_{id}G_b(w))}{zw} \\ &\quad \cdot (Q_i(z,w) - Q_0(z,w)) \\ \Pi_{i,\mu}(z,w) &= \frac{F_{is}^*(\lambda_{iu} - \lambda_{iu}G_b(z))F_{is}^*(\lambda_{id} - \lambda_{id}G_b(w))}{zw} \\ &\quad \cdot \left(\Pi_{i,\mu-1}(z,w) - \Pi_{\mu-1,0}(z,w)\right) \mu = 2\,..\,M \\ Q_i(z,w) &= V_i^*(\lambda_{iu} - \lambda_{iu}G_b(z))V_i^*(\lambda_{id} - \lambda_{id}G_b(w)) \\ &\quad \cdot \left(\sum_{\mu=1}^{M-1}\pi_{0,0}^{i,(\mu)} + q_{0,0}^i + \Pi_{i,M}(z,w)\right) \end{aligned} \tag{3.29}$$

The solution of this last system gives the expression for $Q_i(z,w)$ in the form

$$Q_i(z,w) = V_i^*(\lambda_{iu} - \lambda_{iu}G_b(z))V_i^*(\lambda_{id} - \lambda_{id}G_b(w))z^M w^M \frac{A}{B} \tag{3.30}$$

where

$$\begin{aligned} A &= \sum_{\mu=1}^{M-1}\pi_{0,0}^{i,(\mu)} + q_{0,0}^i - Q_0(z,w)Y^M - \sum_{\mu=1}^{M-1}Y^{M-\mu}\Pi_{\mu,0}(z,w) \\ B &= z^M w^M - V_i^*(\lambda_{iu} - \lambda_{iu}G_b(z))V_i^*(\lambda_{id} - \lambda_{id}G_b(w)) \\ &\quad \cdot F_{is}^*(\lambda_{iu} - \lambda_{iu}G_b(z))^M F_{is}^*(\lambda_{id} - \lambda_{id}G_b(w))^M \\ Y &= \frac{1}{zw}F_{is}^*(\lambda_{iu} - \lambda_{iu}G_b(z))F_{is}^*(\lambda_{id} - \lambda_{id}G_b(w)) \end{aligned} \tag{3.31}$$

However, the solution of (3.30) requires two more elements to be calculated: namely, the LSTs of the frame time and vacation time distributions. Let us find the LST of the channel service time first; this is the time from the moment when the master polls the slave for the first time until either an empty frame has been encountered, or a total of M data frames have been exchanged. (As usual, this time is expressed in time slots T.) This service time can take from one up to M frames. The LSTs of the length of k-th data frame without the POLL/NULL packet pair is

$$\begin{aligned}
F^{*1}(s) &= \frac{\left(Q_i(1,0)+Q_i(0,1)-2q^i_{0,0}\right)}{Q_i(1,1)} G^*_p(s)e^{-s} \\
&+ \frac{(Q_i(1,1)-Q_i(1,0)-Q_i(0,1)+q^i_{0,0})}{Q_i(1,1)} G^*_p(s)^2 \\
F^{*\mu}(s) &= \frac{\left(\Pi_{i,\mu-1}(1,0)+\Pi_{i,\mu-1}(0,1)-2\pi^{i,(\mu-1)}_{0,0}\right)}{\Pi_{i,\mu-1}(1,1)} G^*_p(s)e^{-s} \\
&+ \frac{\left(\Pi_{i,\mu-1}(1,1)-\Pi_{i,\mu-1}(1,0)-\Pi_{i,\mu-1}(0,1)+\pi^{i,(\mu-1)}_{0,0}\right)}{\Pi_{i,\mu-1}(1,1)} G^*_p(s)^2, \\
&\qquad \mu = 2\,..\,M
\end{aligned} \tag{3.32}$$

In the last expression, $Q_i(1,1)$ denotes the probability that a given Markov point corresponds to the end of vacation for slave i. Therefore, the conditional probability that both the uplink and downlink queue are empty at the end of vacation is $\dfrac{q^i_{0,0}}{Q_i(1,1)}$. Next, we observe that the probability that master-to-slave transmission will take k data frames is

$$\begin{aligned}
P_{f,0} &= \frac{q^i_{0,0}}{Q_i(1,1)} \\
P_{f,1} &= \frac{(Q_i(1,1)-q^i_{0,0})}{Q_i(1,1)} \cdot \frac{\pi^{i,(1)}_{0,0}}{\Pi_{i,1}(1,1)} \\
P_{f,k} &= \frac{(Q_i(1,1)-q^i_{0,0})}{Q_i(1,1)} \prod_{\mu=1}^{k-1} \frac{(\Pi_{i,\mu}(1,1)-\pi^{i,(\mu)}_{0,0})}{\Pi_{i,\mu}(1,1)} \frac{\pi^{i,(k)}_{0,0}}{\Pi_{i,k}(1,1)}, \quad k = 2\,..\,M-1 \\
P_{f,M} &= 1 - \sum_{k=0}^{M-1} P_{f,k}
\end{aligned} \tag{3.33}$$

Then, the LST for the master-slave channel service time is

$$S^*_i(s) = \sum_{k=0}^{M-1} P_{f,k} \prod_{\mu=1}^{k} (F^{*\mu}(s))e^{-2s} + P_{f,M} \prod_{\mu=1}^{M} F^{*\mu}(s) \tag{3.34}$$

The uplink and downlink channel service times, $S^*_{i,u}(s)$ and $S^*_{i,d}(s)$, respectively, may be determined in a similar fashion.

Given that piconet has $2 \le m \le 8$ members, one of which is the master, the PGF for the vacation time observed by the slave i is:

$$V^*_i(s) = \prod_{\substack{j=2\\ j\neq i}}^{m} S^*_j(s) \tag{3.35}$$

The first and second moments of vacation time are equal to $\overline{V_i} = V_i'(1)$ and $\overline{V_i^2} = V_i''(1)+V_i'(1)$, respectively. The LST for the cycle time of the piconet, then, becomes

$$C^*(s) = \prod_{i=2}^{m} S_i^*(s) \tag{3.36}$$

A given frame will contain two data packets when the uplink and downlink queues are not empty, one data and one empty (POLL or NULL) packet when one of the queues is empty, or two empty (POLL and NULL) packets when both queues are empty. Given that the LST of a single-slot packet is e^{-s}, the LST for the frame time during the exchange between the master and the slave i is

$$\begin{aligned} F_{is}^*(s) &= PQe^{-2s} \\ &+ (Q_i(0,1)+ Q_i(1,0) + SP - 2PQ)\, G_p^*(s)e^{-2s} \\ &+ (1 - Q_i(0,1) - Q_i(1,0) - SP + PQ)\, (G_p^*(s))^2 \end{aligned} \tag{3.37}$$

where, for simplicity, we have used the notation

$$PQ = \sum_{\mu=1}^{M} \pi_{0,0}^{i,(\mu)} + q_{0,0}^{i}$$

and

$$SP = \sum_{\mu=1}^{M} \left(\Pi_{i,\mu}(0,1) + \Pi_{i,\mu}(1,0)\right)$$

Expressions (3.35) and (3.37) should be substituted in (3.30), which then depends on the packet arrival process for the given slave, as well as on the values $\pi_{0,0}^{i,(\mu)}$ and $q_{0,0}^{i}$. The latter two can be found from the marginal PGF $Q_i(z,1)$ (alternatively, $Q_i(1,w)$ could be used instead), by making use of the fact that $Q_i(z,1)$ must be an analytic function for all $|z| \in (0,1)$. Therefore, its numerator and denominator must have identical roots. The number of roots of the denominator can be determined by Rouché's theorem [Bak and Newman, 1982] and it is equal to M. Obviously, $z_0 = 1$ is one of the roots, while the remaining $M-1$ of them can be determined using Lagrange's theorem [Whittaker and Watson, 1952]:

$$z_j = \sum_{n=1}^{\infty} \frac{e^{2\pi jn\sqrt{-1}/M}}{n!} \cdot \frac{d^{n-1}}{dz^{n-1}} \left(V_i^*(\lambda_{iu} - \lambda_{iu}G_b(z))F_{is}^*(\lambda_{iu} - \lambda_{iu}G_b(z))\right)^{n/M}\Big|_{z=0} \tag{3.38}$$

where $j = 1 \,..\, M-1$ denotes the index of the root in question. In practice it is possible to truncate the sum (3.38) to the first few members only. The solutions thus obtained may contain a small imaginary part which can safely be ignored.

When the $M-1$ roots are substituted in the numerator of (3.30), we obtain a total of $M-1$ equations for the given slave i, with unknowns $q^i_{0,0}$ and $\pi^{i,(k)}_{0,0}$:

$$\left(1-\left(\frac{F^*_{is}(\lambda_{iu}-\lambda_{iu}G_b(z_j))}{z_j}\right)^M\right)q^i_{0,0}+ \sum_{k=1}^{M-1}\left(1-\left(\frac{F^*_{is}(\lambda_{iu}-\lambda_{iu}G_b(z_j))}{z_j}\right)^{M-k}\right)\pi^{i,(k)}_{0,0}=0 \tag{3.39}$$

where $k=1\,..\,M-1$.

The M-th equation is obtained from the condition $Q(1,1)+\sum_{\mu=1}^{M}\Pi_{i,\mu}(1,1)=1$, and it reads:

$$Mq^i_{0,0}+\sum_{k=1}^{M-1}(M-k)\pi^{i,(k)}_{0,0}=\frac{M(1-\lambda_{iu}\overline{F_{is}B})-\lambda_{iu}\overline{BV_i}}{1-\lambda_{iu}\overline{F_{is}B}+\lambda_{iu}\overline{BV_i}} \tag{3.40}$$

Solving the system

As we see, finding the distribution of the number of packets in the uplink or downlink queue upon the master's return from the vacation requires that we know the probability distribution of the vacation time, and vice versa. In order to break the recursion we have applied an iterative approach. Let us introduce marginal mass probabilities of queue lengths as

$$\begin{aligned} q^i_{j_u} &= \sum_{j_d=0}^{\infty} q^i_{j_u,j_d}\\ q^i_{j_d} &= \sum_{j_u=0}^{\infty} q^i_{j_u,j_d}\\ \pi^{i,(j_u)}_{ju} &= \sum_{j_d=0}^{\infty} \pi^{i,(j_u)}_{ju,jd}\\ \pi^{i,(\mu)}_{jd} &= \sum_{j_u=0}^{\infty} \pi^{i,(\mu)}_{ju,jd} \end{aligned} \tag{3.41}$$

The algorithm for solving the system is as follows.

Iteration 0
1. Consider the marginal probability distributions for uplink queues. Assume that $Q^0_i(1,1)=0.9$ and that $q^i_{j_u}=\pi^{i,(j_u)}_{ju}$, $j_u=1\,..\,M-1$. The same process holds for the marginal probability distribution for downlink queue.
2. Calculate the distribution of service time and vacation time for each slave.

3. Solve the system which consists of $M-1$ equations of the type (3.39) and one of the type (3.40) for each queue. (The overall system consists of $2M(m-1)$ equations).
4. Calculate the new values of $Q_i(z,1)$ and find $Q_i(1,1)$. Also, find $q^i_{j_u}$, for $j_u = 0\,..\,M-1$.

Iterations 1 .. k 1. Calculate the new distributions of the service and vacation times, using $Q_i(1,1)$ and values $q^i_{j_u}$ and $j = 0\,..\,M-1$ from the previous iteration.

2. Solve again the system of equations that consists of $2(m-1)$ instances of the system (3.39) and (3.40).
3. With the solutions, calculate $Q_i(z,1)$ and so on.

We have found that good accuracy can be obtained even with a small number of iterations, typically two to three. However, the calculations are rather complex, and solving the entire system of equations may not be the best solution; other ways to estimate the service times of the slaves should be investigated.

Approximating the service times

From (3.34), two observations can be made regarding the interaction among slaves.

First, in case all slaves have identical packet arrival rates (i.e., the piconet load is symmetric), the probabilities that the uplink queue will be empty upon returning from a vacation, $\dfrac{q^i_0}{Q_i(1,1)} = \sum_{j_d=0}^{\infty} q^i_{0,j_d}/Q_i(1,1)$, will be identical. Under asymmetric loads, one might expect the corresponding probabilities to differ: those of slaves with higher arrival rates (where more packets arrive in a given time interval) should be smaller than those of slaves with lower arrival rates.

However, this is not quite the case, because of the feedback effect – the values of $q^i_0/Q_i(1,1)$ depend on the arrival rates of all the other slaves through their service times. For example, consider one of the slaves and its packet burst arrival rate. If the other slaves have smaller arrival rates, their service times will be smaller, which will make the vacation time for the target slave small. This will, so to speak, increase the value of $q^i_0/Q_i(1,1)$ for the target slave beyond its 'expected' value. On the other hand, if the other slaves have larger burst arrival rates, they will also have larger service times, and the vacation time for the target slave will be larger, leading to the decrease in their $q^i_0/Q_i(1,1)$. In fact, for a piconet with two or more slaves ($m \geq 3$), the values of $q^i_0/Q_i(1,1)$ for different slaves ($i = 2\,..\,m$) are approximately equal to each other, even though the corresponding packet burst arrival rates in the uplink direction may be different.

The second observation, which is in part a consequence of the first one, is related to the dependency of $q^i_0/Q_i(1,1)$ on the number of slaves in the piconet. This probability depends on the vacation time observed by the target slave, which in turn depends on the number of transmitted packets and, indirectly, on the arrival rates of the surrounding slaves, rather than their number (which will contribute only to the number of POLL/NULL packets). Therefore, the probability $q_0^i/Q_i(1,1)$ for any slave i will mostly depend on the total load of the piconet, rather than on the packet arrival rates of this individual slave, or its peers.

By extension, the probability $q_0^i/Q_i(1,1)$ mostly depends on the sum of uplink burst arrival rates, rather than the number of slaves and their individual loads. Since the total load on the uplink is the same as one in the downlink, we may also conclude that these observations hold in the downlink direction as well. Similar observations can be made for the other probabilities, i.e., $q_1^i/Q_i(1,1) \ldots q_{M-1}^i/Q_i(1,1)$, which are nearly independent of the load differences among the slaves, and $q_1^i/Q_i(1,1) \ll q_0^i/Q_i(1,1)$.

The property described above has been verified using our Artifex-based simulator; the results of the simulation are shown in Fig. 3.7. The variable slave load is achieved by assigning different (uplink) packet arrival rates to individual slaves. Let λ_{mean} stand for the arithmetic mean of all slaves' uplink packet arrival rates. Then, the load variation of 0 corresponds to symmetric load (all arrival rates are equal, $\lambda_{iu} = \lambda_{mean}$), while the load variation of 0.5 corresponds to uniformly distributed arrival rates in the range $(0.5\lambda_{mean}, 1.5\lambda_{mean})$.

Furthermore, since all the slaves affect each other through their vacation times, it follows that, for $m > 3$, the average cycle time depends on the sum of arrival rates of all the slaves and not on the individual arrival rates of the slaves, provided that differences in arrival rates are not high, say, within 80% of each other, which includes most cases of practical importance.

This property may be verified by observing the dependency of piconet cycle time on the total piconet load, number of slaves, and variation of load among the slaves, obtained by simulations, which is shown in Fig. 3.8.

In order to arrive at more tractable forms of uplink channel service time, the following approximation was found to give a good match for the case of symmetric traffic:

$$\begin{aligned}\overline{S_i^{(s)}} &\approx M\overline{L}\left(1 - \frac{q_0^i}{Q_i(1,1)}\right) + \frac{q_0^i}{Q_i(1,1)} \\ &\quad - \overline{L}\left(1 - \frac{q_0^i}{Q_i(1,1)} - \rho_{tot}^M\right)(G_p^*(\lambda_{iu}))^{2(M-1)}\end{aligned} \tag{3.42}$$

where $\rho_{tot} = 2\overline{L}\,\overline{B}\sum_{i=2}^{m}\lambda_{iu}$ denotes the total piconet load. In case of asymmetric traffic, the approximation has to include an additional correction factor, i.e.,

$$\overline{S_i^{(a)}} \approx \overline{S_i^{(s)}}\sqrt{(m-2)\lambda_{iu}/\sum_{\substack{j=2\\ j\neq i}}^{m}\lambda_{ju}} \tag{3.43}$$

Analogous expressions may be written for the downlink service times.

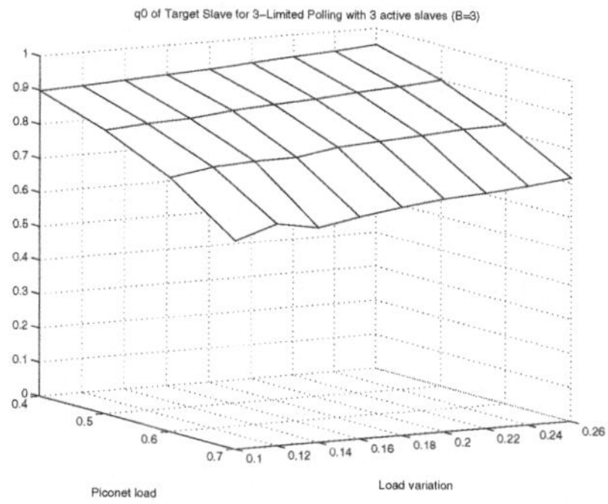

(a) $q_0^i/Q_i(1,1)$ for $m = 4$ (3 active slaves).

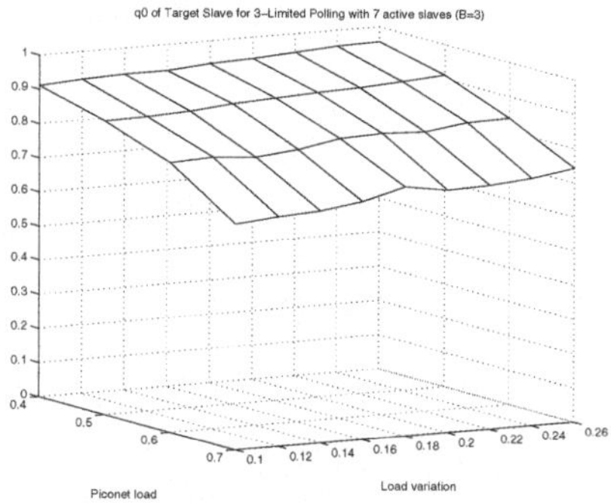

(b) $q_0^i/Q_i(1,1)$ for $m = 8$ (7 active slaves).

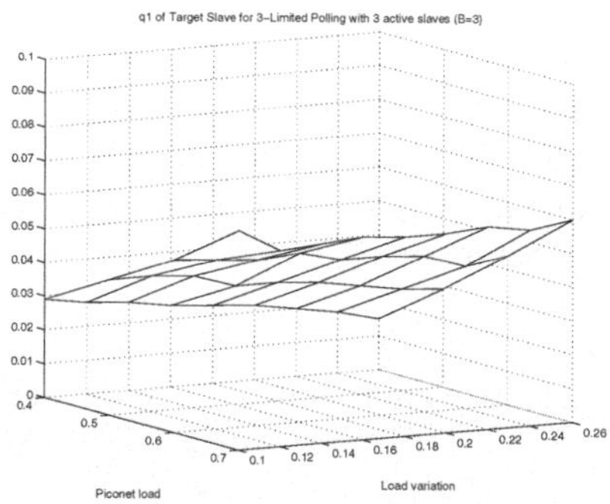

(c) $q_0^i/Q_i(1,1)$ for $m = 4$ (3 active slaves).

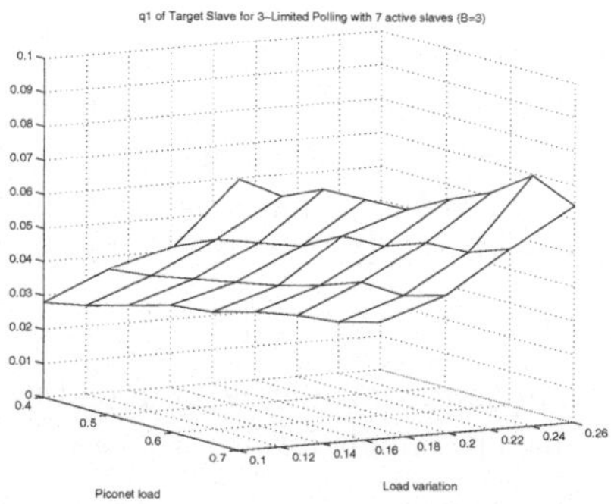

(d) $q_0^i/Q_i(1,1)$ for $m = 8$ (7 active slaves).

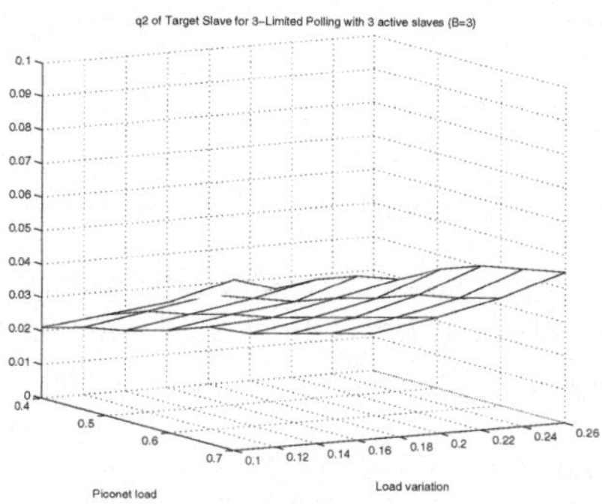

(e) $q_0^i/Q_i(1,1)$ for $m = 4$ (3 active slaves).

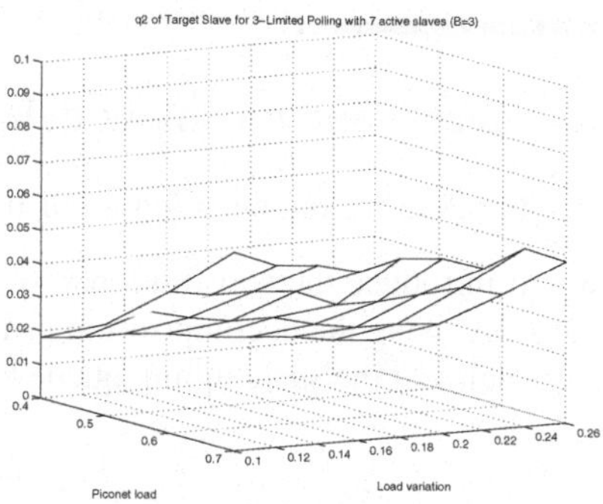

(f) $q_0^i/Q_i(1,1)$ for $m = 8$ (7 active slaves).

FIGURE 3.7

Probabilities that the slave uplink queue contains 0, 1, or 2 packets upon return from the vacation as functions of the total piconet load and variation of load among the slaves; in all cases, $M = 3$, $\overline{B} = 3$. (From J. Mišić, K. L. Chan, and V. B. Mišić, "Admission control in Bluetooth piconets," *IEEE Trans. Veh. Tech.* **53**(3):890–911, © 2004 IEEE. Reprinted with permission.)

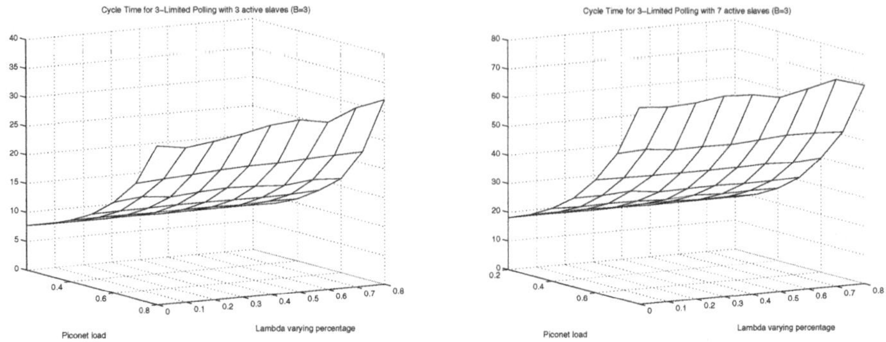

(a) Mean cycle time for $m = 4$ (3 active slaves). (b) Mean cycle time for $m = 8$ (7 active slaves).

FIGURE 3.8
Mean cycle time $\overline{C}$ as a function of the total piconet load and variation of load among the slaves; in all cases, $M = 3$ and $\overline{B} = 3$. (From J. Mišić, K. L. Chan, and V. B. Mišić, "Admission control in Bluetooth piconets," *IEEE Trans. Veh. Tech.* **53**(3):890–911, © 2004 IEEE. Reprinted with permission.)

Both approximations were found to give results within 10 to 15% of the values obtained by simulation.

3.4 Access and downlink delay

Since the PGF for the burst length distribution is geometric, $G_b(z) = \dfrac{z}{\overline{B} + z - z\overline{B}}$, we will introduce the substitution $s = \lambda_{iu} - \lambda_{iu}z/(\overline{B} + z - z\overline{B})$ in the expression (3.30). By using the decomposition principle [Takagi, 1991], the LST for the packet access delay at the slave uplink queue becomes:

$$
\begin{aligned}
W_{ai}^*(s) = & \frac{s(1 - \lambda_{iu}\overline{F_{is}B})}{s - \lambda_{iu} + \lambda_{iu}G_b(F_{is}^*(s))} \cdot \frac{1 - G_b(F_{is}^*(s))}{\overline{B}(1 - F_{is}^*(s))} \\
& \cdot \frac{1 - V_i^*(s)}{s\overline{V_i}} \cdot \frac{Q_i\left(1 - \dfrac{s}{\lambda_{iu}\overline{B} - s\overline{B} + s}, 1\right)}{Q_i(1, 1)V_i^*(s)}
\end{aligned}
\quad (3.44)
$$

The first term in this expression corresponds to the time needed to service the first packet in the burst in the $M^{[x]}/G/1$ system. The second term corresponds to the time needed to service the given target packet in the burst. The third term corresponds to the time needed to service packets which arrive during the vacation, but before the target burst. Finally, the last term corresponds to time needed to service packets which were already in the uplink queue when the vacation has started.

Mean value of the access delay is obtained as $\overline{W_{ai}} = -W_{ai}^{*'}(0)$, which amounts to

$$\overline{W_{ai}} = \frac{\lambda_{iu}\overline{B}(\overline{F_{is}^2})}{(1-\lambda_{iu}\overline{F_{is}B})} + \frac{\overline{B^{(2)}2}}{2\overline{B}(1-\lambda_{iu}\overline{F_{is}B})} + \frac{\overline{V_i^2}}{2\overline{V_i}} - \overline{V_i} + \frac{Q_i^{'}(1,1)}{\lambda_{iu}\overline{B}Q_i(1,1)} \tag{3.45}$$

where $\overline{V_i^2} = V_i^{*''}(0)$. Both analytical and simulation results for the access delay are shown in Fig. 3.9. Two properties of the mean access delay may be deduced from the last expression:

- Under constant offered load $\rho_i = \lambda_{iu}\overline{F_{is}B}$, and under fixed value of M, the access delay will increase with the mean burst size $\overline{B}$. This increase is due to the increase of the second factorial moment of the burst (second term) and the increased number of packets in the uplink queue at the beginning of the vacation (last term).
- Under constant offered load and fixed mean burst size $\overline{B}$, the mean access delay decreases when the value of M increases. This is due to the decreased number of packets in the uplink queue at the beginning of vacation (fourth and fifth terms in the last expression).

Simulation results confirm these observations, as can be seen from Fig. 3.10(a).

The downlink delay is calculated in a similar way. The downlink queue that corresponds to the given slave is actually fed by packets from all other uplink queues, and the exact analysis of its operation is rather involved. Instead, we use a simplified approach in which the downlink queues operate independently of the uplink queues, but their traffic parameters differ from those of the uplink queues.

First, the downlink burst size is smaller than that in the uplink, because the downlink queue may contain interleaved slices of packet bursts from different slaves. We will model this effect by modifying the parameter of the geometric distribution used to model the uplink queue. Let the uplink burst size be determined by the PGF:

$$G_b(z) = \frac{zp_b}{1-z+zp_b} \tag{3.46}$$

where p_b is the parameter of the geometric distribution; the mean burst size in the uplink queue is $\overline{B} = 1/p_b$. In the downlink direction, the p_b parameter will not change if the source slave is the only one which transmits (data packets) in a particular cycle, or if the source slave is the only one that sends packets to the given target slave. In all other cases, the burst will be interleaved with the other burst(s) with the same destination, and the average burst size will change to $\overline{S_{iu}}/\overline{L}$, where $\overline{S_{iu}}$ is the mean service period of all uplink queues which send the traffic toward the given slave.

The probability that the source slave i is the only one to transmit in the current cycle, and the transmission is targeted toward the slave j (where $i, j = 2..m$ and $i \neq j$), is:

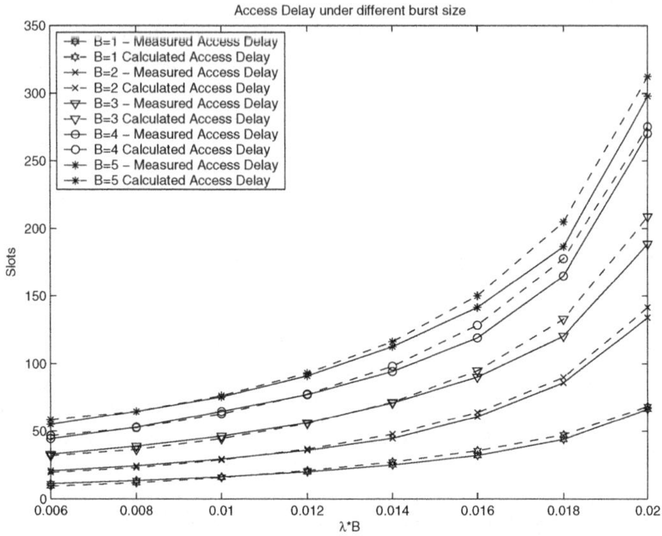

(a) Mean access delay.

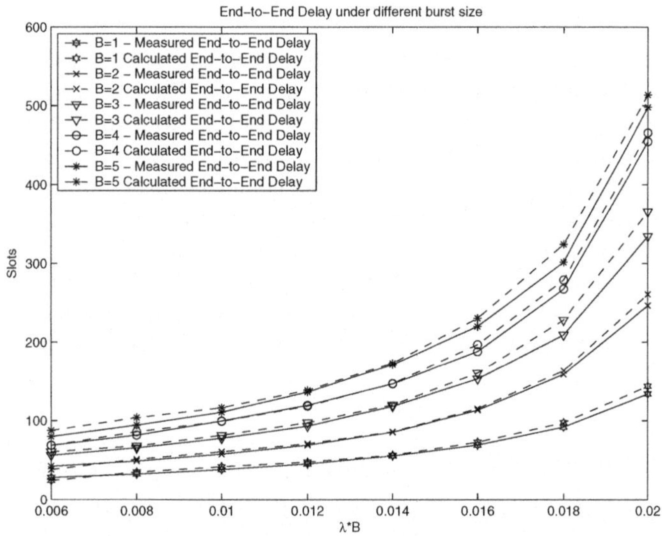

(b) Mean end-to-end delay.

FIGURE 3.9
Analytical and simulation results for access and end-to-end delay as functions of burst arrival rate and mean burst size, for $M = 3$. (From J. Mišić, K. L. Chan, and V. B. Mišić, "Admission control in Bluetooth piconets," *IEEE Trans. Veh. Tech.* **53**(3):890–911, © 2004 IEEE. Reprinted with permission.)

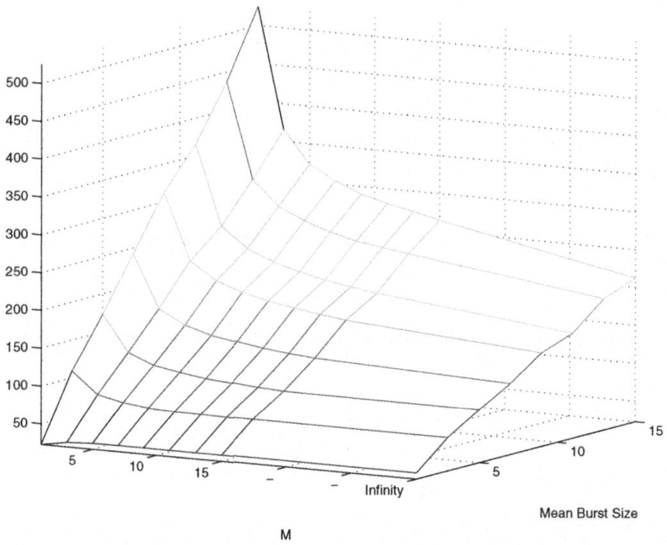

(a) Mean access delay.

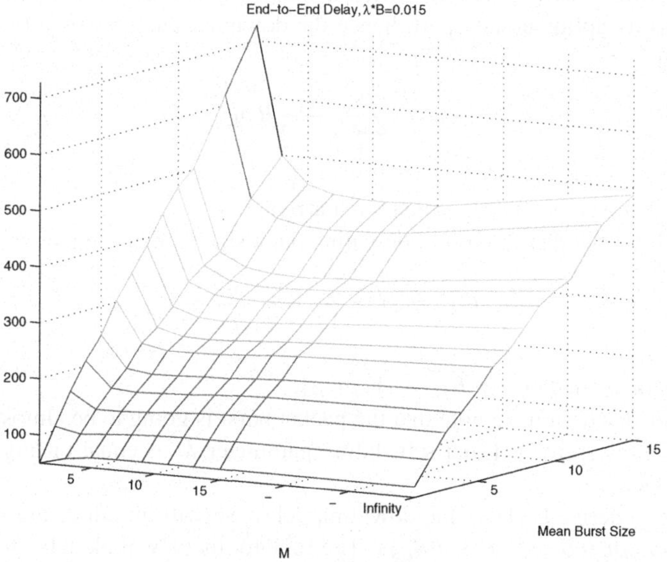

(b) Mean end-to-end delay.

FIGURE 3.10
Packet delays as functions of mean burst size and the polling parameter M. (From J. Mišić, K. L. Chan, and V. B. Mišić, "Admission control in Bluetooth piconets," *IEEE Trans. Veh. Tech.* **53**(3):890–911, © 2004 IEEE. Reprinted with permission.)

$$P_c = \sum_{\substack{i=2 \\ i \neq j}}^{m} \prod_{\substack{k=2 \\ k \neq i,j}}^{m} \frac{q_0^k}{Q_k(1,1)} \frac{(1-q_0^i)}{Q_i(1,1)} \tag{3.47}$$

The probability that slave i is the only one which transmits to the slave j among two or three active slaves in the same cycle is:

$$P_d = \left(1 - \frac{1}{m-3} - \frac{1}{(m-4)^2}\right)^{\left\lceil \frac{B}{M} \right\rceil} \tag{3.48}$$

Then, the parameter of the downlink burst length distribution is given by

$$p_{b,j,d} = \frac{1}{B} P_c + (1 - P_c) \cdot \left(\frac{1}{B} P_d + \frac{\overline{L}}{\overline{S_u}} (1 - P_d)\right) \tag{3.49}$$

where $\overline{S_u} = \sum_{\substack{i=2 \\ i \neq j}}^{m} \overline{S_{iu}} \lambda_{iu} / \sum_{\substack{i=2 \\ i \neq j}}^{m} \lambda_{iu}$.

If the traffic is symmetric, the offered load in the downlink direction must be the same as its uplink counterpart, hence the downlink packet arrival rate has to be adjusted to

$$\lambda_{jd} = \sum_{\substack{i=2 \\ i \neq j}}^{m} \frac{\lambda_{iu}}{m-2} \overline{B} p_{b,j,d} \tag{3.50}$$

where λ_{iu} denotes the uplink packet burst arrival rate.

The PGF for the distribution of downlink burst size then becomes

$$G_{b,j,d}(z) = \frac{z p_{b,j,d}}{1 - z + z p_{b,j,d}} \tag{3.51}$$

and the average burst size is $\overline{B_{j,d}} = 1/p_{b,j,d}$.

After this discussion we note that the packet burst is going to be almost preserved in the downlink, provided that scheduling parameter M is equal or larger than the mean burst size.

In order to derive LST for the downlink delay we will introduce the substitution $u = \lambda_{id} - \lambda_{id} w / (\overline{B_{i,d}} + w - w \overline{B_{i,d}})$. The LST for the downlink delay toward slave i then becomes

$$W_{di}^*(u) = \frac{u(1 - \lambda_{id} \overline{F_{is}}\, \overline{B_{i,d}})}{u - \lambda_{id} + \lambda_{id} G_{b,i,d}(F_{is}^*(u))} \cdot \frac{1 - G_{b,i,d}(F_{is}^*(u))}{\overline{B_{i,d}}(1 - F_{is}^*(u))} \cdot \frac{1 - V_i^*(u)}{u \overline{V_i}} \cdot \frac{Q_i\left(1, 1 - \frac{u}{\lambda_{i,d} \overline{B_{i,d}} - u \overline{B_{i,d}} + u}\right)}{Q_i(1,1) V_i^*(u)} \tag{3.52}$$

and the mean downlink delay is

$$\overline{W_{di}} = \frac{\lambda_{id}\overline{B_{i,d}}\,\overline{F_{is}}}{(1-\lambda_{id}\overline{F_{is}}\,\overline{B_{i,d}})} + \frac{\overline{B_{i,d}^{(2)}}\,\overline{F_{is}}}{2\overline{B_{i,d}}(1-2\lambda_{id}\overline{F_{is}}\,\overline{B_{i,d}})} + \frac{\overline{V_i^2}}{2\overline{V_i}} - \overline{V_i} + \frac{Q_i^{'}(1,1)}{\lambda_{id}\overline{B_{i,d}}Q_i(1,1)} \tag{3.53}$$

The LST for end-to-end delay is equal to $W_{ije}^{*}(s,u) = W_{ai}^{*}(s)W_{dj}^{*}(d)$. Mean end-to-end delay is $\overline{W_{ije}} = \overline{W_{ai}} + \overline{W_{dj}}$. The dependency of end-to-end delay for the piconet with seven identical ACL slaves ($m = 8$) as a function of mean burst size and burst arrival rate is shown in Fig. 3.9.

Upon close inspection, the downlink delay can be seen to behave in a different way from the access delay when the value of M is varied. Namely, the mean burst size in the downlink direction is limited by the interleaving of burst slices under moderate and high burst arrival rates. Therefore, under moderate arrival burst rates, the effective downlink burst size is close to the value of M, and the downlink delay changes in the same direction as does M.

Therefore, when the value of M is varied, the access and downlink delays change in opposing directions, and the end-to-end delay (which is equal to their sum) should exhibit a minimum under relatively small values of M, as shown in Fig. 3.10(b) for the symmetric load case. The location of this minimum can be found by solving the equation

$$\frac{\partial \overline{W_{ai}}}{\partial M} + \frac{\partial \overline{W_{dj}}}{\partial M} = 0 \tag{3.54}$$

which can be accomplished numerically. The resulting dependency of the optimum value of M as the function of $\overline{B}$, in case the aggregate packet arrival rate per slave has a fixed value of $\lambda\overline{B} = 0.015$, is shown in Fig. 3.11.

Roughly speaking, the end-to-end delay is minimized when the value of M is tuned slightly below the value of mean burst size $\overline{B}$, and the minimum delay is below the value obtained with exhaustive scheduling. As the minimum is fairly broad, the choice of M is not too critical; values of $M = 3$ to 5 should give satisfactory results for a wide range of values of $\overline{B}$.

Note that the E-limited service guarantees that each slave will be polled at least once and at most M times within the piconet cycle that cannot exceed $10(m-1)M$ time slots of the Bluetooth clock ($T = 625\mu s$). In this manner, fairness to all slaves is maintained.

As can be seen, both analytical and simulation results show that the E-limited scheme outperforms other two techniques with respect to the end-to-end delay. Second, the E-limited scheme provides fairness by default, since the limit on the number of frames exchanged with any single slave prevents any slave from monopolizing the network. Third, this scheme is efficient since it does not waste bandwidth, save for a single POLL/NULL frame for each poll of the slave with no traffic in either direction – but this inefficiency is present in all other polling schemes.

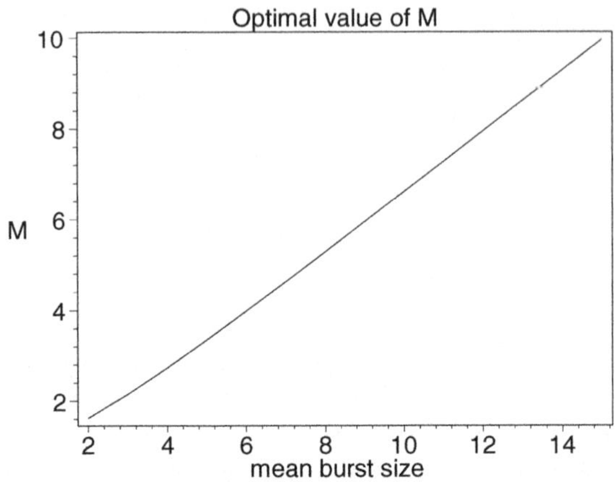

FIGURE 3.11
Optimal value of M as the function of mean burst size $\overline{B}$. (From J. Mišić, K. L. Chan, and V. B. Mišić, "Admission control in Bluetooth piconets," *IEEE Trans. Veh. Tech.* **53**(3):890–911, © 2004 IEEE. Reprinted with permission.)

We have also shown that end-to-end packet delays may be minimized if the value of M (the number of packets to be exchanged in a single visit to a slave) is close to the mean size of the packet bursts coming from the slaves. As these minima are fairly broad, values of M which are close to the optimum can still lead to near-minimum delay values.

4

The impact of finite buffers

Our analysis so far has assumed that both the uplink and downlink queues (or, in other words, the transmit and receive buffers at the baseband level) are of infinite size. This assumption is quite reasonable for stationary devices powered from the AC supply such as printers, desktop computers, and network access points. However, it is not likely to hold for battery-powered mobile devices where chip space is at a premium and, consequently, buffers are made as small as possible. In this chapter, we relax this assumption and analyze the performance of E-limited polling in Bluetooth piconets under bursty traffic and finite buffers.

The impact of finite buffer size at the baseband level on performance is analyzed under the assumption of the so-called total rejection policy [Takagi, 1991]. Namely, the application packets that arrive are segmented into a number of packets that form a packet burst at the baseband level. The baseband packets are arriving in bursts determined by the application. The buffers used to implement the uplink queues have finite size, dictated by the available chip real estate and other design constraints. The incoming packets will be allowed to enter the corresponding buffer only if the entire burst can fit in the buffer; otherwise, the whole burst will be rejected. This is known as the total rejection policy; other policies that utilize partial rejection only are possible but we don't consider them here.

The blocking probability (i.e., the probability that the packet burst will be rejected because of insufficient free space in the corresponding buffer) will be critically dependent on the device buffer size and packet burst arrival rate. Rejected packets will have to be retransmitted under the control of the suitable transport layer protocol, such as TCP; other schemes to deal with packet loss are also possible. Either way, the mean delay and the network throughput – in other words, the performance of the piconet traffic – will be affected.

The problem of packet blocking can be viewed from a practical point of view: namely, how to select the size of hardware buffers so that the packet blocking probability does not exceed a predefined threshold. This is, in fact, the problem considered in this Chapter, where we investigate the impact of buffer size on the performance of Bluetooth piconets expressed in terms of packet blocking probability, access delay, and end to end packet delay.

The Chapter is organized as follows. The basic queueing theoretic model of the piconet is described in Section 4.1, where the probability distributions of queue length sizes in imbedded Markov points are calculated. The probability distribution of queue length in arbitrary time is calculated in Section 4.2, where the burst blocking probability and relevant packet delay distributions are calculated. Section 4.3

explains what these probabilities mean in practice, and outlines the dependencies of blocking probabilities and packet delays on the queue size.

4.1 Queue length distribution in imbedded Markov points

Consider an isolated piconet with the master and $m - 1$ active ACL slaves. The operation of the piconet may be described with a queueing model in which each slave maintains an uplink queue, while the master maintains a number of downlink queues, one per each active slave. Unlike the model used for the analysis in the previous chapter, we assume that the buffers used to implement those queues have finite size, as shown in Fig. 4.1.

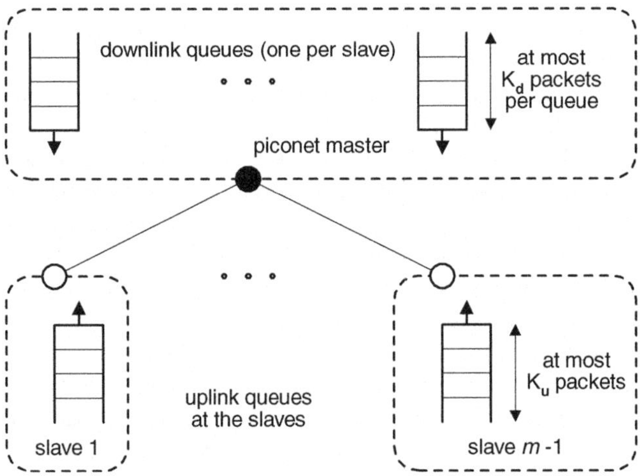

FIGURE 4.1
Queueing model of a single piconet with finite buffers.

Buffer sizes in the network are fixed: the uplink buffer at each slave can take up to K_u packets, while each of the downlink buffers at the master can take up to K_d packets. Other assumptions regarding the traffic are the same as in Chapter 3: i.e., the applications executing on slave devices generate packets which are segmented into a number of baseband packets. Application packet arrivals to the uplink queue of slave i follow a Poisson distribution with the arrival rate λ_{iu}.

Assuming that all slaves use the same segmentation/reassembly mechanism (which is reasonable, as this mechanism is commonly implemented in firmware), the burst length distribution will be the same for all slaves. We assume that the traffic goes from slaves to other slaves only, which simplifies our calculations without undue loss of generality. (Any traffic generated or received by the master can be easily modeled by increasing the packet arrival rates in the downlink queues.) All packets within the given burst will have the same destination node, and the distribution of destinations is assumed to be uniform. In more complex cases, the resulting downlink arrival rates may be calculated from the matrix of probabilities for slave-to-slave communication, without changing the model itself.

The PGF of the burst length probability distribution is

$$G_{bu}(z) = \frac{zp_b}{1 - z + zp_b} \tag{4.1}$$

where p_b is the parameter of the geometric distribution; the mean burst size in the uplink queue is $\overline{B} = G'_{bu}(1) = 1/p_b$. On the downlink, the burst size will be affected by the polling parameter M. This calculation is presented in Chapter 3. One could also expect that burst size may be affected by the finite buffer size. However, we assume that the packet burst will be accepted only if there is sufficient space in the buffer to hold the entire burst, i.e., all the packets from it.

If the traffic is symmetric, the offered load in the downlink direction must be the same as its uplink counterpart, hence the downlink packet arrival rate has to be adjusted to

$$\lambda_{j,d} = \sum_{\substack{i=2 \\ i \neq j}}^{m} \frac{\lambda_{iu}}{m-2} \overline{B} p_{b,d} \tag{4.2}$$

where $p_{b,d}$ is the parameter of the geometric probability distribution of the downlink burst size, as calculated in Chapter 3, and λ_{iu} is the uplink packet burst arrival rate. The PGF for the distribution of downlink burst size is

$$G_{bd}(z) = \frac{zp_{b,d}}{1 - z + zp_{b,d}} \tag{4.3}$$

and the average burst size is $\overline{B_{j,d}} = 1/p_{b,d}$.

The number of packets at the uplink queue of a slave and the corresponding downlink queue at the master can be modeled with a set of imbedded Markov points [Takagi, 1991]. The Markov points correspond to vacation termination times – the times immediately before the master sends a first downlink packet to the slave, and frame service completion times – the times when the polled slave finishes one uplink packet transmission.

Thanks to the symmetry of the piconet with respect to the slaves, it suffices to consider the packet exchange between a single slave, say i, and the master. Let $q^i_{k_u,k_d}$ denote the joint probability that a Markov point in the uplink queue of slave i is a vacation termination time, and that there are $k_u = 0, 1, 2 \ldots K_u$ packets at the uplink queue of slave i and $k_d = 0, 1, 2 \ldots K_d$ packets at the corresponding downlink queue at the master at that time. Also, let $\pi^{i,(\mu)}_{k_u,k_d}$ denote the joint probability that a Markov point is the μ-th frame transmission completion time and that there are k_u packets in the slave's i queue at that time, and k_d packets at the master's queue toward the slave i, where $\mu = 1 \,..\, M$ denotes the index of Markov point and $k_u = 0, 1, 2, \ldots K_u$, $k_d = 0, 1, 2, \ldots K_d$.

Let $f_{is}(x)$ and $v_i(x)$ stand for the pdf-s (probability density functions) of the frame transmission time and vacation time, respectively, at the uplink queue of slave i; their LST transforms will be $F_{is}^*(s)$ and $V_i^*(s)$. We will also make use of the following probabilities:

- the probability of k_u packet arrivals at the slave i's uplink queue during the frame time, denoted with a_{k_u},
- the probability of k_d packet arrivals at the master's downlink queue during the frame time, denoted with a_{k_d},
- the probability of k_u packet arrivals at the slave i's uplink queue during the vacation time (i.e., while the master is servicing other slaves), denoted with f_{k_u},
- and the probability of k_d packet arrivals in the master's downlink during the vacation time, denoted with f_{k_d}.

These probabilities can be calculated as

$$\begin{aligned}
a_{k_u} &= \sum_{l=0}^{\infty} \frac{1}{k_u!} \frac{d^{k_u}}{dz^{k_u}} (G_b(z))^l \Big|_{z=0} \int_0^{\infty} \frac{(\lambda_{iu} x)^l}{l!} e^{-\lambda_{iu} x} f_{is}(x) dx \\
&= \frac{1}{k_u!} \frac{d^{k_u}}{dz^{k_u}} (F_{is}^*(\lambda_{iu} - \lambda_{iu} G_b(z)))\Big|_{z=0} \\
a_{k_d} &= \sum_{l=0}^{\infty} \frac{1}{k_d!} \frac{d^{k_d}}{dz^{k_d}} (G_b(z))^l \Big|_{z=0} \int_0^{\infty} \frac{(\lambda_{id} x)^l}{l!} e^{-\lambda_{id} x} f_{is}(x) dx \\
&= \frac{1}{k_d!} \frac{d^{k_d}}{dz^{k_d}} (F_{is}^*(\lambda_{id} - \lambda_{id} G_b(z)))\Big|_{z=0} \\
f_{k_u} &= \sum_{l=0}^{\infty} \frac{1}{k_u!} \frac{d^{k_u}}{dz^{k_u}} (G_b(z))^l \Big|_{z=0} \int_0^{\infty} \frac{(\lambda_{iu} x)^l}{l!} e^{-\lambda_{iu} x} v_i(x) dx \\
&= \frac{1}{k_u!} \frac{d^{k_u}}{dz^{k_u}} V_i^*(\lambda_{iu} - \lambda_{iu} G_b(z))\Big|_{z=0} \\
f_{k_d} &= \sum_{l=0}^{\infty} \frac{1}{k_d!} \frac{d^{k_d}}{dz^{k_d}} (G_b(z))^l \Big|_{z=0} \int_0^{\infty} \frac{(\lambda_{iu} x)^l}{l!} e^{-\lambda_{iu} x} v_i(x) dx \\
&= \frac{1}{k_d!} \frac{d^{k_d}}{dz^{k_d}} V_i^*(\lambda_{iu} - \lambda_{iu} G_b(z))\Big|_{z=0}
\end{aligned} \tag{4.4}$$

Note that $F_{is}^*(\lambda_{iu} - \lambda_{iu} G_b(z))$ and $V_i^*(\lambda_{iu} - \lambda_{iu} G_b(z))$ denote the PGFs for the number of packet arrivals in the uplink queue during the frame time and vacation time, respectively:

$$\begin{aligned}
A_{F,u}(z) &= \sum_{k_u=0}^{\infty} a_{k_u} z^{k_u} = \sum_{k_u=0}^{\infty} \frac{z_u^k}{k_u!} \frac{d^{k_u}}{dz^{k_u}} F_{is}^*(\lambda_{iu} - \lambda_{iu} G_b(z))\Big|_{z=0} \\
&= F_{is}^*(\lambda_{iu} - \lambda_{iu} G_b(z)) \\
A_{V,u}(z) &= \sum_{k_u=0}^{\infty} f_{k_u} z^{k_u} = \sum_{k_u=0}^{\infty} \frac{z^{k_u}}{k_u!} \frac{d^{k_u}}{dz^{k_u}} V_i^*(\lambda_{iu} - \lambda_{iu} G_b(z))\Big|_{z=0} \\
&= V_i^*(\lambda_{iu} - \lambda_{iu} G_b(z))
\end{aligned} \tag{4.5}$$

Similar expressions hold for the number of arrivals in the downlink queue. The probabilities that the uplink queue of slave i contains k_u packets and that the corresponding downlink queue contains k_d packets in imbedded Markov points satisfy the following equations:

$$\begin{aligned}
\pi_{k_u,k_d}^{i,(1)} &= \sum_{j_u=1}^{k_u+1} \sum_{j_d=1}^{k_d+1} q_{j_u,j_d}^{i} a_{k_u-j_u+1} a_{k_d-j_d+1} + \sum_{j_d=1}^{k_d+1} q_{0,j_d}^{i} a_{k_d-j_d+1} a_{k_u} \\
&\quad + \sum_{j_u=1}^{k_u+1} q_{j_u,0}^{i} a_{k_u-j_u+1} a_{k_d}, \\
&\qquad\qquad 0 \le k_u \le K_u - 2, 0 \le k_d \le K_d - 2 \\
\pi_{K_u-1,k_d}^{i,(1)} &= \sum_{j_u=1}^{K_u} \sum_{j_d=1}^{k_d+1} q_{j_u,j_d}^{i} \left(\sum_{k_u=K_u-j_u}^{\infty} a_{k_u} \right) a_{k_d-j_d+1} \\
&\quad + \sum_{j_d=1}^{k_d+1} q_{0,j_d}^{i} a_{k_d-j_d+1} \sum_{k_u=K_u}^{\infty} a_{k_u} + \sum_{j_u=1}^{k_u+1} q_{j_u,0}^{i} a_{k_d} \sum_{k_u=K_u-j_u}^{\infty} a_{k_u}, \\
&\qquad\qquad 0 \le k_d \le K_d - 2 \\
\pi_{k_u,K_d-1}^{i,(1)} &= \sum_{j_u=1}^{k_u+1} \sum_{j_d=1}^{K_d} q_{j_u,j_d}^{i} a_{k_u-j_u+1} \sum_{k_d=K_d-j_d}^{\infty} a_{k_d} + \sum_{j_d=1}^{K_d} q_{0,j_d}^{i} a_{k_u} \sum_{k_d=K_d-j_d}^{\infty} a_{k_d} \\
&\quad + \sum_{j_u=1}^{k_u+1} q_{0,j_d}^{i} a_{k_u-j_u+1} \sum_{k_d=K_d}^{\infty} a_{k_d}, \\
&\qquad\qquad 0 \le k_u \le K_u - 2 \\
\pi_{k_u,k_d}^{i,(\mu)} &= \sum_{j_u=1}^{k_u+1} \sum_{j_d=1}^{k_d+1} \pi_{j_u,j_d}^{i,(\mu-1)} a_{k_u-j_u+1} a_{k_d-j_d+1} + \sum_{j_d=1}^{k_d+1} \pi_{0,j_d}^{i,(\mu-1)} a_{k_d-j_d+1} a_{k_u} \\
&\quad + \sum_{j_u=1}^{k_u+1} \pi_{j_u,0}^{i,(\mu-1)} a_{k_u-j_u+1} a_{k_d}, \\
&\qquad\qquad 0 \le k_u \le K_u - 2, 0 \le k_d \le K_d - 2, \mu = 2\,..\,M
\end{aligned}$$

continued on next page ...

(4.6)

... continued from previous page

$$\pi^{i,(\mu)}_{K_u-1,k_d} = \sum_{j_u=1}^{K_u}\sum_{j_d=1}^{k_d+1}\pi^{i,(\mu-1)}_{j_u,j_d}\left(\sum_{k_u=K_u-j_u}^{\infty} a_{k_u-j_u+1}\right)a_{k_d-j_d+1}$$
$$+\sum_{j_d=1}^{k_d+1}\pi^{i,(\mu-1)}_{0,j_d}a_{k_d-j_d+1}\sum_{k_u=K_u}^{\infty}a_{k_u}$$
$$+\sum_{j_u=1}^{k_u+1}\pi^{i,(\mu-1)}_{j_u,0}a_{k_d}\sum_{k_u=K_u-j_u}^{\infty}a_{k_u-j_u+1},$$
$$0 \le k_d \le K_d-2,\ \mu=2\,..\,M$$

$$\pi^{i,(\mu)}_{k_u,K_d-1} = \sum_{j_u=1}^{k_u+1}\sum_{j_d=1}^{K_d}\pi^{i,(\mu-1)}_{j_u,j_d}a_{k_u-j_u+1}\sum_{k_d=K_d-j_d}^{\infty}a_{k_d-j_d+1}$$
$$+\sum_{j_d=1}^{K_d}\pi^{i,(\mu-1)}_{0,j_d}a_{k_u}\sum_{k_d=K_d-j_d}^{\infty}a_{k_d}+\sum_{j_u=1}^{k_u+1}\pi^{i,(\mu-1)}_{j_u,0}a_{k_u-j_u+1}\sum_{k_d=K_d}^{\infty}a_{k_d},$$
$$0 \le k_u \le K_u-2,\ \mu=2\,..\,M \tag{4.6}$$

$$q^i_{k_u,k_d} = \left(\sum_{m=1}^{M-1}\pi^{i,(\mu)}_{0,0}+q^i_{0,0}\right)f_{k_u}f_{k_d}+\sum_{j_u=0}^{k_u}\sum_{j_d=0}^{k_d}\pi^{i,(\mu)}_{j_u,j_d}f_{k_u-j_u}f_{k_d-j_d},$$
$$0 \le k_u \le K_u-2,\ 0 \le k_d \le K_d-2$$

$$q^i_{K_u-1,k_d} = \left(\sum_{\mu=1}^{M-1}\pi^{i,(\mu)}_{0,0}+q^i_{0,0}\right)f_{k_d}\sum_{k_u=K_u}^{\infty}f_{k_u}$$
$$+\sum_{j_u=0}^{k_u}\sum_{j_d=0}^{k_d}\pi^{i,(\mu)}_{j_u,j_d}f_{k_d-j_d}\sum_{k_u=K_u-j_u}^{\infty}f_{k_u}$$
$$0 \le k_d \le K_d-2$$

$$q^i_{k_u,K_d-1} = \left(\sum_{\mu=1}^{M-1}\pi^{i,(\mu)}_{0,0}+q^i_{0,0}\right)f_{k_u}\sum_{k_d=K_d}^{\infty}f_{k_d}$$
$$+\sum_{j_u=0}^{k_u}\sum_{j_d=0}^{k_d}\pi^{i,(\mu)}_{j_u,j_d}f_{k_u-j_u}\sum_{k_d=K_d-j_d}^{\infty}f_{k_d},$$
$$0 \le k_u \le K_u-2$$

As always, the sum of all probabilities must be equal to 1, hence

$$\sum_{k_u=0}^{K_u}\sum_{k_d=0}^{K_d}q^i_{k_u,k_d}+\sum_{\mu=1}^{M}\sum_{k_u=0}^{K_u-1}\sum_{k_d=0}^{K_d-1}\pi^{i,(\mu)}_{k_u,k_d}=1 \tag{4.7}$$

Analogous equations should also be set for $(K_u,\ K_d)$ cases (a total of $M+1$), but they are not shown here. This system of equations leads to the probability distribution of uplink and downlink queue lengths the distribution of queue length in Markov points is obtained. However, before solving it, the probability distributions of the frame time and vacation time have to be found as functions of queue length distributions.

The frame time can be determined by taking into account all possible combinations of uplink and downlink queue lengths. A frame will contain NULL/POLL packets only if both queues are empty, it will contain one empty packet and one data packet if one of the queues is empty and the other is non-empty and it will contain two data packets if both queues are non-empty. Therefore, the LST for the frame duration is

$$\begin{aligned} F^*_{is}(s) = & \left(q^i_{0,0} + \sum_{\mu=1}^{M} \pi^{i,(\mu)}_{0,0} \right) e^{-2s} \\ & + \left(\sum_{k_u=1}^{K_u-1} (q^i_{k_u,0} + \sum_{\mu=1}^{M} \pi^{i,(\mu)}_{k_u,0}) + \sum_{k_d=1}^{K_d-1} (q^i_{0,k_d} + \sum_{\mu=1}^{M} \pi^{i,(\mu)}_{0,k_d}) \right) G^*_p(s) e^{-s} \\ & + \sum_{k_u=1}^{K_u-1} \sum_{k_d=1}^{K_d-1} \left(q^i_{k_u,k_d} + \sum_{\mu=1}^{M} \pi^{i,(\mu)}_{k_u,k_d} \right) (G^*_p(s))^2 \end{aligned} \tag{4.8}$$

The probability distribution of the vacation time for a particular slave can be found as a sum of channel service times for other slaves. Channel service time is defined as time elapsed from the moment when master polls the slave for the first time in the piconet cycle until either the empty frame has been encountered or a total of M data frames have been exchanged. Therefore, we need to find the LST of the channel service time for slave i expressed in time slots T. Let us consider frames for $\mu = 1 \,..\, M$ consecutively and find the LST of their durations. The LST for the duration of μ-th frame is

$$\begin{aligned} F^{*1}(s) = & \frac{\sum_{k_u=1}^{K_u-1} q^i_{k_u,0} + \sum_{k_d=1}^{K_d-1} q^i_{0,k_d}}{\sum_{k_u=0}^{K_u-1} \sum_{k_d=0}^{K_d-1} q^i_{k_u,k_d}} G^*_p(s) e^{-s} + \frac{\sum_{k_u=1}^{K_u-1} \sum_{k_d=1}^{K_d-1} q^i_{k_u,k_d}}{\sum_{k_u=0}^{K_u-1} \sum_{k_d=0}^{K_d-1} q^i_{k_u,k_d}} (G^*_p(s))^2 \\ F^{*\mu}(s) = & \frac{\sum_{k_u=1}^{K_u-1} \pi^{i,(\mu)}_{k_u,0} + \sum_{k_d=1}^{K_d-1} \pi^{i,(\mu)}_{0,k_d}}{\sum_{k_u=0}^{K_u-1} \sum_{k_d=0}^{K_d-1} \pi^{i,(\mu)}_{k_u,k_d}} G^*_p(s) e^{-s} + \frac{\sum_{k_u=1}^{K_u-1} \sum_{k_d=1}^{K_d-1} \pi^{i,(\mu)}_{k_u,k_d}}{\sum_{k_u=0}^{K_u-1} \sum_{k_d=0}^{K_d-1} \pi^{i,(\mu)}_{k_u,k_d}} G^*_p(s)^2, \\ & \mu = 2 \,..\, M \end{aligned} \tag{4.9}$$

Now we have to determine the probabilities $P_{f,k}$ (where $k = 0 \,..\, M$) that the channel service time will take exactly k data frames:

$$
\begin{aligned}
P_{f,0} &= \frac{q^i_{0,0}}{\sum\limits_{k_u=0}^{K_u-1}\sum\limits_{k_d=0}^{K_d-1} q^i_{k_u,k_d}} \\
P_{f,1} &= \left(1 - \frac{q^i_{0,0}}{\sum\limits_{k_u=0}^{K_u-1}\sum\limits_{k_d=0}^{K_d-1} q^i_{k_u,k_d}}\right) \frac{\pi^{i,(1)}_{0,0}}{\sum\limits_{k_u=0}^{K_u-1}\sum\limits_{k_d=0}^{K_d-1} \pi^{i,(1)}_{k_u,k_d}} \\
P_{f,k} &= \left(1 - \frac{q^i_{0,0}}{\sum\limits_{k_u=0}^{K_u-1}\sum\limits_{k_d=0}^{K_d-1} q^i_{k_u,k_d}}\right) \\
&\cdot \prod_{\mu=1}^{k-1}\left(1 - \frac{\pi^{i,(\mu)}_{0,0}}{\sum\limits_{k_u=0}^{K_u-1}\sum\limits_{k_d=0}^{K_d-1} \pi^{i,(\mu)}_{k_u,k_d}}\right) \cdot \frac{\pi^{i,(k)}_{0,0}}{\sum\limits_{k_u=0}^{K_u-1}\sum\limits_{k_d=0}^{K_d-1} \pi^{i,(k)}_{k_u,k_d}}, \\
&\qquad k = 2\,..\,M-1 \\
P_{f,M} &= 1 - \sum_{k=0}^{M-1} P_{f,k}
\end{aligned}
\tag{4.10}
$$

Then, the LST for the channel service time is

$$
S^*_i(i) = \sum_{k=0}^{M-1} P_{f,k} \prod_{\mu=1}^{k} (F^{*\mu}(s))e^{-2s} + P_{f,M} \prod_{\mu=1}^{M} F^{*\mu}(s) \tag{4.11}
$$

and, assuming that the slaves have indices $i = 1\,..\,m-1$, the LST of the vacation time for slave i is

$$
V^*_i(s) = \prod_{\substack{j=1 \\ j\neq i}}^{m-1} S^*_j \tag{4.12}
$$

When (4.8) and (4.12) are substituted in the system (4.6), the system can be solved for unknown mass probabilities of joint uplink-downlink queue length probability distribution.

Let us denote the probability that vacation starts after the uplink transmission as $h_i = \sum_{\mu=1}^{M-1} \pi^{i,(\mu)}_{0,0} + \sum_{k_u=0}^{K_u-1}\sum_{k_d=0}^{K_d-1} \pi^{i,(M)}_{k_u,k_d}$. The probability that the vacation will start after an arbitrary Markov point is $q^i_{0,0} + h_i$, while the average time interval between two consecutive Markov points at slave i is

$$\eta_i = (q^i_{0,0} + h_i)\overline{V_i} + (1 - q^i_{0,0} - h_i)\overline{F_{is}} \tag{4.13}$$

4.2 Uplink queue length distribution

By using the probability distribution of the uplink queue length at Markov points, we will derive the probability distribution of the uplink queue length at arbitrary time between two Markov points. We can also derive the PDF of the remaining vacation time (if the previous Markov point was the start of vacation) or the PDF of the remaining frame service time (if the previous Markov point was the start of the packet service). Let us introduce the following variables:

- the pdf of the vacation time, $v_i(x)$, and its PDF, $V_i(x)$,
- uplink and downlink queue length at arbitrary time, $L_{q,i}$,
- elapsed vacation time, $V_{\_,i}$,
- remaining vacation time, as $V_{+,i}$,
- the number of packet arrivals (results of packet burst arrivals) in the slave uplink queue during the elapsed vacation time, $A_u(V_\_)$, and the corresponding number of arrivals in the downlink queue, $A_d(V_\_)$,
- the pdf of the frame service time, $f_{is}(x)$, and its PDF, $F_{is}(x)$,
- elapsed frame service time, $X_{\_,i}$,
- remaining frame service time, as $X_{+,i}$,
- the number of packet arrivals (as results of packet burst arrivals) in the slave uplink queue during the elapsed frame service time, $A_u(X_\_)$, and the corresponding number of arrivals in the downlink queue, $A_d(X_\_)$.

We will use results from the renewal theory [Kleinrock, 1972; Takagi, 1991] which determine the pdf's of elapsed frame service time and vacation time. In the case of frame service time, the pdf of the elapsed vacation time is $\dfrac{1 - V_i(x)}{\overline{V_i}}$, and the pdf of the remaining vacation time is $\dfrac{v_i(x)}{1 - V_i(x)}$.

For the time between the start of the vacation and end of vacation, we define the joint probability of the uplink/downlink queue lengths and remaining vacation time as

$$\Omega^*_{k_u,k_d,i}(s) = \int_0^\infty e^{-sy}\mathcal{P}[L_{q,i} = (k_u,k_d),\, y < V_{+,i} < y+dy], \qquad 0 \le k_u \le K_u,\, 0 \le k_d \le K_d \tag{4.14}$$

where $\mathcal{P}[x]$ denotes the probability of event x.

For the time between the start and end of the frame service for phase $1 \le \mu \le M$, we define the joint probability of the uplink/downlink queue lengths and remaining frame service time as

$$\Pi^*_{k_u,k_d,\mu,i}(s) = \int_0^\infty e^{-sy}\mathcal{P}[L_{q,i} = (k_u,k_d),\, y < X_{+,i} < y+dy], \qquad 1 \le k_u \le K_u,\, 1 \le k_d \le K_d,\, 1 \le \mu \le M \tag{4.15}$$

Using the probabilities of the uplink queue state in the previous Markov point, we obtain

$$\begin{aligned}
\Omega^*_{k_u,k_d,i}(s) &= \frac{\overline{V_i}}{\eta_i}\left(q^i_{0,0} + \sum_{\mu=1}^{M-1}\pi^{i,(\mu)}_{0,0}\right) E[e^{-sV_{+,i,u}}|A_u(V_{-,i}) = k_u,\, A_d(V_{-,i}) = k_d] \\
&\quad \cdot\mathcal{P}[A_u(V_{-,i}) = k_u]\mathcal{P}[A_d(V_{-,i}) = k_d] \\
&\quad + \frac{\overline{V_i}}{\eta_i}\sum_{j_u=0}^{k_u}\sum_{j_d=0}^{k_d}\pi^{i,(M)}_{j_u,j_d} E[e^{-sV_{+,i,u}}|A_u(V_{-,i}) = k_u - j_u,\, A_d(V_{-,i}) = k_d - j_d] \\
&\quad \cdot\mathcal{P}[A_u(V_{-,i}) = k_u - j_u]\cdot\mathcal{P}[A_d(V_{-,i}) = k_d - j_d], \\
&\qquad 0 \le k_u \le K_u - 1,\, 0 \le k_d \le K_d - 1 \\
\Omega^*_{K_u,k_d,i}(s) &= \\
&\frac{\overline{V_i}}{\eta_i}\left(q^i_{0,0} + \sum_{\mu=1}^{M-1}\pi^{i,(\mu)}_{0,0}\right)\sum_{k_u=K_u}^{\infty} E[e^{-sV_{+,i}}|A_u(V_{-,i}) = k_u,\, A_d(V_{-,i}) = k_d] \\
&\quad \cdot\mathcal{P}[A_u(V_{-,i}) = k_u]\mathcal{P}[A_d(V_{-,i}) = k_d] \\
&\quad + \frac{\overline{V_i}}{\eta_i}\sum_{j_u=0}^{k_u}\sum_{j_d=0}^{k_d}\pi^{i,(M)}_{j_u,j_d}\sum_{k_u=K_u-j_u}^{\infty} E[e^{-sV_{+,i}}|A_u(V_{-,i}) = k_u,\, A_d(V_{-,i}) = k_d - j_d] \\
&\quad \cdot\mathcal{P}[A_u(V_{-,i}) = k_u]\mathcal{P}[A_d(V_{-,i}) = k_d - j_d], \\
&\qquad 0 \le k_d \le K_d - 1
\end{aligned}$$

continued on next page . . .

(4.16)

... continued from previous page

$$
\begin{aligned}
&\Omega^*_{k_u,K_d,i}(s) = \\
&\quad \frac{\overline{V_i}}{\eta_i}\left(q^i_{0,0} + \sum_{\mu=1}^{M-1} \pi^{i,(\mu)}_{0,0}\right) \sum_{k_d=K_d}^{\infty} E[e^{-sV_{+,i}} | A_u(V_{-,i}) = k_u, A_d(V_{-,i}) = k_d] \\
&\qquad\qquad \cdot \mathcal{P}[A_u(V_{-,i}) = k_u]\mathcal{P}[A_d(V_{-,i}) = k_d] \\
&\quad + \frac{\overline{V_i}}{\eta_i} \sum_{j_u=0}^{k_u} \sum_{j_d=0}^{k_d} \pi^{i,(M)}_{j_u,j_d} \sum_{k_d=K_d-j_d}^{\infty} E[e^{-sV_{+,i}} | A_u(V_{-,i}) = k_u - j_u, A_d(V_{-,i}) = k_d] \\
&\qquad\qquad \cdot \mathcal{P}[A_u(V_{-,i}) = k_u - j_u]\mathcal{P}[A_d(V_{-,i}) = k_d], \\
&\qquad\qquad\qquad 0 \le k_u \le K_u - 1 \\
&\Pi^*_{k_u,k_d,1,i}(s) = \\
&\quad \frac{\overline{F_{is}}}{\eta_i} \sum_{j_u=1}^{k_u} \sum_{j_d=1}^{k_d} q^i_{j_u,j_d} E[e^{-sX_{+,i}} | A_u(X_{-,i}) = k_u - j_u, A_d(X_{-,i}) = k_d - j_d] \\
&\qquad\qquad \cdot \mathcal{P}[A_u(X_{-,i}) = k_u - j_u]\mathcal{P}[A_d(X_{-,i}) = k_d - j_d] \\
&\qquad\qquad\qquad 1 \le k_u \le K_u - 1, 1 \le k_d \le K_d - 1 \\
&\Pi^*_{K_u,k_d,1,i}(s) = \\
&\quad \frac{\overline{F_{is}}}{\eta_i} \sum_{j_u=1}^{K_u} \sum_{j_d=1}^{k_d} q^i_{j_u,j_d} \sum_{k_u=K_u-j_u}^{\infty} E[e^{-sX_{+,i}} | A_u(X_{-,i}) = k_u, A_d(X_{-,i}) = k_d - j_d] \\
&\qquad\qquad \cdot \mathcal{P}[A_u(X_{-,i}) = k_u]\mathcal{P}[A_d(X_{-,i}) = k_d - j_d], \\
&\qquad\qquad\qquad 1 \le k_d \le K_d - 1 \\
&\Pi^*_{k_u,K_d,1,i}(s) = \\
&\quad \frac{\overline{F_{is}}}{\eta_i} \sum_{j_u=1}^{k_u} \sum_{j_d=1}^{K_d} q^i_{j_u,j_d} \sum_{k_d=K_d-j_d}^{\infty} E[e^{-sX_{+,i}} | A_u(X_{-,i}) = k_u - j_u, A_d(X_{-,i}) = k_d] \\
&\qquad\qquad \cdot \mathcal{P}[A_u(X_{-,i}) = k_u - j_u]\mathcal{P}[A_d(X_{-,i}) = k_d] \\
&\qquad\qquad\qquad 1 \le k_u \le K_u - 1 \\
&\Pi^*_{k_u,k_d,\mu,i}(s) = \\
&\quad \frac{\overline{F_{is}}}{\eta_i} \sum_{j_u=1}^{k_u} \sum_{j_d=1}^{k_d} \pi^{i,(\mu)}_{j_u,j_d} E[e^{-sX_{+,i}} | A_u(X_{-,i}) = k_u - j_u, A_d(X_{-,i}) = k_d - j_d] \\
&\qquad\qquad \cdot \mathcal{P}[A_u(X_{-,i}) = k_u - j_u]\mathcal{P}[A_d(X_{-,i}) = k_d - j_d], \\
&\qquad\qquad\qquad 1 \le k_u \le K_u - 1, 1 \le k_d \le K_d - 1, 2 \le \mu \le M \\
&\Pi^*_{K_u,k_d,\mu,i}(s) = \\
&\quad \frac{\overline{F_{is}}}{\eta_i} \sum_{j_u=1}^{K_u} \sum_{j_d=1}^{k_d} \pi^{i,(\mu)}_{j_u,j_d} \sum_{k_u=K_u-j_u}^{\infty} E[e^{-sX_{+,i}} | A_u(X_{-,i}) = k_u, A_d(X_{-,i}) = k_d - j_d] \\
&\qquad\qquad \cdot \mathcal{P}[A_u(X_{-,i}) = k_u]\mathcal{P}[A_d(X_{-,i}) = k_d - j_d], \\
&\qquad\qquad\qquad 2 \le \mu \le M, 1 \le k_d \le K_d - 1
\end{aligned}
$$

continued on next page ...

(4.16)

... continued from previous page

$$\Pi^*_{k_u,K_d,\mu,i}(s) = \frac{\overline{F_{is}}}{\eta_i} \sum_{j_u=1}^{k_u} \sum_{j_d=1}^{K_d} \pi^{i,(\mu)}_{j_u,j_d} \sum_{k_d=K_d-j_d}^{\infty} E[e^{-sX_{+,i}} | A_u(X_{-,i}) = k_u - j_u, A_d(X_{-,i}) = k_d]$$
$$\cdot \mathcal{P}[A_u(X_{-,i}) = k_u - j_u] \mathcal{P}[A_d(X_{-,i}) = k_d],$$
$$2 \le \mu \le M, 1 \le k_u \le K_d \tag{4.16}$$

For brevity, we have not shown the additional $M+1$ equations that should also be set for (K_u, K_d) cases.

The system (4.16) can be simplified as follows. We have initially assumed that the uplink and downlink burst sizes should be the same when the values of M_s and M_b are larger than $\overline{B}$. We will relax that assumption and calculate them separately, using the notation $G_{b_u}(z)$ and $G_{b_d}(z)$ for the PGFs for burst sizes in the uplink and downlink, respectively.

$$\begin{aligned}\phi^*_{k_u,k_d}(s) &= E[e^{-sV_{+,i}} | A_u(V_{-,i}) = k_u, A_d(V_{-,i}) = k_d] \\ &\quad \cdot \mathcal{P}[A_u(V_{-,i}) = k_u] \mathcal{P}[A_d(V_{-,i}) = k_d] \\ &= \sum_{l_u=0}^{\infty} \frac{1}{k_u!} \frac{d^{k_u}}{dz_u^{k_u}} \left(G_{b_u}(z_u)\right)^{l_u} \Big|_{z_u=0} \\ &\quad \cdot \sum_{l_d=0}^{\infty} \frac{1}{k_d!} \frac{d^{k_d}}{dz_d^{k_d}} \left(G_{b_d}(z_d)\right)^{l_d} \Big|_{z_d=0} \\ &\quad \cdot \int_0^{\infty} \frac{(\lambda_{iu}x)^{l_u}}{l_u!} e^{-\lambda_{iu}x} \frac{(\lambda_{id}x)^{l_d}}{l_d!} e^{-\lambda_{id}x} \frac{1-V_i(x)}{\overline{V_i}} dx \\ &\quad \cdot \int_0^{\infty} e^{-sy} \frac{v_i(x+y)}{1-V_i(x)} dy \end{aligned} \tag{4.17}$$

If we introduce the substitution $u = x + y$ and change the order and limits of integration of variables x and u, we obtain

$$\phi^*_{k_u,k_d}(s) = \frac{1}{\overline{V_i} k_u! k_d!} \frac{d^{k_u}}{dz_u^{k_u}} \frac{d^{k_d}}{dz_d^{k_d}} \left(\frac{V_i^*(-\lambda_{iu} G_{bu}(z_u) + \lambda_{iu} - \lambda_{id} G_{bd}(z_d) + \lambda_{id}) - V_i^*(s)}{\lambda_{iu} G_{bu}(z_u) + \lambda_{id} G_{bd}(z_d) - \lambda_{iu} - \lambda_{id} + s} \right) \Bigg|_{z_u=0, z_d=0} \tag{4.18}$$

In an analogous fashion, starting from

$$\begin{aligned}\psi^*_{k_u,k_d}(s) &= E[e^{-sX_{+,i}} | A_u(X_{-,i}) = k_u, A_d(X_{-,i}) = k_d] \\ &\quad \cdot \mathcal{P}[A_u(X_{-,i}) = k_u] \mathcal{P}[A_d(X_{-,i}) = k_d] \end{aligned} \tag{4.19}$$

we obtain

$$\psi^*_{k_u,k_d}(s) = \frac{1}{\overline{F_{is}}k_u!k_d!} \cdot \frac{d^{k_u}}{dz_u^{k_u}} \frac{d^{k_d}}{dz_d^{k_d}} \left(\frac{F^*_{is}(-\lambda_{iu}G_{bu}(z_u) + \lambda_{iu} - \lambda_{id}G_{bd}(z_d) + \lambda_{id}) - F^*_{is}(s)}{\lambda_{iu}G_{bu}(z_u) + \lambda_{id}G_{bd}(z_d) - \lambda_{iu} - \lambda_{id} + s} \right)\Bigg|_{z_u=0,z_d=0} \tag{4.20}$$

Then the system (4.16) can be transformed to

$$\begin{aligned}
\Omega^*_{k_u,k_d,i}(s) &= \frac{\overline{V_i}}{\eta_i}\left(q^i_{0,0} + \sum_{\mu=1}^{M-1}\pi^{i,(\mu)}_{0,0}\right)\phi^*_{k_u,k_d}(s) \\
&\quad + \frac{\overline{V_i}}{\eta_i}\sum_{j_u=0}^{k_u}\sum_{j_d=0}^{k_d}\pi^{i,(M)}_{j_u,j_d}\phi^*_{k_u-j_u,k_d-j_d}, \\
&\qquad\qquad 1 \le k_u \le K_u - 1, 1 \le k_d \le K_d - 1 \\
\Omega^*_{K_u,k_d,i}(s) &= \frac{\overline{V_i}}{\eta_i}\left(q^i_{0,0} + \sum_{\mu=1}^{M-1}\pi^{i,(\mu)}_{0,0}\right)\sum_{k_u=K_u}^{\infty}\phi^*_{k_u,k_d}(s) \\
&\quad + \frac{\overline{V_i}}{\eta_i}\sum_{j_u=0}^{K_u}\sum_{j_d=0}^{k_d}\pi^{i,(M)}_{0,0}\sum_{k_u=K_u-j_u}^{\infty}\phi^*_{k_u,k_d-j_d}(s),\ 1 \le k_d \le K_d - 1 \\
\Omega^*_{K_u,K_d,i}(s) &= \frac{\overline{V_i}}{\eta_i}\left(q^i_{0,0} + \sum_{\mu=1}^{M-1}\pi^{i,(\mu)}_{0,0}\right)\sum_{k_u=K_u}^{\infty}\sum_{k_d=K_d}^{\infty}\phi^*_{k_u,k_d}(s) \\
&\quad + \frac{\overline{V_i}}{\eta_i}\sum_{j_u=0}^{K_u}\sum_{j_d=0}^{K_d}\pi^{i,(M)}_{0,0}\sum_{k_u=K_u-j_u}^{\infty}\sum_{k_d=K_d-j_d}^{\infty}\phi^*_{k_u,k_d}(s) \\
\Pi^*_{k_u,k_d,1,i}(s) &= \frac{\overline{F_{is}}}{\eta_i}\sum_{j_u=1}^{k_u}\sum_{j_d=1}^{k_d}q^i_{j_u,j_d}\psi^*_{k_u-j_u,k_d-j_d}(s), \\
&\qquad\qquad 1 \le k_u \le K_u - 1, 1 \le k_d \le K_d - 1 \\
\Pi^*_{K_u,k_d,1,i}(s) &= \frac{\overline{F_{is}}}{\eta_i}\sum_{j_u=1}^{K_u}\sum_{j_d=1}^{k_d}q^i_{j_u,j_d}\sum_{k_u=K_u-j_u}^{\infty}\psi^*_{k_u,k_d-j_d}(s),\ 1 \le k_d \le K_d - 1 \\
\Pi^*_{K_u,K_d,1,i}(s) &= \frac{\overline{F_{is}}}{\eta_i}\sum_{j_u=1}^{K_u}\sum_{j_d=1}^{K_d}q^i_{j_u,j_d}\sum_{k_u=K_u-j_u}^{\infty}\sum_{k_d=K_d-j_d}^{\infty}\psi^*_{k_u,k_d}(s) \\
\Pi^*_{k_u,k_d,m,i}(s) &= \frac{\overline{F_{is}}}{\eta_i}\sum_{j_u=1}^{k_u}\sum_{j_d=1}^{k_d}\pi^{i,(\mu)}_{j_u,j_d}\psi^*_{k_u-j_u,k_d-j_d}(s), \\
&\qquad\qquad 1 \le k_u \le K_u - 1, 1 \le k_d \le K_d - 1, 2 \le \mu \le M
\end{aligned}$$

continued on next page ...

(4.21)

... continued from previous page

$$\Pi^*_{K_u,k_d,\mu,i}(s) = \frac{\overline{F_{is}}}{\eta_i} \sum_{j_u=1}^{K_u} \sum_{j_d=1}^{k_d} \pi^{i,(\mu)}_{j_u,j_d} \sum_{k_u=K_u-j_u}^{\infty} \psi^*_{k_u,k_d-j_d}(s),$$
$$1 \le k_d \le K_d - 1, 2 \le \mu \le M$$
$$\Pi^*_{K_u,K_d,\mu,i}(s) = \frac{\overline{F_{is}}}{\eta_i} \sum_{j_u=1}^{K_u} \sum_{j_d=1}^{K_d} \pi^{i,(\mu)}_{j_u,j_d} \sum_{k_u=K_u-j_u}^{\infty} \sum_{k_d=K_d-j_u}^{\infty} \psi^*_{k_u,k_d}(s), 2 \le \mu \le M \tag{4.21}$$

Equations for $\Omega^*_{k_u,K_d,i}(s))$, $\Pi^*_{k_u,K_d,1,i}(s)$, and $\Pi^*_{k_u,K_d,\mu,i}(s))$ are analogous to the second, fifth, and eighth of the equations above, respectively.

The joint distribution of the uplink and downlink queue size toward the slave at arbitrary time is

$$\mathcal{P}[L_{q,i} = (0,0)] = \Omega^*_{0,0,i}(0)$$
$$\mathcal{P}[L_{q,i} = (0,k_d)] = \Omega^*_{0,k_d,i}(0) + \sum_{\mu=1}^{M} \Pi^*_{0,k_d,\mu,i}(0), \quad 1 \le k_d \le K_d$$
$$\mathcal{P}[L_{q,i} = (k_u,0)] = \Omega^*_{k_u,0,i}(0) + \sum_{\mu=1}^{M} \Pi^*_{k_u,0\mu,i}(0), \quad 1 \le k_u \le K_u$$
$$\mathcal{P}[L_{q,i} = (k_u,k_d)] = \Omega^*_{k_u,k_d,i}(0) + \sum_{\mu=1}^{M} \Pi^*_{k_u,k_d,\mu,i}(0), \tag{4.22}$$
$$1 \le k_u \le K_u - 1, 1 \le k_d \le K_d - 1$$
$$\mathcal{P}[L_{q,i} = (K_u,k_d)] = \Omega^*_{K_u,k_d,i}(0) + \sum_{\mu=1}^{M} \Pi^*_{K_u,k_d,\mu,i}(0), 1 \le k_d \le K_d$$
$$\mathcal{P}[L_{q,i} = (k_u,K_d)] = \Omega^*_{k_u,K_d,i}(0) + \sum_{\mu=1}^{M} \Pi^*_{k_u,K_d,\mu,i}(0), 1 \le k_u \le K_u$$

We can now calculate the burst blocking probability, under total rejection policy, in the uplink queue of slave i at arbitrary times. (Note that the corresponding blocking probability can be calculated for the downlink queues at the master as well.) To that end, let us denote the mass probability of burst size being equal to l packets as $g_l = \frac{1}{l!}\frac{d^l}{dz^l} G_{bu}(z)|_{z=0}$. Then we obtain

$$\begin{aligned} P_{B,i,u} &= \sum_{k_u=0}^{K_u} \sum_{k_d=0}^{K_d} \mathcal{P}[L_{q,i} = (k_u,k_d)]\mathcal{P}[\text{burst} > K_u - k_u] \\ &= \sum_{k_u=0}^{K_u} \sum_{k_d=0}^{K_d} \mathcal{P}[L_{q,i} = (k_u,k_d)] \sum_{l=K_u-k_u}^{\infty} g_l \end{aligned} \tag{4.23}$$

The joint queue length distributions contain information about remaining service/vacation time. This enables us to determine the LST for the access delay for the first packet in the burst. Let us denote the blocking probability for the first packet of the burst in the uplink queue as $P^g_{B,i,u} = \sum_{k_d=0}^{K_d} \left(\Omega^*_{K_u,k_d,u}(0) + \sum_{\mu=1}^{M} \Pi^*_{K_u,k_d,\mu,i}(0) \right)$. Then the access delay of the first packet in the burst becomes

$$W^*_{g,ai}(s) = \frac{1}{1-P^g_{B,i,u}} \left(\sum_{k_d=0}^{K_d} \sum_{k_u=0}^{K_u-1} \Omega^*_{k_u,k_d,i}(s) F^*_{is}(s)^{k_u} V^*_i(s)^{\lfloor k/M \rfloor} \right. \\ \left. + \sum_{k_d=0}^{K_d} \sum_{k_u=1}^{K_u-1} \sum_{\mu=1}^{M} \Pi^*_{k_u,k_d,\mu,i}(s) F^*_{is}(s)^{(k_u-1)} V^*_i(s)^{\lfloor (k_u+\mu-1)/M \rfloor} \right) \tag{4.24}$$

The mean access delay for the first packet in the burst is determined as

$$\overline{W_{g,ai}} = -\frac{d}{ds} W^*_{g,ai}(s)\Big|_{s=0} \\ = \frac{1}{\lambda^2_{iu}(1-P^g_{B,i,u})\eta_i} \sum_{k_d=0}^{K_d} \sum_{k_u=1}^{K_u-1} k_u \left(\sum_{\mu=1}^{M} \pi^{i,(\mu)}_{k_u,k_d} \right) + \frac{K_u}{\lambda_{iu}} \frac{P^g_{B,i,u}}{1-P^g_{B,i,u}} - \overline{F_{is}} \tag{4.25}$$

In order to calculate the delay of the arbitrary packet in the burst, we need the distribution of the distance of the arbitrary packet from the first packet in the burst. The probability that there are k packets before the arbitrary packet in the burst is $g^-_k = \frac{1}{\overline{B}} \sum_{j=k+1}^{\infty} g_j$. Then, the LST for the delay of the arbitrary packet in the accepted burst becomes

$$W^*_{ai}(s) = \frac{\sum_{k_d=0}^{K_d} \sum_{k_u=0}^{K_u-1} \Omega^*_{k_u,k_d,i}(s) \sum_{j=1}^{K_u-k_u} g_j \sum_{w=1}^{j-1} g^-_w F^*_{is}(s)^{(k_u+w)} V^*_i(s)^{\lfloor (k_u+w)/M \rfloor}}{1-P_{B,i,u}} \\ + \frac{\sum_{k_d=0}^{K_d} \sum_{k_u=1}^{K_u-1} \sum_{\mu=1}^{M} \Pi^*_{k_u,k_d,\mu,i}(s) \sum_{j=1}^{K_u-k_u} g_j \sum_{w=1}^{j-1} g^-_w F^*_{is}(s)^{(k_u-1)} V^*_i(s)^{\lfloor (k_u+\mu+w-1)/M \rfloor}}{1-P_{B,i,u}} \tag{4.26}$$

4.3 Experimental results

In order to validate the results of the queueing theoretic analysis, we have modified our Bluetooth piconet simulator by introducing buffers with variable (and adjustable) size. The uplink buffers at each slave had the same size. The size of the master queue has been obtained for all downlink queues taken together, as this was the more realistic scenario. All buffers were operated under the total rejection policy, i.e., all packets from a packet burst were rejected if the available buffer space could not accommodate the entire burst.

A series of experiments has been conducted to measure the blocking probability and packet delays (both access delay and end-to-end delay) in the piconet. We have used the E-limited polling scheme described in Chapter 3, with the parameter M set as specified below.

The first experiment deals with the blocking probability at the master queue under varying offered load ρ; the slave buffer size has been fixed at 10KBytes. The results are shown in Fig. 4.2; values of blocking probability less than 10^{-5} are not shown for clarity. As can be seen, the blocking probability increases with offered load and decreases with the buffer size. Master buffer size of a few kilobytes or more can guarantee the blocking probability less than 10^{-4} at loads up to $\rho = 0.5$, which is probably more than will be encountered in practice. Since packet losses are small, there was no significant difference in packet delays.

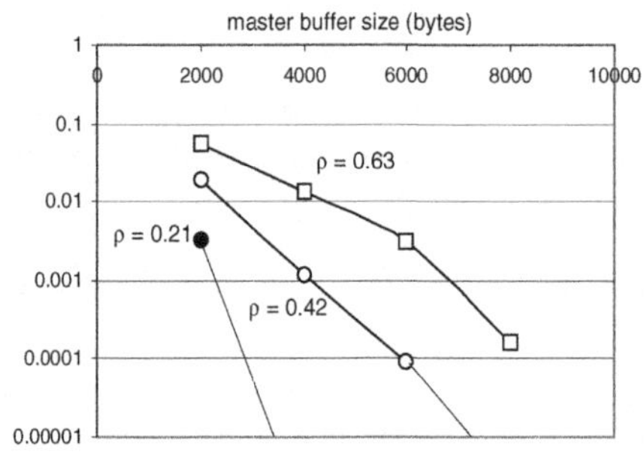

FIGURE 4.2
Blocking probability at the master buffer as the function of piconet load ρ.

The situation is slightly different at the slave uplink queues, as can be seen from Fig. 4.3. In this case, the master buffer size was fixed at 10KBytes. Again, larger buffers lead to lower blocking probability, but for smaller buffers the offered load does not affect this probability too much. At the same time, the packet delays are lower, but the reason for this decrease is actually the decrease in the number of packets that are allowed to get through, rather than a performance improvement.

The next experiment investigates the effect of packet burstiness, as expressed by varying the mean burst size while keeping the overall packet arrival rate constant. The performance under varying slave uplink buffer size is shown in Fig. 4.4(a); in all cases, the polling parameter has been fixed at $M = 3$, while the offered load was kept at $\rho = 0.42$. As can be seen, more bursty traffic corresponds to higher blocking

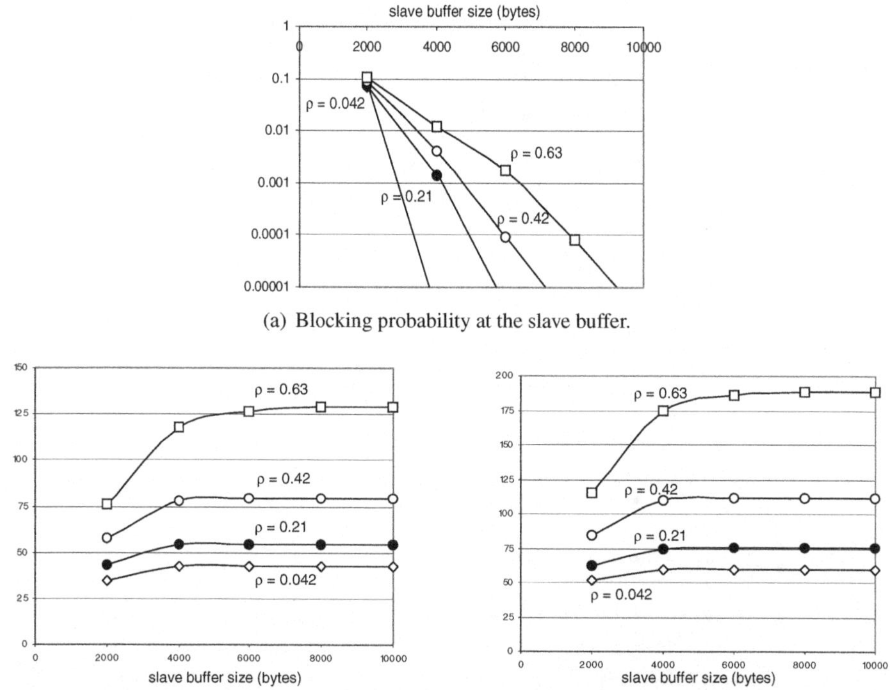

(a) Blocking probability at the slave buffer.

(b) Access delay.

(c) End-to-end packet delay.

FIGURE 4.3
Performance of the finite slave uplink buffer as the function of piconet load ρ.

probability, but also to lower packet delays. The explanation of this phenomenon is similar to the previous one: for packet bursts with sizes up to the polling parameter, $\overline{B} \leq M$, the master is able to grab the entire burst from the slave, and therefore the delays are not much affected. At the same time, the blocking probability is small even for small uplink buffers, as the burst is quickly transferred to the master.

When the mean burst size is higher, the master cannot transfer the entire burst during a single visit to the slave. As the consequence, the probability that the burst will not fit in the uplink buffer increases. For very bursty traffic and small buffer sizes, the blocking probability may well exceed 0.1, which makes the piconet hardly usable for data traffic. (As explained above, a reliable transport protocol such as TCP will probably take care of packet losses, but at the expense of enormous increase in packet delays.)

The behavior is slightly different at the master, as can be seen from Fig. 4.5. In this case, the blocking probability at small buffer sizes – although high – does not change much for bursts of five or more packets. This may be explained by the fact that large bursts arrive infrequently, and the master is still able to handle them with (relative) ease. The difference in packet delays was small and it's not shown.

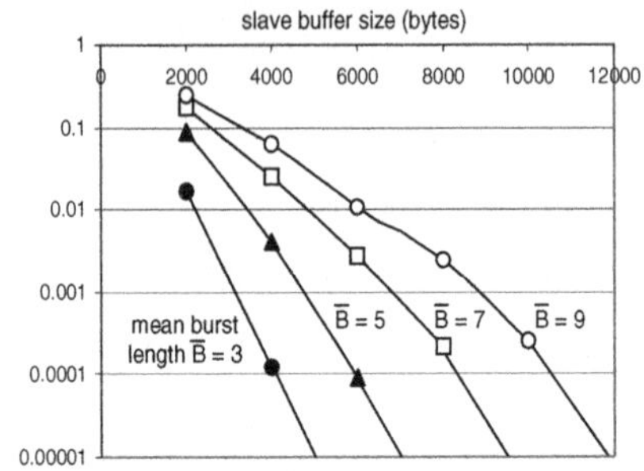

(a) Blocking probability at the slave buffer.

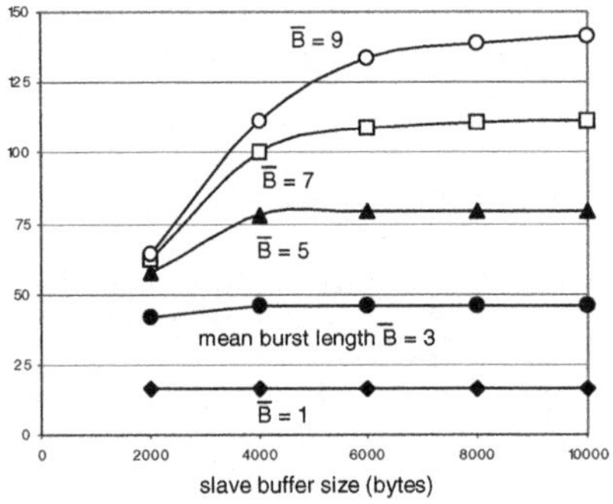

(b) Access delay.

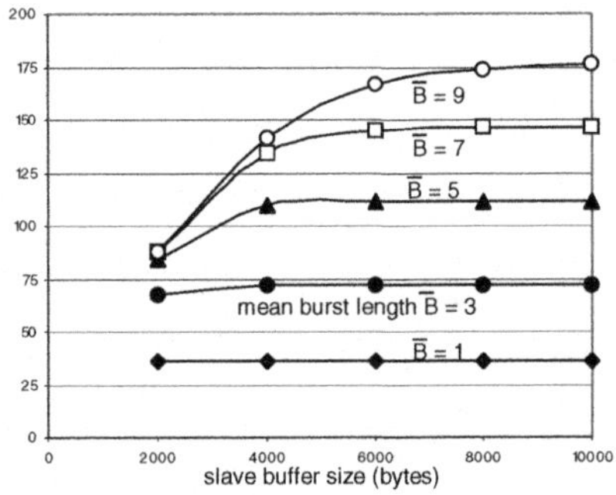

(c) End-to-end packet delay.

FIGURE 4.4
Performance of the finite slave uplink buffer as the function of mean burst size $\overline{B}$.

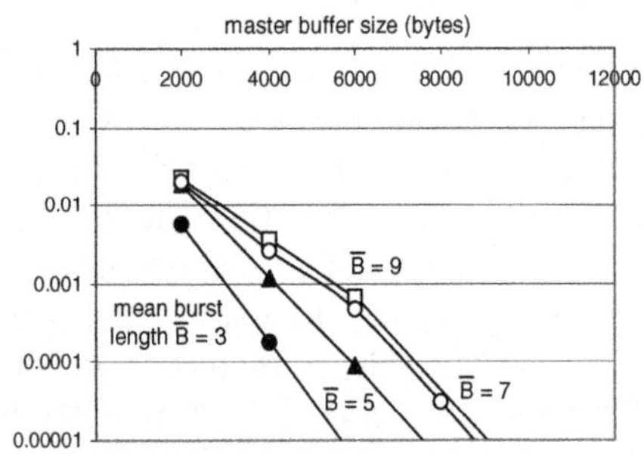

FIGURE 4.5
Blocking probability at the master buffer as the function of mean burst size $\overline{B}$.

We have also measured the performance of the piconet under varying polling parameter M and varying buffer sizes. Blocking probability and end-to-end packet delays with limited slave buffer size are shown in Fig. 4.6. As can be seen, the

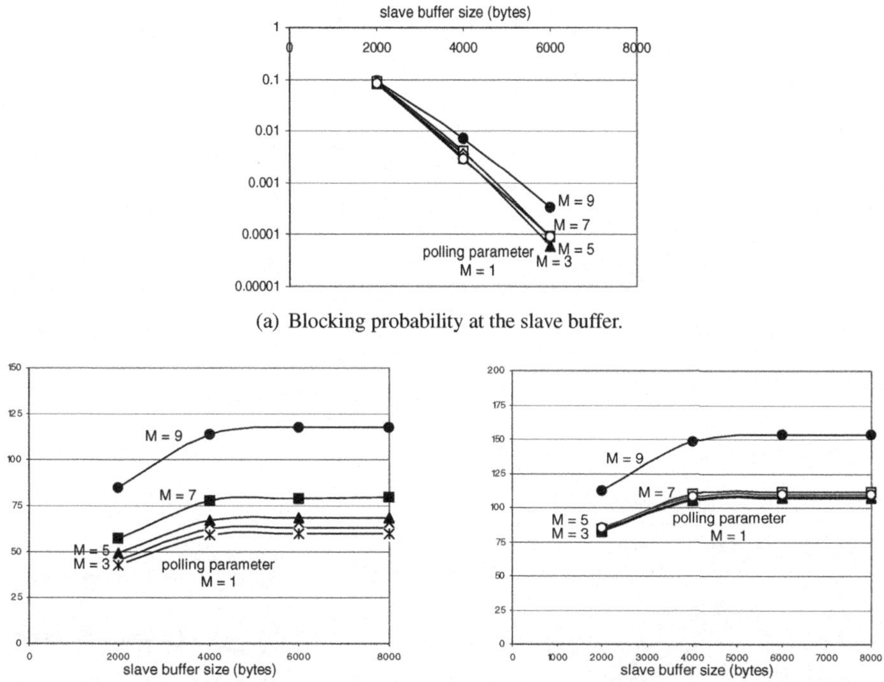

(a) Blocking probability at the slave buffer.

(b) Access delay.

(c) End-to-end packet delay.

FIGURE 4.6
Performance of finite slave uplink buffer as the function of the polling parameter M.

increase in M leads to a decrease in blocking probability at the slave, as does the increase in slave uplink buffer size – but this becomes more pronounced only for higher values of M. At the same time, the end-to-end packet delay is consistently decreasing with M, while the access delay decreases slightly, the largest jump being the one that corresponds to the increase of M from 1 to 3. (This is consistent with earlier results in [Mišić and Mišić, 2003*b*].)

In case of finite buffer size at the master, Fig. 4.7 shows the blocking probability and the end-to-end packet delays for values of M from 1 to 9. The behavior is quite similar to that at the slave uplink queue.

In all cases, the buffer sizes of a few Kbytes – say, 6 to 10 – are sufficient to ensure that the resulting blocking probability at the device, be it master or slave, will remain lower than 10^{-5} under typical values of piconet load.

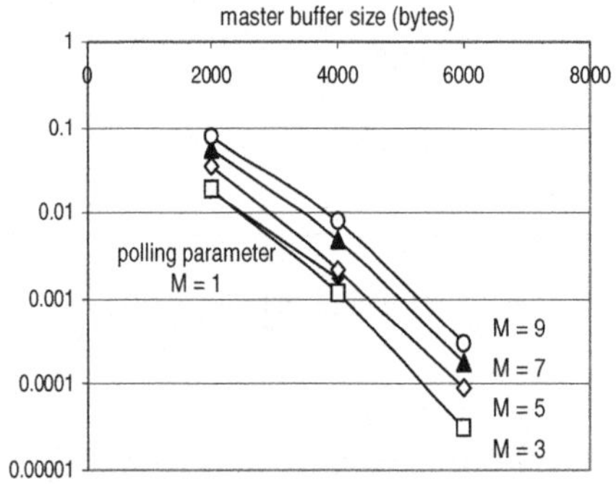

(a) Blocking probability at the master buffer.

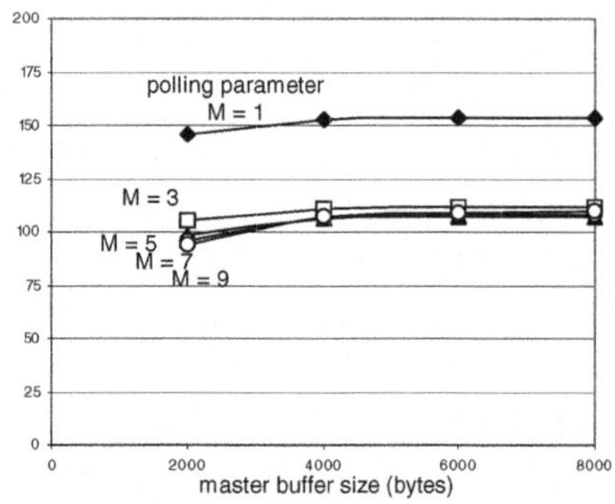

(b) End-to-end packet delay.

FIGURE 4.7
Performance of finite master buffer as the function of the polling parameter M.

5

Admission control

If the performance of Bluetooth networks is to remain within acceptable limits, the piconet master must carefully manage all aspects of piconet operation. Earlier chapters have discussed a number of polling algorithms and the choice of their parameters, which may be considered to belong to 'tactical' and 'operational' levels of management, respectively. However, management also involves decisions that have to be made at the time the piconet is formed: whether to invite (and subsequently admit) another device or not, and what would be the impact of this admission on the performance of the piconet. This chapter is devoted to issues of admission control in Bluetooth piconets, and it shows that a range of algorithms can be utilized to exercise such control in order to maintain the performance indicators within predefined bounds. As shown in Chapter 3, the critical performance indicators mainly depend on the total piconet load, rather than on the traffic load of individual slaves. This observation will be used as the foundation upon which three algorithms for admission control are designed.

The simplest algorithm is based on queue stability or, in case the buffer size is limited, the blocking probability. Either way, its low computational complexity makes it suitable for battery power-limited master nodes. The second algorithm estimates the access delay of the slave upon admission using the estimated first and second moments of the vacation times, and makes the admission decision on the basis of predefined access delay bounds. The third algorithm is based on predefined cycle time bounds, which makes it suitable for applications that generate Constant-Bit Rate data flows.

Either of these admission control schemes could easily be incorporated in the formation algorithms for both piconets and scatternets. In the inquiry phase, the piconet master may use a particular IAC code [Bluetooth SIG, 2003*a*] to advertise its use of the chosen admission policy. The slaves responding to the inquiry would provide information about their anticipated packet arrival rate and, possibly, about the mean burst sizes. Alternatively, if the segmentation and reassembly process is done by the same systems software layer, such as BNEP [Bluetooth SIG, 2001*a*], all devices may be assumed to generate packet bursts with the same mean size, and this information need not be supplied. The master can then decide to page, and subsequently admit, only those slaves for which the appropriate admission criterion will be satisfied, and thus enable the piconet to operate within the desired limits of quality of service.

In all cases, we assume that the mean flow rate of each slave must be known at the time of admission; the same assumption has been adopted by other authors who discuss admission control in Bluetooth piconets [Ait Yaiz and Heijenk, 2001;

Lapeyrie and Turletti, 2003]. In practice, intelligent devices can provide the actual or estimated flow rates of their respective applications, while simpler devices, such as mice or keyboards, could be admitted using the default rates for that particular type of device.

The three admission control algorithms are presented in Sections 5.1, 5.2, and 5.3, respectively.

5.1 Admission control based on queue stability

For traffic which is not delay sensitive, it suffices to check the stability condition of uplink (and, possibly, downlink) queues upon the admission of new traffic flow. (We will refer to this algorithm as QS, for Queue Stability.) If the stability conditions are met for all the queues, the new slave can be admitted to the piconet. If one or more of the stability conditions are not met, the admission of a new slave will cause one or more uplink queues to become unstable. The existence of an unstable queue, or several of them, means that the access delays will increase indefinitely (first in those queues, and then in others as well) and buffer overflows will occur.

In case of uniform probability of all destinations, the downlink queues are fed in by the $(m-1)$-th part of the sum of all uplink burst arrival rates, and the downlink traffic will be symmetric. Consequently, if all uplink queues are stable, the downlink queues will be stable as well. In case the distribution of downlink traffic is not uniform, the algorithm should check the stability of the downlink queues as well.

The QS algorithm operates as follows. Assume that $m-1 \leq 6$ traffic flows are already admitted, and a new slave x is about to be admitted to the network. (We assume that the inquiry phase has been completed, and that the new slave has to be paged and accepted to the network.) The admission process requires that the new slave has to specify the uplink burst arrival rate λ_{xu} and, optionally, the mean burst size $\overline{B_{x,u}}$. (We assume that all slaves will have the same uplink burst size distribution.)

The piconet master, then, has to verify that the admission of data traffic to and from the new slave will not violate the stability of the uplink queues of the slaves that are already members of the piconet. This may be accomplished by calculating the new vacation times, i.e., if the master solves the system of equations from Section 3.3.

However, solving the system, even in a simplified form, can be time- and energy-consuming. If the piconet master operates under limited power conditions, the same task may be accomplished in a simpler and faster manner. We assume that the master is keeping a table that numerically describes the dependency of these probabilities on the total load in the piconet – essentially, a two-dimensional representation of the data shown in Fig. 3.7. In cases where power limitations do not exist, an algorithmic approximation might be used as well.

First, the master calculates the new total offered load, using the packet burst arrival

TABLE 5.1
Slave load and activation sequence in the simulation of the QS admission scheme. (From J. Mišić, K. L. Chan, and V. B. Mišić, "Admission control in Bluetooth piconets," *IEEE Trans. Veh. Tech.* **53**(3):890–911, © 2004 IEEE. Reprinted with permission.)

Slave number	Starts at (s)	With load	Giving total load ρ
1	0	0.06	0.06
2	0	0.12	0.18
3	300	0.18	0.36
4	600	0.24	0.60
5	900	0.30	0.90
6	1200	0.36	1.26

rates λ_{iu} and mean burst sizes $\overline{B_{iu}}$.

Second, using the tabular dependency kept in memory, the master obtains the value for $\overline{C}$ that corresponds to the new total load. As discussed in Chapter 3, these values will be nearly equal for all slaves, regardless of their individual packet arrival rates.

In the third step, the master verifies the stability condition for the queues of all slaves. The stability condition for the uplink queue i is based on the fact that the average number of packets that arrive in the slave queue during its average cycle time must be less than M:

$$\lambda_{iu}\overline{BC} < M \tag{5.1}$$

Assuming that the destinations are uniformly distributed among the slaves, the stability condition for downlink queues does not need to be checked. However, in cases where the assumption does not hold, the stability of downlink queues has to be checked using the steps outlined above.

To illustrate the operation of the QS admission scheme, we have modified our Bluetooth simulator so that new slaves are introduced at predefined times, thus simulating admission. Each of the slaves starts transmitting data at a predefined time and contributes a predefined portion of the total load. The number of slaves, their loads (relative to the rated load of the piconet), total offered load and the time of their admission, are shown in Table 5.1. Note that the rated load of the piconet is actually exceeded in the presence of all six slaves. In all measurements, we have used $\overline{B} = M = 3$.

Then, we have calculated the stability conditions and measured the mean access delay for each uplink queue over a period of $1400s$, or slightly more than 23 minutes, of simulated time. The resulting diagrams of the total offered load and calculated stability conditions are shown together on Fig. 5.1(a), while Fig. 5.1(b) shows the measured access delays, averaged over $10s$ intervals.

As can be seen, the delays increase with each added slave. With two and three slaves, mean access delay is well below $50T$. The admission of the fourth slave

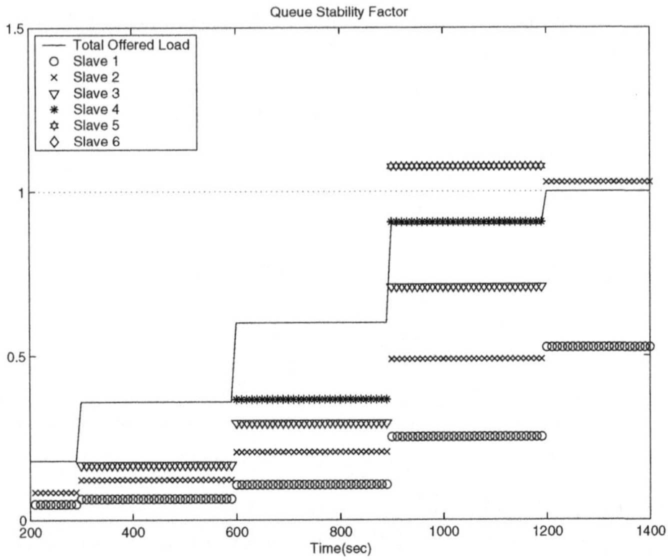

(a) Offered load and queue stability conditions.

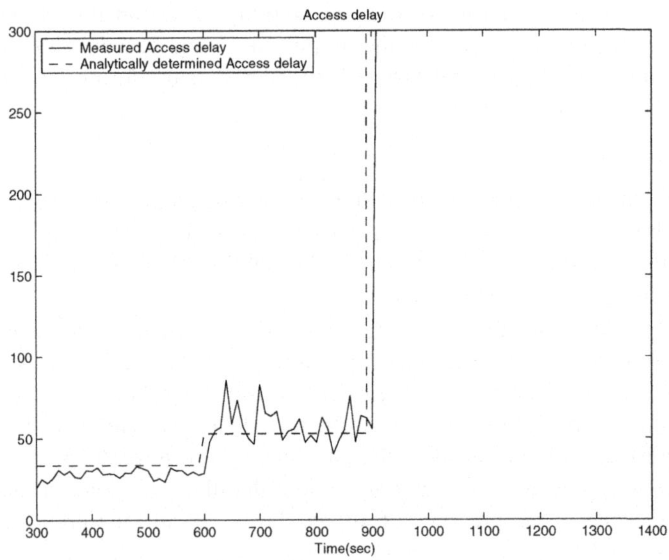

(b) Mean access delay at the slave uplink queues.

FIGURE 5.1
Pertaining to the operation of the QS admission control scheme. (From J. Mišić, K. L. Chan, and V. B. Mišić, "Admission control in Bluetooth piconets," *IEEE Trans. Veh. Tech.* **53**(3):890–911, © 2004 IEEE. Reprinted with permission.)

brings the total load to $\rho = 0.6$, and the mean delay rises to about $60T$. (The noise spikes are the consequence of bursty traffic.)

However, when the fifth slave is admitted, the stability condition for its uplink queue will be violated, even though the overall piconet load is still $\rho = 0.9$. Obviously, the QS algorithm would have rejected the sixth slave.

In our experiment, however, this slave did get admitted. As a consequence, the access delay has risen to about 900 to $1000T$, well beyond the range shown in Fig. 5.1(b), and thereafter increases indefinitely – the queues can never be emptied. Note that $800T$ is equal to $0.5s$, and this is access delay only; end-to-end delays can be much longer, possibly even twice as long. If the sixth slave is admitted, the stability conditions will be violated for five uplink queues, and the piconet will effectively cease to function.

For increased safety, the stability threshold could be set to a value smaller than 1, such as 0.9 or even 0.8.

In case of finite buffers, the blocking probability should be checked upon new flow arrival instead of the stability condition. However, the analysis from [Takagi, 1993*a*] shows that stability condition and blocking probability are correlated. Since the stability condition is easier to compute than the estimate of blocking probability, we believe it is more suitable for use in admission control.

5.2 Admission control based on access delay

In cases where traffic is delay sensitive and access delay is the critical parameter, a more accurate (but also more complex) admission control scheme may be utilized. Under this scheme, when the new slave requests to be admitted, the master has to recalculate mean access delays for all existing slaves and calculate the access delay for the new slave. (This will also limit the end-to-end delay, which, in general, does not exceed twice the value of the access delay.) If any of these delays exceeds the agreed-upon limits, the new slave will not be admitted.

The mean access delay under E-limited polling was calculated in Chapter 3; we repeat the expression here for convenience:

$$\overline{W_{ia}} = \frac{\lambda_{iu}\overline{B}(\overline{F_{is}^2})}{(1 - \lambda_{iu}\overline{F_{is}B})} + \frac{\overline{B^{(2)}2}}{2\overline{B}(1 - \lambda_{iu}\overline{F_{is}B})} + \frac{\overline{V_i^2}}{2\overline{V_i}} - \overline{V_i} + \frac{Q_i'(1,1)}{\lambda_{iu}\overline{B}Q_i(1,1)} \quad (5.2)$$

We use $\rho_{iu} = \lambda_{iu}\overline{BF_{is}}$ to denote the portion of the uplink load contributed by the slave i, while the value of $Q_i'(1, 1)$ is obtained as

$$
\begin{aligned}
Q_i'(1,1) &= Q_i(1,1)(\lambda_{iu}\overline{B}\overline{V_i} + M) \\
&- \frac{\left(\rho_{iu}^2/2 + 2\overline{L^2}(\lambda_{iu}\overline{B})^2 + 2\lambda_{iu}\overline{L}\overline{B^{(2)}} - 3\rho_{iu} + 2\right)}{2((1-\rho_{iu}) + \lambda_{iu}\overline{B}\overline{V_i})} \\
&- \frac{q_{0,0}^i M(M-1)\left((1-\rho_{iu})^2 - \rho_{iu}\right) + \Pi_{i,0}''(1,1)(1-\rho_{iu})^2}{2(M(1-\rho_{iu}) - \lambda_{iu}\overline{B}\,\overline{V_i})} \\
&+ \frac{Q_i(1,1)\left(-M(M-1) + \overline{V_i^2}(\lambda_{iu}\overline{B})^2 + \overline{V_i}\lambda_{iu}\overline{B^{(2)}} + 4M\overline{V_i}\,\overline{L}(\lambda_{iu}\overline{B})^2\right)}{2(M(1-\rho_{iu}) - \lambda_{iu}\overline{B}\overline{V_i})} \\
&+ \frac{Q_i(1,1)\left(M(2M-1)\frac{\rho_{iu}^2}{2} + 2M\overline{L^2}(\lambda_{iu}\overline{B})^2 + M\lambda_{iu}\overline{F_{is}}\,\overline{B^{(2)}}\right)}{2(M(1-\rho_{iu}) - \lambda_{iu}\overline{B}\overline{V_i})}
\end{aligned}
\tag{5.3}
$$

where $\Pi_{i,0}(z,w) = \sum_{\mu=1}^{M-1} \pi_{0,0}^{i,(\mu)} z^{M-\mu} w^{M-\mu}$.

It may be interesting to note that the denominators of the last three components actually include the stability conditions utilized in the QS scheme. Therefore, this scheme is more restrictive than the QS scheme described above.

The operation of the AD admission control scheme will be illustrated with an example. Let us assume that the individual slaves are admitted to the piconet according to the data shown in Table 5.2, and that the access delay threshold is set to $150T$. For reference, we have shown the total offered load and queue stability conditions in Fig. 5.2(a), while the estimated and measured values of access delay are shown in Fig. 5.2(b).

TABLE 5.2
Slave load and activation sequence in the simulation data of the AD and CT admission schemes. (From J. Mišić, K. L. Chan, and V. B. Mišić, "Admission control in Bluetooth piconets," *IEEE Trans. Veh. Tech.* **53**(3):890–911, © 2004 IEEE. Reprinted with permission.)

Slave number	Starts at (s)	With load	Giving total load ρ
1	0	0.04	0.04
2	0	0.08	0.12
3	300	0.12	0.24
4	600	0.16	0.40
5	900	0.20	0.50
6	1200	0.24	0.84

As can be seen, the admission of the fifth slave at $t = 900s$ leads to an increase in access delay, but its mean values are well below $150T$. However, when the sixth slave is about to be added at $t = 1200s$, the AD algorithm estimates that the access delay will rise to about $400T$, or $0.25s$. As this value exceeds the predefined threshold of $150T$, this slave would be rejected. In case it is admitted, the delays indeed increase well above the threshold. Of course, a threshold set at a sufficiently high value might have allowed the admission of the sixth slave, while a lower threshold might not have admitted the fifth, or even the fourth one. Note that the total offered load is below the rated piconet load, with the maximum of $\rho = 0.84$; hence, the QS admission criteria would be satisfied by all six slaves in this case.

We note that the AD admission control scheme requires more information to be kept by the master than its QS counterpart, even though we retain the fitted dependencies of service and vacation times on the marginal probabilities $\sum_{kd=0}^{\infty} q^i_{0,kd}/Q_i(1,1)$ and $\sum_{ku=0}^{\infty} q^i_{ku,0}/Q_i(1,1)$. First, it requires estimation of the second moment of vacation time; this may be accomplished by using the new estimated average vacation time together with the variance of the previous cycle time. Second, it requires estimation of the probabilities that the slave uplink queues are empty after the corresponding service periods, i.e., $\sum_{kd=0}^{\infty} \pi^{i,(m)}_{0,kd}$, $m = 1..M-1$; these probabilities can be estimated in the same manner as $\sum_{kd=0}^{\infty} q^i_{0,kd}/Q_i(1,1)$. For simplicity, the term $\Pi''_{i,0}(1,1)(1-\rho_{iu})^2$ can be neglected, which leads to a slight overestimation of the access delay. We also assume that both first and second moments of the packet length distribution and burst length distribution are known to the master; as mentioned above, this information may be available if all devices use the same segmentation/reassembly algorithm.

5.3 Admission control based on cycle time

In some cases, the data flow from or to a slave might have a specified bandwidth; such is the case with audio and video streams, as well as with other types of traffic commonly referred to as CBR traffic. The Bluetooth specification indeed provides rudimentary support for CBR traffic at the LMP level [Bluetooth SIG, 2001*b*]. The master and an ACL slave may set up the maximum polling interval T_{poll}, i.e., the maximum time between subsequent transmissions. This polling interval is guaranteed in the active mode, except when there are collisions with page, page scan, inquiry, and inquiry scan. The value of the polling interval may be negotiated between

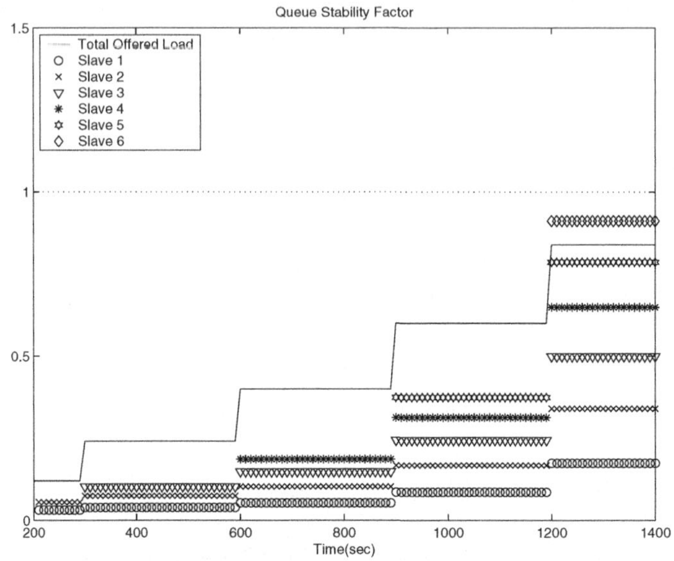

(a) Offered load and queue stability conditions.

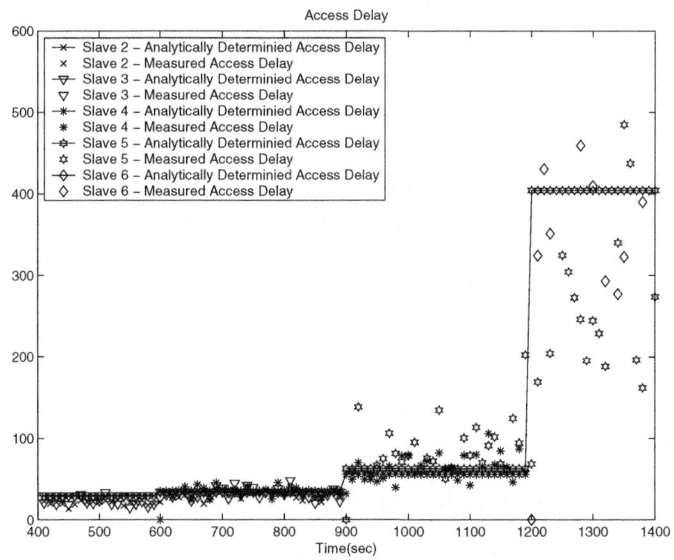

(b) Mean access delay at the slave uplink queues.

FIGURE 5.2
Pertaining to the operation of AD admission control scheme. (From J. Mišić, K. L. Chan, and V. B. Mišić, "Admission control in Bluetooth piconets," *IEEE Trans. Veh. Tech.* **53**(3):890–911, © 2004 IEEE. Reprinted with permission.)

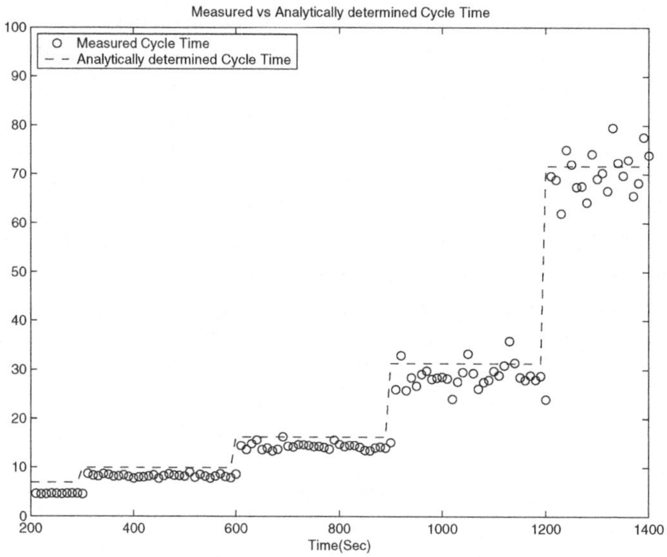

FIGURE 5.3
Pertaining to the operation of CT admission control scheme. (From J. Mišić, K. L. Chan, and V. B. Mišić, "Admission control in Bluetooth piconets," *IEEE Trans. Veh. Tech.* **53**(3):890–911, © 2004 IEEE. Reprinted with permission.)

the master and the slave, and the constraint is set up only when both participants accept it.

However, if the bandwidth requirements are known in advance, as is often the case, it may be advantageous to perform the check and/or negotiation before the slave is admitted to the piconet, rather than afterward. Fortunately, the AD admission control scheme may easily be adapted to cater for this case. Instead of targeting the mean access delay, the master should use Equations (3.42), (3.43), and (13.42) to obtain an estimate for mean cycle time. Note that, in this case, the variances of service times are not needed, hence the computational requirements will be somewhat lower than under the AD admission control scheme. If the estimated mean cycle time exceeds the limits required by either the new slave, or one or more of the slaves already admitted to the piconet, the new slave will be rejected.

To illustrate this admission control scheme, let us consider again the example from Table 5.2. Let us assume that the traffic of the fifth slave is actually CBR traffic with approximately $112Kbps$ bandwidth. (It might be a video broadcast coming from the Internet, with the master as the access point to a wired LAN.) Assuming that the broadcast uses $B = 3$ packets of DH3 type with 183 bytes each, the maximum cycle time to achieve the required bandwidth is about $63T$. (We assume that the broadcast is unidirectional, and that the slave responds with single slot packet most of the time.)

Now, consider the estimated and measured values of cycle time, as in Fig. 5.3.

When the fifth slave requests admission at $t = 900s$, the piconet master will estimate the new cycle time. Since it is below the required value of $63T$, the fifth slave will get admitted. As can be seen, the new cycle time is indeed below that value. However, at $t = 1200s$, the sixth slave requests to be admitted; the master calculates the new estimate for the cycle time, which turns out to be about $70T$. As this violates the previous agreement with the fifth slave, the sixth slave will be rejected.

As before, the diagram shows the measured values of mean cycle time if this slave were admitted after all, averaged over $10s$ intervals. As can be seen, the average value of the cycle time is rather close to the previously calculated estimate.

Finally, we note that the analytical approach used in Chapters 3 and 4 actually gives the probability distribution for service times. Therefore, it is possible to calculate the maximum piconet cycle time that will not be exceeded with a given probability, rather than just the mean cycle time. However, such calculation is likely to be prohibitively expensive, and it is questionable whether the increased accuracy would be worth the effort.

6

Performance of TCP traffic

Initial use of Bluetooth was simple cable replacement, rather than general networking tasks. However, the number of possible uses has grown to include various networking tasks between computers and computer-controlled devices such as PDAs, mobile phones, smart peripherals, and the like. As a consequence, the majority of the traffic over Bluetooth networks will belong to different flavors of the ubiquitous TCP/IP family of protocols. Even though a number of early papers have analyzed the performance of Bluetooth networks under TCP traffic, they have generally adopted a somewhat simplistic approach that does not account for interactions between the TCP congestion control and the L2CAP flow control and baseband polling mechanisms. In this case, TCP segments are embedded into IP datagrams and sent to the L2CAP layer, where they are segmented into Bluetooth packets and placed in infinite-size buffers; the baseband layer then takes individual packets and sends them over the radio link. The TCP congestion window and transmission rate will increase until time-out events occur due to high end-to-end delays. However, these delays will be caused by very high loads, more or less independently of the polling scheme utilized at the baseband layer. In this manner, it is hard, or even impossible, to gain in-depth knowledge of the performance of TCP traffic in Bluetooth networks.

Furthermore, the adoption of version 1.2 of the Bluetooth specification [Bluetooth SIG, 2003*a*] has introduced significant changes in the manner in which the Bluetooth protocol stack operates. According to version 1.2, each L2CAP channel may operate in either Basic L2CAP mode (essentially identical to the L2CAP mode supported by the previous version of the specification [Bluetooth SIG, 2001*b*]), Flow Control, or Retransmission mode [Bluetooth SIG, 2003*d*]. All three modes offer segmentation, but only the latter two control buffering through protocol data unit (PDU) sequence numbers, and control the flow rate by limiting the required buffer space. Additionally, the Retransmission Mode uses a go-back-n repeat mechanism with an appropriate timer to retransmit missing or damaged PDUs as necessary. The architectural blocks of the L2CAP layer are schematically shown in Fig. 6.1(a).

It is clear that complex interactions between the TCP congestion control and the L2CAP flow control and baseband polling mechanisms may be expected in a Bluetooth network carrying TCP traffic. The main focus of this chapter is to investigate those interactions and thus provide insight into the performance of TCP data traffic in Bluetooth piconets. We present the analytical model for the segment loss probability, probability distribution of TCP round trip time (including the L2CAP round trip time) and TCP sending rate. We also discuss the dimensioning of various parameters of the architecture with regard to the tradeoff between end-to-end delay and

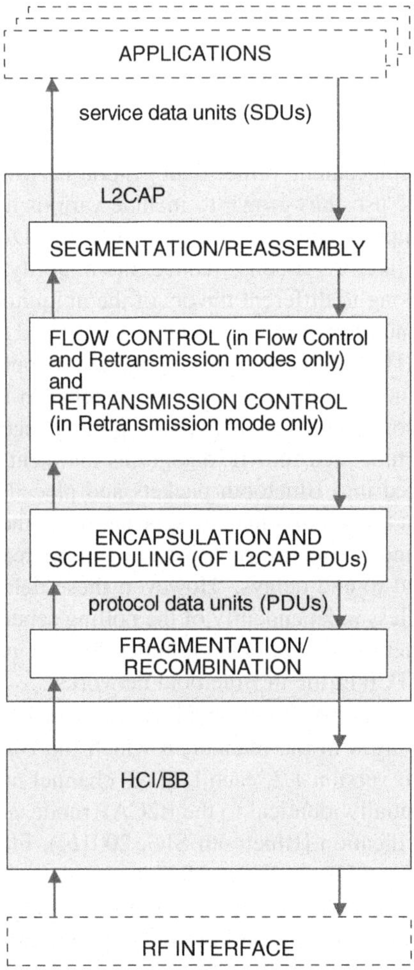

(a) Original specification from (Bluetooth SIG, 2003*c*).

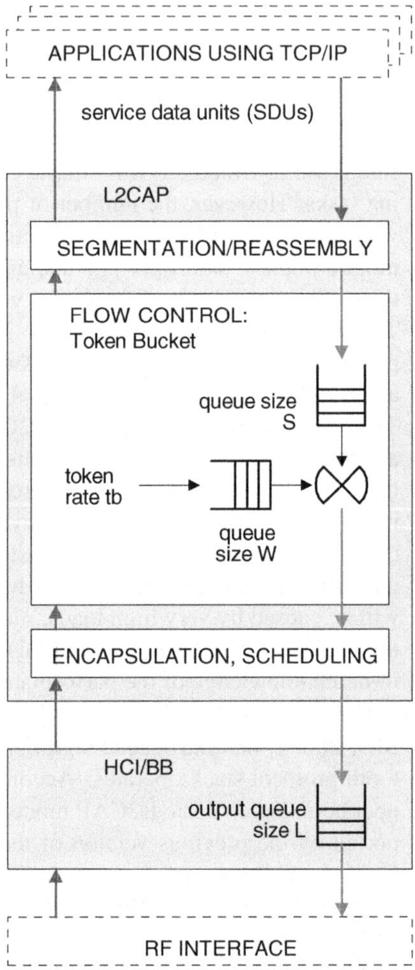

(b) Our test setup of the Flow Control mode.

FIGURE 6.1
Architectural blocks of the Bluetooth L2CAP layer (control paths not shown for clarity).

achievable throughput under multiple TCP connections from different slave devices.

The Chapter is organized as follows. Section 6.1 describes the system model and the basic assumptions about the piconet operation under TCP traffic, and discusses some recent related work in the area of performance modeling and analysis of TCP traffic. Section 6.2 discusses the segment loss probability and the probability distribution of the congestion window size, and presents the analytical results. Derivations of the probability density functions for the TCP segment and acknowledgment delay, as well as the blocking probability through the token bucket filter and the output queue serviced by the E-limited poller, are presented in Sections 6.3 and 6.4. Finally, Section 6.5 presents simulation results for the performance of piconet with slave-to-slave TCP connections.

6.1 System model and related work

We consider a piconet in which the slaves create TCP connections with each other, so that each TCP connection will traverse two hops in the network. There are no TCP connections starting or ending at the master. We focus on TCP Reno [Jacobson, 1990*b*], which is a widely used variant of TCP at the moment of this writing. We assume that the application layer at slave i (where $i = 1 \mathinner{..} m-1$) sends messages of 1460 bytes at a rate of λ_i. This message will be sent within a single TCP segment, provided the TCP congestion window is not full. The TCP segment will be encapsulated in an IP packet with the appropriate header; this packet is then passed on to the L2CAP layer. We assume that the segmentation algorithm produces the minimum number of Bluetooth baseband packets [Kalia, Bansal and Shorey, 2000], i.e., four DH5 packets and one DH3 packet per TCP segment [Bluetooth SIG, 2003*c*]. We also assume that each TCP segment will be acknowledged with a single, empty TCP segment carrying only the TCP ACK bit; due to the size of the TCP header, such acknowledgment requires one three-slot (e.g., DH3) baseband packet. Therefore, the total throughput per piconet can reach a theoretical maximum of $0.5\frac{1460\cdot 8}{(4\cdot 5+2\cdot 3)\cdot 625ms} \approx 360$ kbps. As the piconet contains $m-1 \leq 7$ slaves with identical traffic, each slave can achieve a goodput of only about $\dfrac{360}{m-1}$ kbps.

Our analysis setup implements the L2CAP Flow Control mode through a token bucket [Bertsekas and Gallager, 1991] with the data queue of size S and a token queue of size W, wherein tokens arrive at a constant rate of tb. Furthermore, there is an outgoing queue for baseband data packets of size L, from which the data packets are serviced by the poller utilizing the chosen intra-piconet polling scheme. Note that the uplink and downlink queues at the baseband level are still present. This implementation is shown in Fig. 6.1(b).

Some of the parameters of this architecture, such as the sizes of the queues and the token rate in the token bucket, may be adjusted in order to achieve delay-throughput trade-off for each slave. This scheme can limit the throughput of the slave through the token rate, while simultaneously limiting the length of the burst of baseband packets submitted to the network. Depending on the token buffer size and traffic intensity, overflows of the token buffer can be detected through the TCP loss events such as three duplicate acknowledgments or time-outs. In the former case, the size of the congestion window will be halved and TCP will continue working in its Additive Increase-Multiplicative Decrease (AIMD) phase. In the latter case, the congestion window will shrink to one, and TCP will enter its slow-start routine.

At the same time, the performance of Bluetooth networks is also affected by the intra-piconet polling scheme. We will assume that the E-polling scheme is used, as it has been shown to offer better performance than either limited or exhaustive service (see Chapter 3). In addition, it provides fairness by default, since the limit on the number of frames exchanged with any single slave prevents the slaves from monopolizing the network. It should be noted that the TCP protocol itself provides mechanisms to regulate fairness among TCP connections originating from one device, but the throughput that can be achieved in Bluetooth networks is too small to warrant a detailed analysis in this direction.

The performance of TCP traffic, in particular the steady-state send rate of the bulk-transfer TCP flows, has recently been assessed as the function of segment loss rate and round trip time (RTT) [Padhye, Firoiu, Towsley and Kurose, 2000]. The system is modeled at the ends of rounds that are equal to the round trip times and during which a number of segments equal to the current size of the congestion window is sent. The model assumes that both the RTT and loss probability can be obtained by measurement, and derives average value of congestion window size.

The same basic model, but with improved accuracy with regard to the latency and steady state transmission rate for TCP Tahoe, Reno, and SACK variants, has been used in [Sikdar, Kalyanaraman and Vastola, 2003]. The authors have also studied the impact of correlated segment losses under three TCP versions.

In our approach, we will model the system at the moments of acknowledgement arrivals and time-out events, instead of at the ends of successive rounds as in both papers mentioned above. In this manner, we are able to obtain more accurate information about performance and to derive the TCP congestion window size and throughput in both non-saturated and saturated operating regimes. In addition, the blocking probabilities of all the queues along the path are known and each packet loss can be treated independently.

We note that both of the papers mentioned above assume that RTT is a constant due to the large number of hops. In the model utilized in this discussion there are two hops only, and the RTT must be modeled as a random variable which is dependent on the current congestion window size.

Recently, the Adaptive Increase-Multiplicative Decrease (AIMD) technique was applied, using the concept similar to the token bucket filter, to control the flow over the combination of wireless and wireline paths [Cai, Shen and Mark, 2003]. It is assumed that packets can be lost over the wireless link only, and that all packet transmission times take exactly one slot. The system is modeled at the ends of the packet transmission times. This approach is orthogonal to ours, since we assume that the wireless channel errors will be handled through Forward Error Correction (FEC) and focus on finite queue losses at the data-link layer instead.

It should be noted that the performance of TCP/IP traffic in Bluetooth has been the topic of several research papers. The TCP Vegas variant has been shown to provide satisfactory performance in the presence of losses in the radio channel [Johansson, Kihl and Körner, 2000]. The estimation of channel state has been used to guide the segmentation/reassembly at the L2CAP level, as well as the packet format, so as to optimize the performance of TCP traffic [Balatti, Marzegalli and Vitello, 2001]. A similar approach has been proposed in [Chen, Kapoor, Sanadidi and Gerla, 2004], where the baseband packet size is varied according the channel state information to enhance TCP performance. The main focus of all these papers is the state of the wireless channel, rather than the values of TCP parameters and the size of finite device buffers, which are investigated in detail in this Chapter.

6.2 TCP window size

Under the assumptions outlined in the previous Section, the probability generating function (PGF) for the burst size of data packets at the baseband layer is $G_{bd}(z) = z^5$. The corresponding PGF for the acknowledgment packet burst size is $G_{ba}(z) = z$. The PGF for the size distribution of baseband data packets is $G_{pd}(z) = 0.8z^5 + 0.2z^3$, while the PGF for the size distribution of acknowledgment packets is $G_{pa}(z) = z^3$. Then, the PGF for the packet size and the mean packet size can be written as $G_p(z) = 0.5G_{bd}(G_{pd}(z)) + 0.5G_{ba}(G_{pa}(z))$ and $\overline{L_{ad}} = 0.5G'_{bd}(G_{pd}(z))|_{z=1} + 0.5G'_{ba}(G_{pa}(z))|_{z=1} = 3.8$, respectively. (All time variables are expressed in units of time slots of the Bluetooth clock, $T = 625\mu s$.) We will also assume that the receiver advertised window is larger than the congestion window at all times.

The paths traversed by the TCP segment sent from slave i to slave j and the corresponding acknowledgment sent in the opposite direction, are shown schematically in Fig. 6.2. A TCP segment or acknowledgment can be lost if any of the buffers along the path is full (and, consequently, blocks the reception of the packet in question). In subsequent discussion, we will calculate the probability distributions of token bucket queue lengths and outgoing buffer queue lengths at arbitrary times, as well as the corresponding blocking probabilities for TCP segments (P_{Bd}, P_{BdL}) and TCP acknowledgments (P_{Ba}, P_{BaL}). Segments/acknowledgments are also passing through the outgoing downlink queue at the master. However, we assume that this queue is much longer than the corresponding queues at the slaves, so that the blocking probabilities P_{BMd} and P_{BMa} are much smaller and may safely be ignored in calculations. Then, the total probability p of losing a TCP segment or its acknowledgment is

$$p = 1 - (1 - P_{Bd})(1 - P_{BdL})(1 - P_{BMd})(1 - P_{Ba})(1 - P_{BaL})(1 - P_{BMa}) \quad (6.1)$$

We will model the length of the TCP window at the moments of acknowledgments

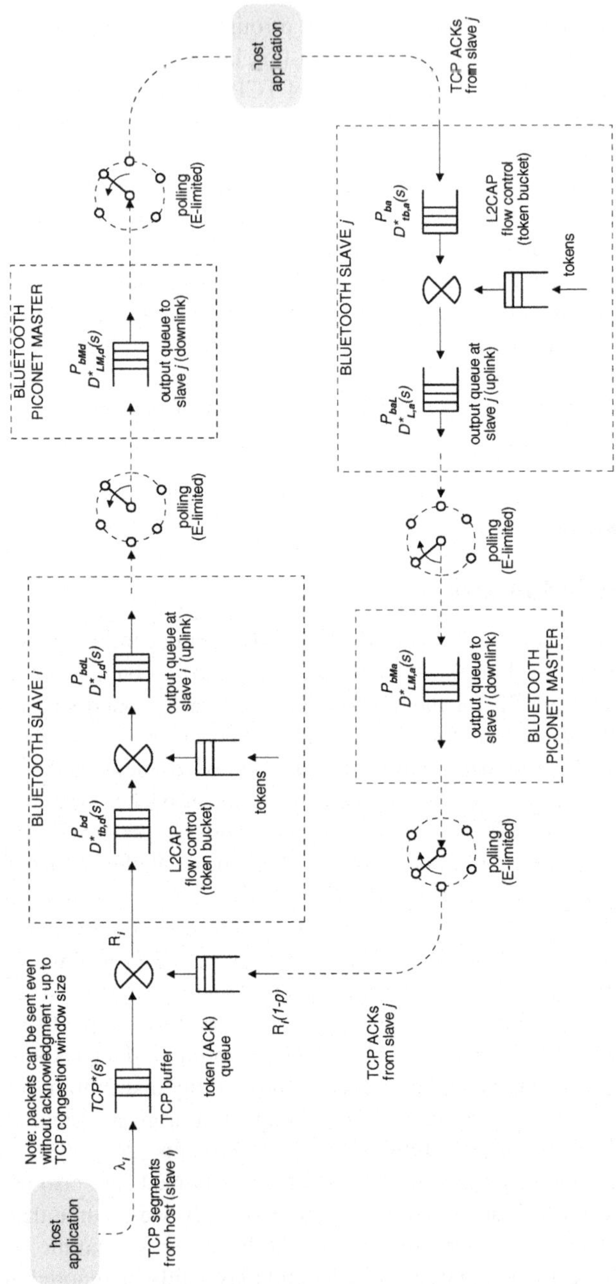

FIGURE 6.2
The path of the TCP segment and its acknowledgment, together with the blocking probabilities and LSTs of the delays in respective queues.

arrivals and segment loss events. Since TCP window of size w grows by $1/w$ after the successful acknowledgment, it is not possible to model the system using the probability generating functions, and we have to resort to finding the corresponding pdf (probability density function) instead. This probability distribution is a hybrid function represented by the mass probability w_1 of window size being 1, and by the continuous probability density function $w(x)$ for window sizes from 2 to ∞.

We also need to determine the pdf of the congestion window threshold $t(x)$ at the moments of acknowledgement arrival or packet loss. Let the probability of the time-out event be denoted with $P_{to} = p(1-(1-p)^3)$, and the probability of loss by three duplicate acknowledgements with $P_{td} = p(1-p)^3$. In order to simplify the derivation, we assume that window size immediately after the change of the state is x. If the event that changed the state was a positive acknowledgement, we assume that the window size before this event was $x - \frac{1}{x}$. Then the pdf of congestion window size can be approximated as $w(x-\frac{1}{x}) \approx w(x) - w'(x)\frac{1}{x}$. The pdfs can be described by

$$\begin{aligned} w_1 &= P_{to} + P_{td}\int_{x=2}^{3} w(x)dx, & w &= 1 \\ t(x) &= t(x)(1-p) + w(2x)p & w &\geq 2 \\ w(x) &= \left(w(x) - w'(x)\frac{1}{x}\right)(1-p)\int_0^{x-1/x} t(y)dy \\ &\quad + w(0.5x)(1-p)\int_{0.5x}^{\infty} t(y)dy + w(2x)P_{td} \end{aligned} \tag{6.2}$$

where $\int_0^{x-1/x} t(y)dy$ denotes the probability that the congestion window size before the acknowledgement, $x - 1/x$, is above the threshold size (i.e., that TCP is working in the AIMD mode), and the probability that the current size of the congestion window, $0.5x$, is lower than the threshold (i.e., that the system is in the slow start mode) is given by $\int_{0.5x}^{\infty} t(y)dy$.

The system (6.2) of integro-differential equations could be solved numerically, but sufficient accuracy may be obtained through the approximation

$$w(x) = C_1 e^{-0.25px^2(1+3p-3p^2+p^3)/(1-p)} \tag{6.3}$$

where the normalization constant C_1 is determined from the condition $1 - w_1 = \int_2^{\infty} w(x)dx$. The pdf of window size for various values of TCP window size and segment loss probability is shown in Fig. 6.3(a).

The mean TCP window size $\overline{w}$ and mean threshold size $\overline{t}$ are calculated as

$$\begin{aligned} \overline{w} &= w_1 + \int_2^{\infty} xw(x)dx \\ \overline{t} &= \int_1^{\infty} xw(2x)dx \end{aligned} \tag{6.4}$$

since $t(x) = w(2x)$. The dependency of mean window size on TCP window size and segment loss probability p is shown in Fig. 6.3(b).

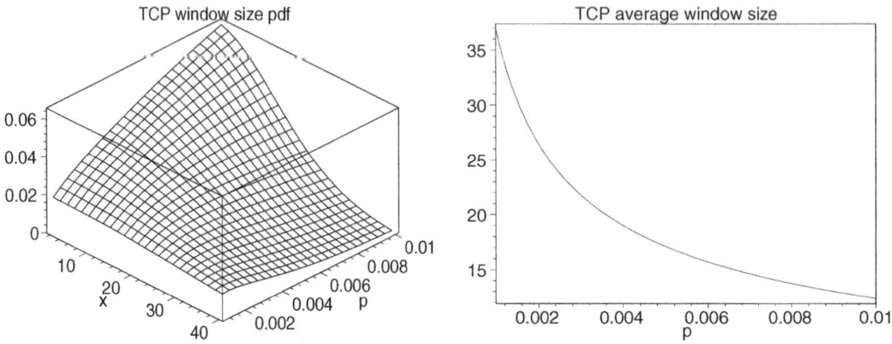

(a) pdf of window size versus window size and segment loss probability p.

(b) Mean value of window size $\overline{w}$ versus segment loss probability p.

FIGURE 6.3
Characteristics of TCP window size.

As will be shown in Sections 6.3 and 6.4, we can find the probability distribution of delays through all the token bucket filters and the baseband buffers along the path of the TCP segment and its acknowledgment. The delay is calculated from the moment when the TCP segment enters the token bucket filter at the source device until the acknowledgment is received by that same device. This delay is equal to the sum of delays in all the buffers from Bluetooth protocol stack along the path. We will refer to this delay as the L2CAP round trip delay, and its probability distribution can be described through the corresponding Laplace-Stieltjes transform:

$$D^*_{L2CAP}(s) = D^*_{tb,d}(s)D^*_{L,d}(s)D^*_{LM,d}(s)D^*_{tb,a}(s)D^*_{L,a}(s)D^*_{LM,a}(s) \tag{6.5}$$

We can also calculate the probability distribution of the number of outstanding (unacknowledged) segments at the moments of segment acknowledgement arrivals. This number for an arbitrary TCP segment is equal to the number of segment arrivals during the L2CAP round trip time of that segment. The PGF for the number of segment arrivals during the L2CAP round trip time can be calculated, using the approach from [Takagi, 1991], as $A(z) = D^*_{L2CAP}(\lambda_i - z\lambda_i)$. The probability of k segment arrivals during the RTT time is equal to $a_k = \frac{d^k}{dz^k} D^*_{L2CAP}(\lambda_i - z\lambda_i)|_{z=0}$.

Using the PASTA property (Poisson Arrivals See Time Averages), we conclude that the arriving segment will see the same probability distribution of outstanding segments as the incoming acknowledgement. Therefore, the probability P_t that the arriving segment will find a free token and leave the TCP buffer immediately is

$$P_t = \sum_{k=1}^{\infty} a_k \int_{k+1}^{\infty} w(x)dx \tag{6.6}$$

and the probability that the segments will be stored in the TCP buffer is $1 - P_t$. Then, the rate at which TCP sends the segments to the token bucket filter, which will be referred to as the TCP send rate R_i, has two components. One of these is contributed by the segments that find acknowledgments waiting in the token queue, and thus can leave the TCP buffer immediately; another one comes from the segments that have to wait in the TCP buffer until an acknowledgement (for an earlier segment) arrives. The expression for TCP send rate, then, becomes $R_i = P_t\lambda_i + R_i(1-p)(1-P_t)$, which may be simplified to

$$R_i = \lambda_i \frac{P_t}{P_t + p - pP_t} \tag{6.7}$$

As both components of the previous expression can be modeled as Poisson processes, the TCP sending process can be modeled as Poisson process as well.

In order to calculate the delay through the TCP buffer, we need to determine the probability distribution of the number of segments buffered by the TCP due to the insufficient size of the congestion window. (We assume that the TCP buffer has infinite capacity.) The probability that k segments are buffered is

$$q_k = \sum_{i=1}^{\infty} a_{i+k} \int_i^{i+1} w(x)dx + a_{k+1}w_1 \tag{6.8}$$

and the LST for the delay through the TCP buffer is

$$D^*_{TCP}(s) = \sum_{k=0}^{\infty} q_k e^{-sk} \tag{6.9}$$

The entire round trip time of the TCP segment, then, becomes

$$RTT^*(s) = D^*_{TCP}(s) D^*_{L2CAP}(s) \tag{6.10}$$

6.3 Queueing analysis of the token bucket filter

Let us now consider the token bucket (TB) filter, the queueing model of which is shown in Fig. 6.4(a). We assume that the TCP segments arrive at the TCP sending rate R_i calculated in (6.7), while the TCP acknowledgements arrive at the rate of $R_i(1-p)$. The token arrival rate will be denoted as tb, which means that the token arrival period will be $T_b = 1/tb$. The queue which holds tokens has length of W baseband packet tokens. Packets leave the queue at max_rate, which is much larger than the token arrival rate. Using this model, we first derive the probability distribution function (PDF) of the number of the packets in the token bucket queue at the moments of token arrival, and then proceed to calculate that same PDF at arbitrary time.

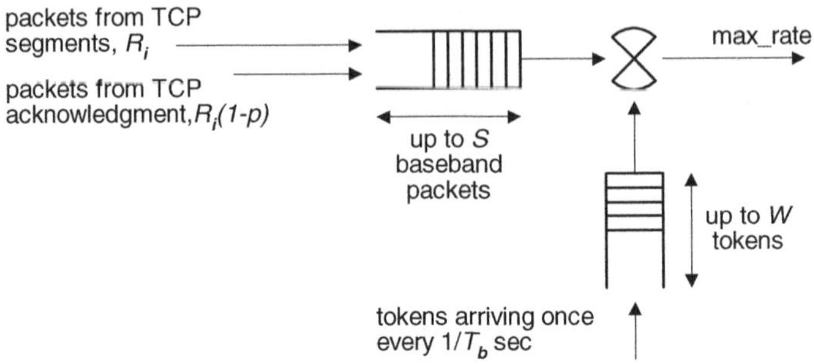

(a) Queueing model of the token bucket with finite capacity, accepting two types of packets.

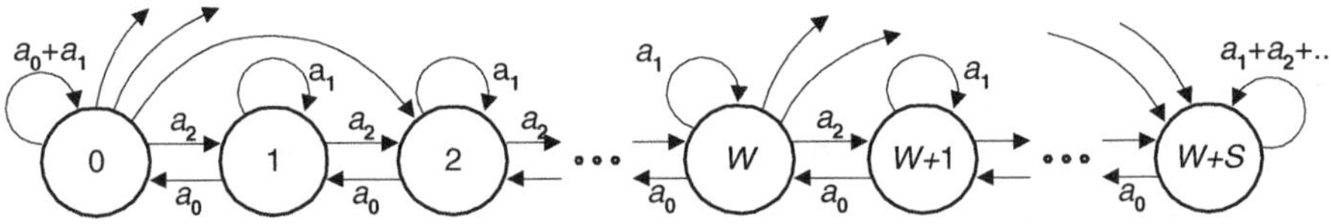

(b) Token bucket may be represented as a discrete Markov chain.

FIGURE 6.4
Pertaining to the analysis of the token bucket filter.

The probability of a_k baseband packet arrivals in the TB queue is equal to the sum of probabilities of data and acknowledgment packet arrivals:

$$\begin{aligned} a_k &= \sum_{i=1}^{k} \left[\sum_{l=0}^{\infty} \frac{1}{i!} \frac{d^i}{dz^i} G_{bd}(z)^l \Big|_{z=0} e^{-R_i T_b} \frac{(R_i T_b)^l}{l!} \right. \\ &+ \left. \sum_{l=0}^{\infty} \frac{1}{(k-i)!} \frac{d^{k-i}}{dz^{k-i}} G_{ba}(z)^l \Big|_{z=0} e^{-R_i T_b} \frac{(R_i T_b)^l}{l!} \right] \\ &= \sum_{i=1}^{k} \left[\frac{1}{i!} \frac{d^i}{dz^i} e^{-R_i T_b (1-G_{bd}(z))} \Big|_{z=0} + \frac{1}{(k-i)!} \frac{d^{k-i}}{dz^{k-i}} e^{-R_i T_b (1-G_{ba}(z))} \Big|_{z=0} \right] \end{aligned} \tag{6.11}$$

The TB can be modeled as a discrete-time Markov chain, in which the state i (when $0 \le i \le W$) corresponds to the situation when $W-i$ tokens are available in the token buffer, but no data packets are present in the data queue. The remaining states (from $W+1$ to $W+S$) correspond to the situation where there are data packets in the data queue, but the token queue is empty. Since both queues have finite lengths, the Markov chain, which is shown in Fig. 6.4(b), is finite as well. The balance equations for this Markov chain are

$$\begin{aligned}
\pi_0 &= a_0\pi_1 + (a_0 + a_1)\pi_0 \\
\pi_i &= \sum_{j=0}^{i+1} a_{i-j+1}\pi_j, \quad 0 < i < W + S - 1 \\
\pi_{W+S-1} &= \sum_{j=0}^{S-1} \pi_j \sum_{k=S-j}^{\infty} a_k
\end{aligned} \tag{6.12}$$

The PDF for the queue lengths at the moments of token arrivals can be found by solving this system with the condition $\sum_{k=0}^{S-1} \pi_k = 1$.

By using the probability distribution of the TB queue length at the moments of token arrivals, we can derive the joint probability distribution of the TB queue length and the remaining time before the token arrival. In this manner we are able to derive the blocking probability which in turn determines the segment loss probability at the TCP level. We will introduce the following variables:

- The total queue length (including both token queue and data queue), L_q.
- The elapsed token time – from a given token arrival to the arbitrary time before the arrival of the next token, $T_{b\_}$.
- The remaining token time – from the arbitrary time between two successive token arrivals till the next token arrival, T_{b+}.
- The number of packet arrivals (results of burst arrivals) during the elapsed token time, $A(T_{b\_})$.
- The blocking probability at arbitrary time, P_B. Since the burst represents a TCP segment, we will adopt the total rejection policy for calculating the blocking probability, i.e., either the burst (TCP segment) will fit into the baseband buffer in its entirety, or will be rejected.
- Finally, the probability distribution function $T_b(x)$ of the token inter-arrival time, and the corresponding probability density function $t_b(x)$.

For the time between two successive token arrivals, the joint probability distribution of the TB queue length and remaining token time is

$$\Pi_k^*(s) = \int_0^\infty e^{-sy}\mathcal{P}[L_q = k, y < T_{b+} < y + dy], 1 \le k \le W + S \tag{6.13}$$

where, as in Chapter 4, $\mathcal{P}[\S]$ denotes the probability of event x.

By using the TB queue length distribution at the time of arrival of the previous token, we obtain

$$
\begin{aligned}
\Pi_k^*(s) &= R_i(\overline{G_{bd}} + \overline{G_{ba}})T_b(1 - P_B) \\
&\quad \cdot \sum_{j=0}^{k} \pi_j E[e^{-sT_{b+}} | A(T_{b\_}) = k - j] \\
&\quad \cdot \mathcal{P}[A(T_{b\_}) = k - j], 1 \leq k \leq W + S - 1 \\
\Pi_{W+S}^*(s) &= R_i(\overline{G_{bd}} + \overline{G_{ba}})T_b(1 - P_B) \cdot \\
&\quad \sum_{j=0}^{W+S-1} \pi_j \sum_{k=S-j}^{\infty} E[e^{-sT_{b+}} | A(T_{b\_}) = k - j] \\
&\quad \cdot \mathcal{P}[A(T_{b\_}) = k - j]
\end{aligned}
\tag{6.14}
$$

The expression

$$
\begin{aligned}
\psi_k^*(s) &= E[e^{-sT_{b+}} | A(T_{b\_}) = k] \mathcal{P}[A(T_{b\_}) = k] \\
&= \sum_{l=0}^{\infty} \sum_{i=1}^{k} \left(\frac{1}{i!} \frac{d^i}{dz^i} G_{bd}(z)^l + \frac{1}{(k-i)!} \frac{d^{k-i}}{dz^{k-i}} G_{ba}(z)^l \right) \Bigg|_{z=0} \\
&\quad \cdot \int_0^{\infty} \frac{(R_i x)^l}{l!} e^{-R_i x} \frac{1 - T_b(x)}{T_b} dx \int_0^{\infty} e^{-sy} \frac{t_b(x+y)}{1 - T_b(x)} dy
\end{aligned}
\tag{6.15}
$$

may be simplified by introducing the substitution $u = x + y$ and by changing the order of integration:

$$
\begin{aligned}
\psi_k^*(s) &= \frac{1}{T_b} \int_0^{\infty} \sum_{l=0}^{\infty} \sum_{i=1}^{k} \left(\frac{1}{i!} \frac{d^i}{dz^i} \frac{(R_i x G_{bd}(z))^l}{l!} \right. \\
&\qquad \left. + \frac{1}{(k-i)!} \frac{d^{k-i}}{dz^{k-i}} \frac{(R_i x G_{ba}(z))^l}{l!} \right) \Bigg|_{z=0} \\
&\qquad \cdot e^{-R_i x} dx \int_0^{\infty} e^{-sy} t_b(x+y) dy
\end{aligned}
\tag{6.16}
$$

$$
\begin{aligned}
&= \frac{1}{T_b} \int_0^{\infty} \sum_{i=1}^{k} \left(\frac{1}{i!} \frac{d^i}{dz^i} e^{R_i G_{bd}(z) x} + \frac{1}{(k-i)!} \frac{d^{k-i}}{dz^{k-i}} e^{R_i G_{ba}(z) x} \right) \Bigg|_{z=0} \\
&\quad \cdot \cdot e^{(s - R_i) x} dx \int_x^{\infty} e^{-su} t_b(u) du
\end{aligned}
$$

continued on next page ...

... continued from previous page

$$= \frac{1}{T_b} \int_0^{\infty} e^{-su} t_b(u) du \int_0^{u} \sum_{i=1}^{k} \left(\frac{1}{i!} \frac{d^i}{dz^i} e^{(R_i G_{bd}(z)+s-R_i)x} + \frac{1}{(k-i)!} \frac{d^{k-i}}{dz^{k-i}} e^{(R_i G_{ba}(z)+s-R_i)x} \right) \Bigg|_{z=0} dx$$

$$= \frac{1}{T_b} \int_0^{\infty} e^{-su} \sum_{i=1}^{k} \left(\frac{1}{i!} \frac{d^i}{dz^i} \frac{e^{(R_i G_{bd}(z)+s-R_i)u} - 1}{R_i G_{bd}(z) + s - R_i} + \frac{1}{(k-i)!} \frac{d^{k-i}}{dz^{k-i}} \frac{e^{(R_i G_{ba}(z)+s-R_i)u} - 1}{R_i G_{ba}(z) + s - R_i} \right) \Bigg|_{z=0} t_b(u) du$$

$$= \frac{1}{T_b} \sum_{i=1}^{k} \left(\frac{1}{i!} \frac{d^i}{dz^i} \frac{e^{(R_i G_{bd}(z)-R_i)T_b} - e^{-sTb}}{R_i G_{bd}(z) + s - R_i} + \frac{1}{(k-i)!} \frac{d^{k-i}}{dz^{k-i}} \frac{e^{(R_i G_{pa}(z)-R_i)T_b} - e^{-sT_b}}{R_i G_{pa}(z) + s - R_i} \right) \Bigg|_{z=0} \tag{6.16}$$

Now the system (6.14) can be written as

$$\Pi_k^*(s) = R_i(\overline{G_{bd}} + \overline{G_{ba}}) T_b (1 - P_B) \sum_{j=0}^{k} \pi_j \psi_{k-j}^*(s), \quad 1 \le k \le W + S - 1 \tag{6.17}$$

$$\Pi_{W+S}^*(s) = R_i(\overline{G_{bd}} + \overline{G_{ba}}) T_b (1 - P_B) \sum_{j=0}^{W+S-1} \pi_j \sum_{k=S-j}^{\infty} \psi_{k-j}^*(s)$$

The probabilities of having exactly k packets in the TB filter are simply $P_k = \mathcal{P}[L_q = k] = \Pi_k^*(0)$. Also, we note that the probability of the empty queue is equal to $P_0 = 1 - R_i(\overline{G_{bd}} + \overline{G_{ba}}) T_b (1 - P_B)$. It should be observed that all queue state probabilities at arbitrary time are functions of P_B; therefore, we need to express the blocking probability as the function of the queue state probabilities, and then solve this equation for P_B. The blocking probability under two types of packet bursts in the TB filter is

$$P_B = \sum_{k=0}^{W+S-1} P_k \sum_{j=W+S-k+1}^{\infty} 0.5(g_{bd,j} + g_{ba,j}) + P_{W+S} \tag{6.18}$$

where $g_{bd,j} = \frac{1}{j!} \frac{d^j}{dz^j} G_{bd}(z)|_{z=0}$ and $g_{ba,j} = \frac{1}{j!} \frac{d^j}{dz^j} G_{ba}(z)|_{z=0}$ are mass probabilities of the burst size probability distribution for the TCP data segment and the TCP acknowledgment segment, respectively. Expression (6.18) can be rearranged to find the blocking probability P_B, which leads to the queue length distribution. After that, the individual blocking probabilities for each traffic type can be found as

$$\begin{aligned} P_{Bd} &= \sum_{k=0}^{W+S-1} P_k \sum_{j=W+S-k+1}^{\infty} g_{bd,j} + P_{W+S} \\ P_{Ba} &= \sum_{k=0}^{W+S-1} P_k \sum_{j=W+S-k+1}^{\infty} g_{ba,j} + P_{W+S} \end{aligned} \tag{6.19}$$

The delay through the token bucket filter should be calculated separately for the data segments and for the acknowledgments. The queueing delay of the entire TCP segment is equal to the queueing delay of the first baseband packet from the burst representing that TCP segment. The LST of the delay for the data segment is given with

$$D^*_{tb,d}(s) = \frac{\left(P_0 + \sum_{l=1}^{W} \Pi^*_k(0)\right) \sum_{k=1}^{W+S-l} g_{bd,k} + \sum_{k=W}^{W+S-1} \Pi^*_k(s)[G^*_{pd}]^{k-1} \sum_{j=1}^{W+S-k} g_{bd,j}}{1 - P_{Bd}} \tag{6.20}$$

The LST for the delay of the acknowledgment, $D^*_{tb,d}(s)$, can be determined in an analogous manner.

6.4 The outgoing queue at the baseband level

We will now derive the expressions for the delay and blocking probability for the other queue (buffer): the outgoing buffer at the baseband level. This buffer has finite length of L baseband packets, and it is fed by packets that pass through the token buffer filter. We assume that the TCP segment packets arrive at the rate of $\lambda_{i,d} = R_i(1 - P_{Bd})$, and that the TCP acknowledgment packets arrive at the rate of $\lambda_{i,a} = R_i(1-p)(1 - P_{Ba})$. We again assume that each buffer has the total rejection policy, i.e., the entire burst is rejected if it cannot fit into the buffer.

The baseband queue is serviced using the E-limited polling scheme from Chapter 3. Let us begin by determining the probability distribution of slave queue lengths in imbedded Markov points that correspond to vacation termination times and uplink transmission completion times. Then, we will determine the PDF for the slave queue length at arbitrary point of time, and use it to derive the access delay and the blocking probability of the burst.

Outgoing queue length distribution in Markov points

Let $q_{k,i,u}$ denote the joint probability that a Markov point in the uplink queue of slave i is a vacation termination time and that there are $k = 0, 1, 2\ldots$ packets at the outgoing queue of the slave i at that time. Also, let $\pi^{(\mu)}_{k,i,u}$ denote the joint probability that a Markov point in the uplink queue i is the μ-th uplink transmission completion time and that there are k packets in the queue, where $\mu = 1\,..\,M$ and $k = 0\,..\,L-1$. The analogous probabilities for the corresponding downlink queue are denoted with $q_{k,i,d}$ and $\pi^{(\mu)}_{k,i,d}$. Note that the model that decouples the uplink and downlink queues, which was presented in Chapter 4, is more accurate – but it is also computationally complex. The extension of the model in this Subsection along the lines of Chapter 4 is straightforward, and it is left as an exercise to the reader.

Let $g_p(x)$ and $v_i(x)$ denote the probability density functions of the packet transmission time and vacation time, respectively, at the uplink queue of slave i; their LST transforms will be $G_p^*(s)$ and $V_i^*(s)$. The LST transform of the probability distribution function for the length of a frame is $G_p^*(s)^2$. (Remember that the frame consists of a downlink packet and the subsequent uplink frame.) Let us also denote the probability of k packet arrivals at the uplink queue of slave i during the frame time as $a_{k,i,u}$, and the probability of k packet arrivals during the vacation time (i.e., while master is serving other slaves) as $f_{k,i,u}$. These probabilities may be calculated as

$$\begin{aligned}
a_{k,i,u} &= \sum_{i=1}^{k}\left[\sum_{l=0}^{\infty}\frac{1}{i!}\frac{d^i}{dz^i}\left(G_{bd}(z)\right)^l\Big|_{z=0}\int_0^{\infty}\frac{(\lambda_{i,d}x)^l}{l!}e^{-\lambda_{i,d}x}g_p * g_p(x)dx\right.\\
&\quad\left.+\sum_{l=0}^{\infty}\frac{1}{(k-i)!}\frac{d^{k-i}}{dz^{k-i}}\left(G_{ba}(z)\right)^l\Big|_{z=0}\int_0^{\infty}\frac{(\lambda_{i,a}x)^l}{l!}e^{-\lambda_{i,a}x}g_p * g_p(x)dx\right]\\
&= \sum_{i=1}^{k}\left[\frac{1}{i!}\frac{d^i}{dz^i}\left(G_p^*(\lambda_{i,d}-\lambda_{i,d}G_{bd}(z))\right)^2\Big|_{z=0}\right.\\
&\quad\left.+\frac{1}{(k-i)!}\frac{d^{k-i}}{dz^{k-i}}\left(G_p^*(\lambda_{i,a}-\lambda_{i,a}G_{ba}(z))\right)^2\Big|_{z=0}\right]\\
f_{k,i,u} &= \sum_{i=1}^{k}\left[\sum_{l=0}^{\infty}\frac{1}{i!}\frac{d^i}{dz^i}left.\left(G_{bd}(z)\right)^l\Big|_{z=0}\int_0^{\infty}\frac{(\lambda_{i,d}x)^l}{l!}e^{-\lambda_{i,d}x}v_i(x)dx\right.\\
&\quad\left.+\sum_{l=0}^{\infty}\frac{1}{(k-i)!}\frac{d^{k-i}}{dz^{k-i}}\left(G_{ba}(z)\right)^l\Big|_{z=0}\int_0^{\infty}\frac{(\lambda_{i,a}x)^l}{l!}e^{-\lambda_{i,a}x}v_i(x)dx\right]\\
&= \sum_{i=1}^{k}\left[\frac{1}{i!}\frac{d^i}{dz^i}\left(V_i^*(\lambda_{i,d}-\lambda_{i,d}G_{bd}(z))\right)^2\Big|_{z=0}\right.\\
&\quad\left.+\frac{1}{(k-i)!}\frac{d^{k-i}}{dz^{k-i}}\left(V_i^*(\lambda_{i,a}-\lambda_{i,a}G_{ba}(z))\right)^2\Big|_{z=0}\right]
\end{aligned} \tag{6.21}$$

where $g_p * g_p(x)$ denotes the convolution of $g_p(x)$ with itself. Note that $(G_p^*(\lambda_{i,d} - \lambda_{i,d}G_{bd}(z)))^2$ and $G_p^*(\lambda_{i,a} - \lambda_{i,a}G_{ba}(z))$ denote the PGFs for the number of packet arrivals in the uplink queue from TCP data segments and acknowledgments, respectively, during the frame time. Also, the terms $(V_i^*(\lambda_{i,d} - \lambda_{i,d}G_{bd}(z)))^2$ and

$V_i^*(\lambda_{i,a} - \lambda_{i,a}G_{ba}(z))$ denote the corresponding PGFs for the number of packet arrivals in the uplink queue, but during the vacation time.

For the packet departure times when the master polls the slaves, we note that the buffer occupancy can be between 0 and $L-1$. The probabilities that the uplink queue contains k packets in Markov points are given by

$$\begin{aligned}
\pi_{k,i,u}^{(1)} &= \sum_{j=1}^{k+1} q_{j,i,u} a_{k-j+1,i,u}, && 0 \le k \le L-2 \\
\pi_{L-1,i,u}^{(1)} &= \sum_{j=1}^{L} q_{j,i,u} \sum_{k=L-j}^{\infty} a_{k,i,u} \\
\pi_{k,i,u}^{(\mu)} &= \sum_{j=1}^{k+1} \pi_{j,i,u}^{(\mu-1)} a_{k-j+1,i,u}, && 0 \le k \le L-2,\ \mu = 2\,..\,M \\
\pi_{L-1,i,u}^{(\mu)} &= \sum_{j=1}^{L-1} \pi_{j,i,u}^{(\mu-1)} \sum_{k=L-j}^{\infty} a_{k,i,u}, && \mu = 2\,..\,M \\
q_{k,i,u} &= \left(\sum_{m=1}^{M-1} \pi_{0,i,u}^{(\mu)} + q_{0,i,u} \right) f_{k,i,u} + \sum_{j=0}^{k} \pi_{j,i,u}^{(M)} f_{k-j,i,u}, \\
& && 0 \le k \le L-1 \\
q_{K,i,u} &= \left(\sum_{m=1}^{M-1} \pi_{0,i,u}^{(\mu)} + q_{0,i,u} \right) \sum_{k=L}^{\infty} f_{k,i,u} + \sum_{j=0}^{L-1} \pi_{j,i,u}^{(M)} \sum_{k=L-j}^{\infty} f_{k,i,u}
\end{aligned} \tag{6.22}$$

Also, $\sum_{k=0}^{L} q_{k,i,u} + \sum_{\mu=1}^{M} \sum_{k=0}^{L-1} \pi_{k,i,u}^{(\mu)} = 1.$

The distribution of the queue length in Markov points is obtained when the system of these linear equations is solved for the mass probabilities of queue lengths.

Let us denote the probability that the vacation starts after the uplink transmission as $h_{i,u} = \sum_{\mu=1}^{M-1} \pi_{0,i,u}^{(\mu)} + \sum_{k=0}^{L-1} \pi_{k,i,u}^{(M)}$. Then, the probability that the vacation will start after an arbitrary Markov point is $q_{0,i,u} + h_{i,u}$. The average distance in time between two consecutive Markov points at slave i is

$$\eta_{i,u} = (q_{0,i,u} + h_{i,u})\overline{V_i} + (1 - q_{0,i,u} - h_{i,u})2\overline{L_{ad}} \tag{6.23}$$

Outgoing queue length distribution at arbitrary time

By using the probability distribution of the uplink queue length in Markov points, we can derive the probability distribution of this queue length at arbitrary time between two Markov points, together with the PDF of the remaining vacation time (if the previous Markov point was the start of a vacation) or the PDF of the remaining frame service time (if the previous Markov point was the start of a packet service). We will introduce the following variables:

- The probability density function of the vacation time, $v_i(x)$, and its PDF, $V_i(x)$.
- The queue length at an arbitrary time, $L_{q,i,u}$.
- The elapsed vacation time, $V_{\_,i}$.
- The remaining vacation time, $V_{+,i}$.
- The number of packet arrivals resulting from packet burst arrivals in the elapsed vacation time, $A(V_{\_,i})$.
- The probability density function of the frame service time, $g_p * g_p(x)$, and its PDF, $F_s(x)$ (where $*$ denotes the convolution operator).
- The elapsed frame service time, $X_{\_,i}$.
- The remaining frame service time, $X_{+,i}$.
- The number of packet arrivals resulting from packet burst arrivals in the elapsed frame service time, $A(X_{\_,i})$.

For the time between the start and end of vacation, we define the joint probability of the queue length and the remaining vacation time as

$$\Omega^*_{k,i,u}(s) = \int_0^\infty e^{-sy}\mathcal{P}[L_{q,i,u} = k, y < V_{+,i} < y + dy], \quad 0 \le k \le L \tag{6.24}$$

For the time between the start and end of the frame service, for the frame $1 \le \mu \le M$, we define the joint probability of the queue length and remaining frame service time as

$$\Pi^*_{k,\mu,i,u}(s) = \int_0^\infty e^{-sy}\mathcal{P}[L_{q,i,u} = k, y < X_{+,i} < y + dy], \quad 1 \le k \le L, \le \mu \le M \tag{6.25}$$

Using the corresponding probabilities in the previous Markov point, we obtain

$$\begin{aligned}\Omega^*_{k,i,u}(s) &= \frac{\overline{V_i}}{\eta_{i,u}}\Big(q_{0,i,u} + \sum_{m=1}^{M-1} \pi^{(\mu)}_{0,i,u}\Big)E[e^{-sV_{+,i}}|A(V_{\_,i}) = k]\mathcal{P}[A(V_{\_,i}) = k] \\ &+ \frac{\overline{V_i}}{\eta_{i,u}}\sum_{j=0}^{k}\pi^{(M)}_{j,i,u}E[e^{-sV_{+,i}}|A(V_{\_,i}) = k - j]\mathcal{P}[A(V_{\_,i}) = k - j], \\ &\qquad 0 \le k \le L - 1\end{aligned}$$

continued on next page ...

(6.26)

... continued from previous page

$$\Omega^*_{L,i,u}(s) = \frac{\overline{V_i}}{\eta_{i,u}}\Big(q_{0,i,u} + \sum_{m=1}^{M-1} \pi^{(\mu)}_{0,i,u}\Big) \sum_{k=L}^{\infty} E[e^{-sV_{+,i}}|A(V_{-,i})=k]\mathcal{P}[A(V_{-,i})=k]$$
$$+ \frac{\overline{V_i}}{\eta_{i,u}} \sum_{j=0}^{k} \pi^{(M)}_{j,i,u} \sum_{k=L-j}^{\infty} E[e^{-sV_{+,i}}|A(V_{-,i}) = k]\mathcal{P}[A(V_{-,i}) = k]$$

$$\Pi^*_{k,1,i,u}(s) = \frac{2\overline{L_{ad}}}{\eta_{i,u}} \sum_{j=1}^{k} q_{j,i,u} E[e^{-sX_{+,i}}|A(X_{-,i}) = k-j]\mathcal{P}[A(X_{-,i}) = k-j],$$
$$1 \le k \le L-1$$

$$\Pi^*_{L,1,i,u}(s) = \frac{2\overline{L_{ad}}}{\eta_{i,u}} \sum_{j=1}^{L} q_{j,i,u} \sum_{k=K-j}^{\infty} E[e^{-sX_{+,i}}|A(X_{-,i}) = k]\mathcal{P}[A(X_{-,i}) = k],$$
$$1 \le k \le L-1$$

$$\Pi^*_{k,\mu,i,u}(s) = \frac{2\overline{L_{ad}}}{\eta_{i,u}} \sum_{j=1}^{k} \pi^{(\mu)}_{j,i,u} E[e^{-sX_{+,i}}|A(X_{-,i}) = k-j]\mathcal{P}[A(X_{-,i}) = k-j],$$
$$1 \le k \le L-1, 2 \le \mu \le M$$

$$\Pi^*_{L,\mu,i,u}(s) = \frac{2\overline{L_{ad}}}{\eta_{i,u}} \sum_{j=1}^{L} \pi^{(\mu)}_{j,i,u} \sum_{k=K-j}^{\infty} E[e^{-sX_{+,i}}|A(X_{-,i}) = k]\mathcal{P}[A(X_{-,i}) = k],$$
$$2 \le \mu \le M \qquad (6.26)$$

Using the expressions

$$\phi_k(s) = E[e^{-sV_{+,i}}|A(V_{-,i}) = k]\mathcal{P}[A(V_{-,i}) = k]$$
$$= \sum_{i=1}^{k}\Bigg[\frac{1}{\overline{V_i} i!}\frac{d^i}{dz^i}\Big|_{z=0}\Big(\frac{V_i^*(-\lambda_{i,d}G_{bd}(z)+\lambda_{i,d}) - V_i^*(s)}{\lambda_{i,d}G_{bd}(z)+s+\lambda_{i,d}}\Big)$$
$$+ \frac{1}{\overline{V_i}(k-i)!}\frac{d^{k-i}}{dz^{k-i}}\Big|_{z=0}\Big(\frac{V_i^*(-\lambda_i G_{ba}(z)+\lambda_{i,a}) - V_i^*(s)}{\lambda_{i,a}G_{ba}(z)+s+\lambda_{i,a}}\Big)\Bigg]$$
$$\psi^*_k(s) = E[e^{-sX_{+,i}}|A(X_{-,i}) = k]\mathcal{P}[A(X_{-,i}) = k]$$
$$= \sum_{i=1}^{k}\Bigg[\frac{1}{2\overline{L_{ad}} i!}\frac{d^i}{dz^i}\Big|_{z=0}\Big(\frac{G_p^*(-\lambda_{i,d}G_{bd}(z)+\lambda_{i,d})^2 - G_p^*(s)^2}{\lambda_{i,d}G_{bd}(z)+s+\lambda_{i,d}}\Big)$$
$$+ \frac{1}{2\overline{L_{ad}}(k-i)!}\frac{d^{k-i}}{dz^{k-i}}\Big|_{z=0}\Big(\frac{G_p^*(-\lambda_{i,a}G_{ba}(z)+\lambda_{i,a})^2 - G_p^*(s)^2}{\lambda_{i,a}G_{ba}(z)+s+\lambda_{i,a}}\Big)\Bigg] \qquad (6.27)$$

the system (6.26) can be transformed to

$$
\begin{aligned}
\Omega_k^*(s) &= \frac{\overline{V_i}}{\eta_{i,u}}(q_{0,i,u} + \sum_{m=1}^{M-1} \pi_{0,i,u}^{(\mu)})\phi_k^*(s) + \frac{\overline{V_i}}{\eta_{i,u}} \sum_{j=0}^{k} \pi_{j,i,u}^{(M)} \phi_{k-j}^*(s) \\
\Omega_{L,i,u}^*(s) &= \frac{\overline{V_i}}{\eta_{i,u}}(q_{0,i,u} + \sum_{\mu=1}^{M-1} \pi_{0,i,u}^{(\mu)}) \sum_{k=K}^{\infty} \phi_k^*(s) + \frac{\overline{V_i}}{\eta_{i,u}} \sum_{j=0}^{k} \pi_{j,i,u}^{(M)} \sum_{k=K-j}^{\infty} \phi_k^*(s) \\
\Pi_{k,1,i,u}^*(s) &= \frac{2\overline{L_{ad}}}{\eta_{i,u}} \sum_{j=1}^{k} q_{j,i,u} \psi_k^*(s), & 1 \le k \le L-1 \\
\Pi_{L,1,i,u}^*(s) &= \frac{2\overline{L_{ad}}}{\eta_{i,u}} \sum_{j=1}^{L} q_{j,i,u} \sum_{k=K-j}^{\infty} \psi_k^*(s), & 1 \le k \le L-1 \\
\Pi_{k,\mu,i,u}^*(s) &= \frac{2\overline{L_{ad}}}{\eta_{i,u}} \sum_{j=1}^{k} \pi_{j,i,u}^{(\mu)} \psi_{k-j}^*(s), & 1 \le k \le L-1, 2 \le \mu \le M \\
\Pi_{L,\mu,i,u}^*(s) &= \frac{2\overline{L_{ad}}}{\eta_{i,u}} \sum_{j=1}^{L} \pi_{j,i,u}^{(\mu)} \sum_{k=L-j}^{\infty} \psi_k^*(s), & 1 \le k \le L-1, 2 \le \mu \le M
\end{aligned}
\tag{6.28}
$$

The distribution of the size of the uplink queue at arbitrary time is given by

$$
\begin{aligned}
\mathcal{P}[L_{q,i,u} = 0] &= \Omega_0^*(0) \\
\mathcal{P}[L_{q,i,u} = k] &= \Omega_k^*(0) + \sum_{m=1}^{M} \Pi_{k,m,i,u}^*(0) \\
\mathcal{P}[L_{q,i,u} = L] &= \Omega_L^*(0) + \sum_{\mu=1}^{M} \Pi_{L,\mu,i,u}^*(0)
\end{aligned}
\tag{6.29}
$$

which allows us to calculate the burst blocking probability in the uplink queue at an arbitrary time. To that end, let us denote the mass probability of the burst size being exactly l packets as $g_l = \frac{1}{l!}\frac{d^l}{dz^l} G_b(z)|_{z=0}$. Then, we may write

$$
P_{B,i,L} = \sum_{k=0}^{L} \mathcal{P}[L_{q,i,u} = k]\mathcal{P}[\text{burst} > L-k] \tag{6.30}
$$

and the blocking probabilities for TCP segments and acknowledgments become

$$
\begin{aligned}
P_{BdL} &= \sum_{k=0}^{L} \mathcal{P}[L_{q,i,u} = k] \sum_{l=L-k}^{\infty} g_{bd,l} \\
P_{BaL} &= \sum_{k=0}^{L} \mathcal{P}[L_{q,i,u} = k] \sum_{l=L-k}^{\infty} g_{ba,l}
\end{aligned}
\tag{6.31}
$$

The delay of the entire TCP segment is equal to the delay of the first baseband packet from that segment:

$$D^*_{L,d}(s) = \frac{1}{1 - P_{BdL}} \left[\sum_{k=0}^{K-1} \Omega^*_{k,i,u}(s) G^*_{pd}(s)^{2k} V^*_i(s)^{\lfloor k/M \rfloor} + \sum_{k=1}^{L-1} \sum_{\mu=1}^{M} \Pi^*_{k,\mu,i,u}(s) G^*_{pd}(s)^{2(k-1)} V^*_{i,u}(s)^{\lfloor (k+\mu-1)/M \rfloor} \right] \tag{6.32}$$

Note that this delay is not equal to the delay for the acknowledgment segment, $D^*_{L,a}(s)$, even though the expressions used to derive them are similar.

6.5 Performance assessment

We first consider a piconet with two slaves only, having two simultaneous but independent TCP connections: one from slave 1 to slave 2, the other from slave 2 to slave 1. In the first set of experiments, we have varied the token buffer queue size S and offered load per connection, while other parameters were fixed. The value of the polling parameter was $M = 5$, so that the entire TCP segment can be sent in one piconet cycle, i.e., during a single visit of the master to the source slave. The token rate was fixed to $tb = 250kbps$, the token buffer capacity was $W = 3$KB, the output rate of the token bucket queue was set to $max\_rate = 1Mbps$, and the outgoing (uplink) queue size was $L = 20$.

The values of TCP goodput and RTT obtained by using the ns-2 simulator [ns2, 2003] with Bluehoc extension [Bluehoc, 2003], are shown in Figs. 6.5(a) and 6.5(b), respectively. The size of token bucket buffer varied from 5 to 25 baseband packets, while the offered load varied from 80 to 240kbps. We observe that only for the largest token bucket buffer, at $S = 25$, does the goodput reach the physical limit (for the given segment size) of approximately 180kbps per connection. At the same time, the round trip delay reaches $400ms$.

The mean congestion window size and mean slow start threshold are shown in Figs. 6.5(c) and 6.5(d), respectively, when the buffer size S varies from 5 baseband packets to 25 and the offered load varies from 80 to 240 kbps. The mean congestion window size grows with S, which is expected since larger S means lower buffer loss. Furthermore, the congestion window grows with the segment arrival rate under very low loads, since there are not enough packets to expand the window size. For moderate and high offered loads, the average window size experiences a sharp decrease with the offered load. This is consistent with analytical results, since the packet loss probability is directly proportional to the offered load.

Finally, Figs. 6.5(e) and 6.5(f) show the rate of TCP time-outs and fast retransmissions as functions of offered load and buffer size S.

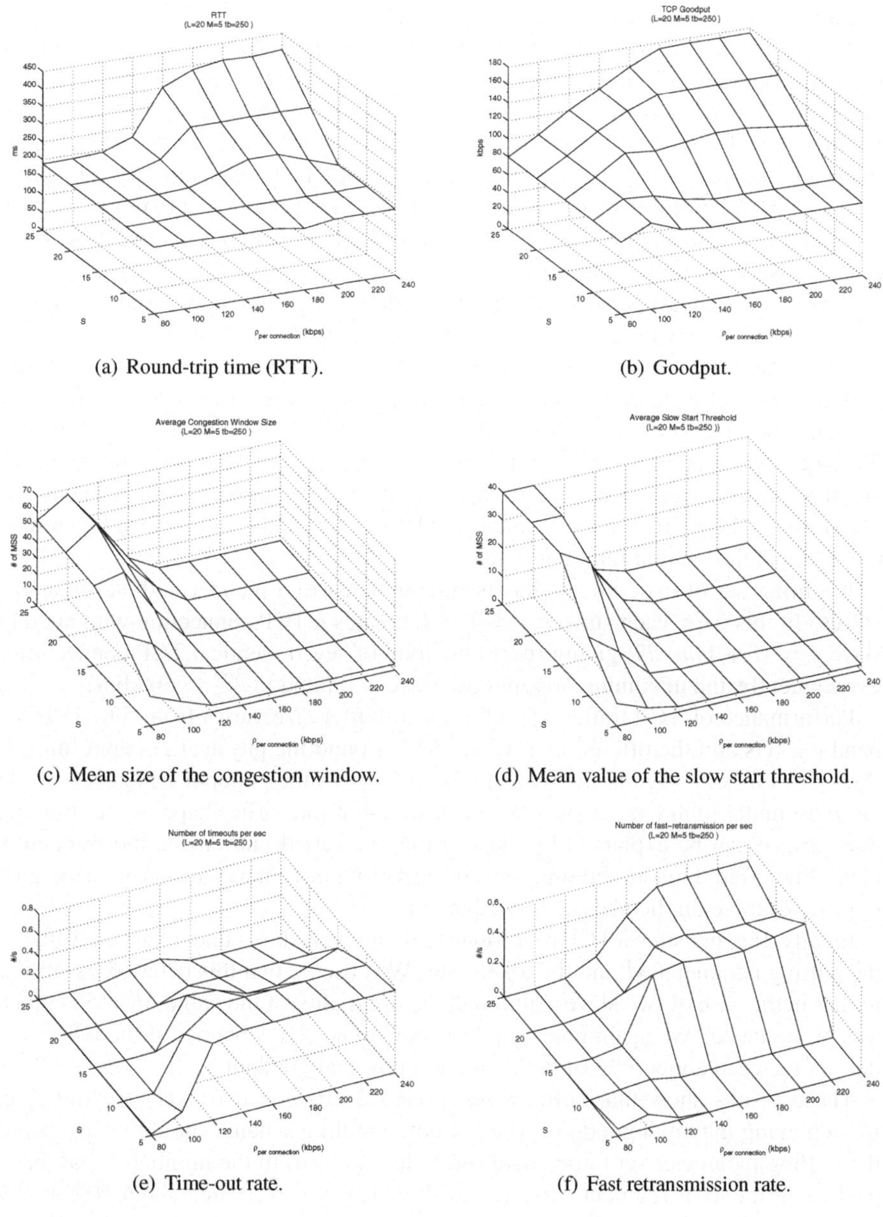

(a) Round-trip time (RTT).

(b) Goodput.

(c) Mean size of the congestion window.

(d) Mean value of the slow start threshold.

(e) Time-out rate.

(f) Fast retransmission rate.

FIGURE 6.5
TCP performance as the function of the buffer size S, in the piconet with two slaves.

Together, the dependencies shown in Fig. 6.5 hint that the value of $S = 25$ leads to maximum achievable throughput and negligible time-out rate for the offered load equal to the maximum achievable goodput (180kbps).

The next set of experiments considered TCP performance as a function of the polling parameter M, with constant values of $S = 25$ and $L = 20$. The resulting diagrams are shown in Fig. 6.6. We observe that the value of $M = 5$, which is sufficient to carry a TCP segment of five baseband packets, gives maximum goodput, minimum time-out rate and maximum fast-retransmission rate, when compared to larger values of M. Therefore, the minimal value of M which is sufficient to transfer a TCP segment in one piconet cycle appears also to be optimal with respect to goodput and other measures of performance.

We have also investigated TCP behavior with varying offered load and varying token rate; the corresponding diagrams are shown in Fig. 6.7. We observe that increasing the token rate over the maximal achievable goodput per connection can result only in marginal increase of mean congestion window size and goodput, despite the fact that the blocking probability at the token bucket filter is decreased. However, the blocking probability at the outgoing buffer at the baseband will still increase, and this increase leads to increased time-out rate and increased overall segment loss probability. Therefore, the token rate should not be set to the value much larger than the physical throughput limit per slave.

A similar set of experiments has been conducted in a piconet with seven active slaves. In this case, each slave i, $i = 1 \,..\, 7$, creates a TCP connection with another slave $j = (i + 1) mod 7$, giving rise to the total of seven identical TCP connections. Consequently, the maximum goodput per slave is limited to $\frac{360}{7} \approx 50$ kbps.

Performance of TCP traffic when the token buffer size varies from 5 to 25 baseband packets and the offered load varies 50% around the physical goodput limit, is shown in Fig. 6.8. Again, the value of $S = 25$ gives the goodput which is close to the limit under high load, as well as a low time-out rate. The shape of the time-out rate surface can be explained by the fact that packets do not arrive too frequently under low offered loads, and time-outs occur before three new packets are generated to provoke three duplicated acknowledgments.

Finally, Fig. 6.9 shows TCP performance with seven slaves under varying value of the polling parameter M and the token rate. We observe that this behavior is similar to that in the case of two slaves, although the optimality of the value $M = 5$ is much less pronounced. We again note that good performance is obtained if the token rate does not exceed about 50% of the maximum physical goodput.

These results show that buffer sizes around 25 baseband packets are sufficient for achieving maximal goodput. The E-limited polling scheme should be used with the polling parameter set to be equal (or, at least, close) to the number of baseband packets in the TCP segment. Token rate should be set to a value about 50% larger than the portion of the total piconet throughput dedicated to a particular slave.

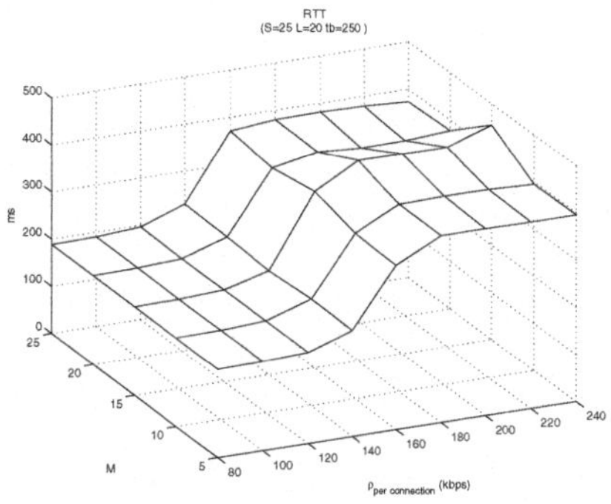

(a) Round-trip time (RTT).

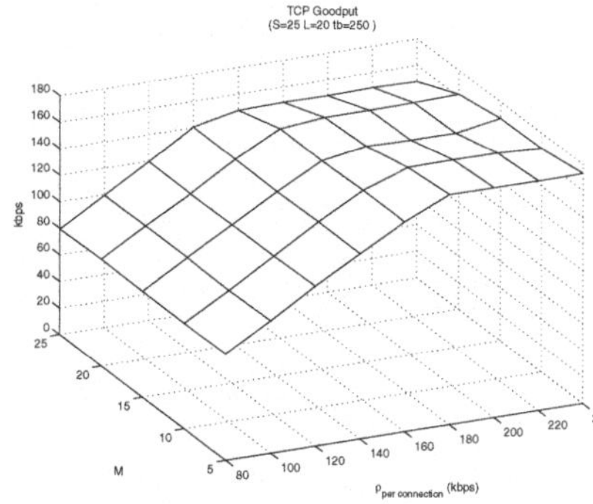

(b) Goodput.

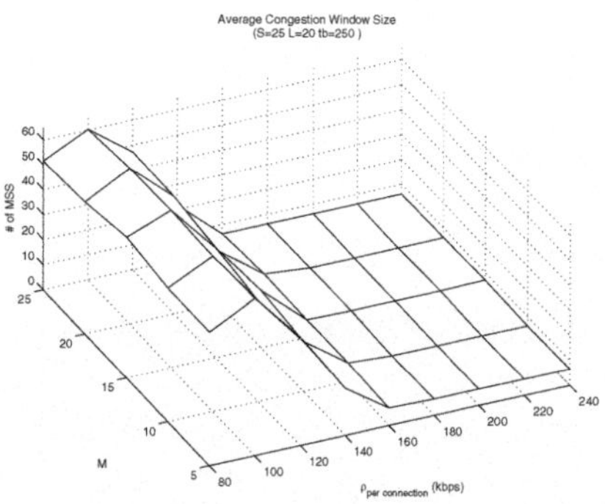

(c) Mean size of the congestion window.

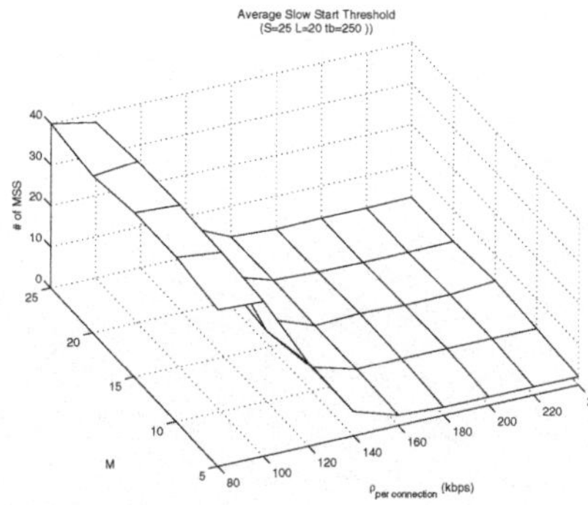

(d) Mean value of the slow start threshold.

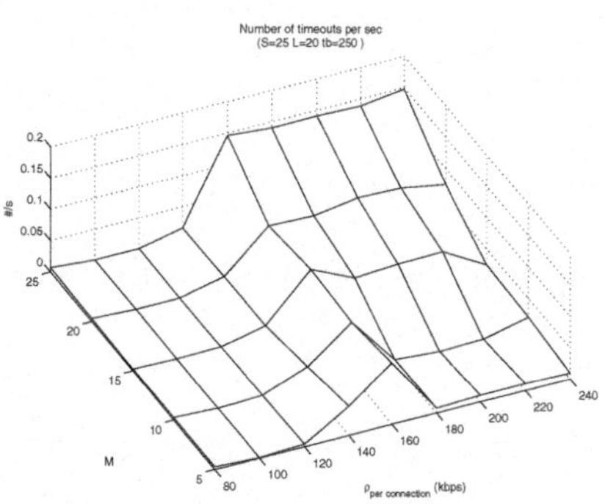

(e) Time-out rate.

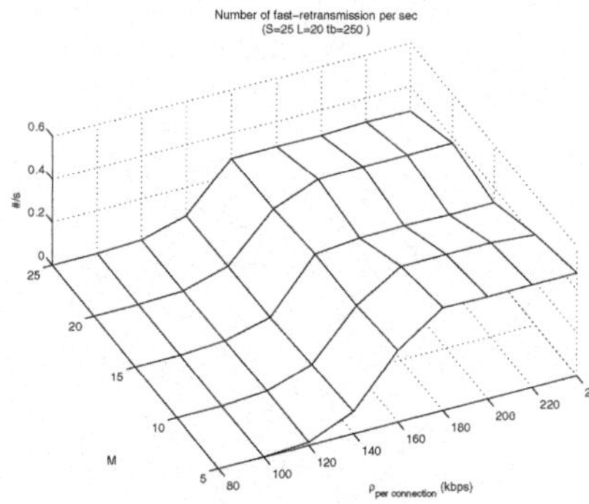

(f) Fast retransmission rate.

FIGURE 6.6
TCP performance as the function of the polling parameter M, in the piconet with two slaves.

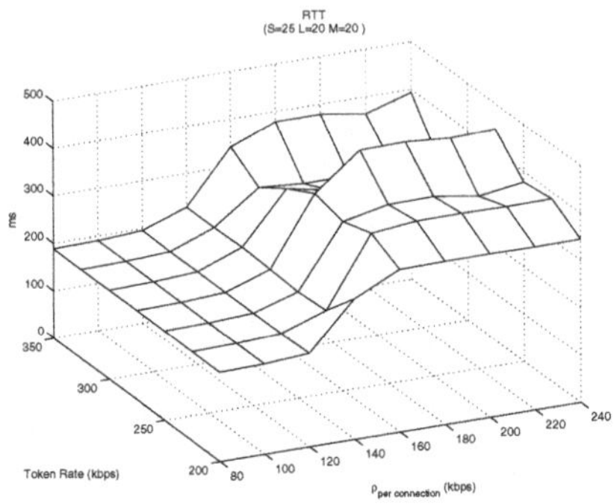

(a) Round-trip time (RTT).

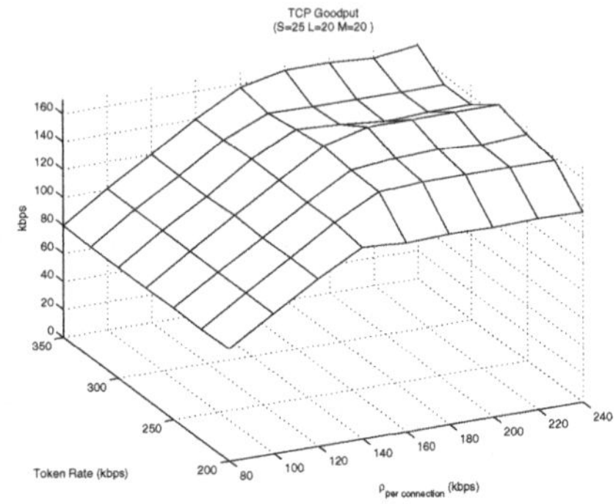

(b) Goodput.

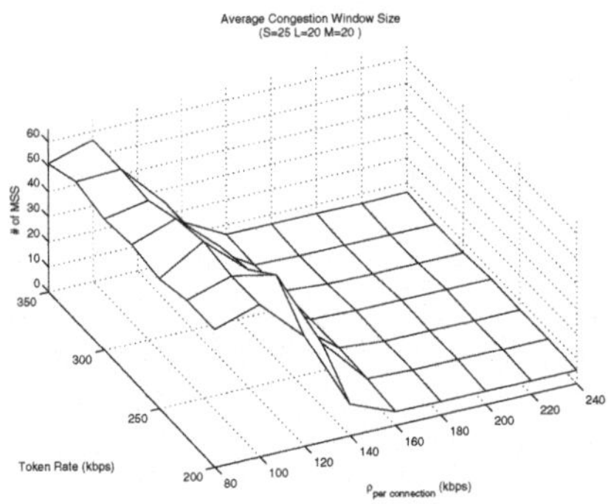

(c) mean size of the congestion window.

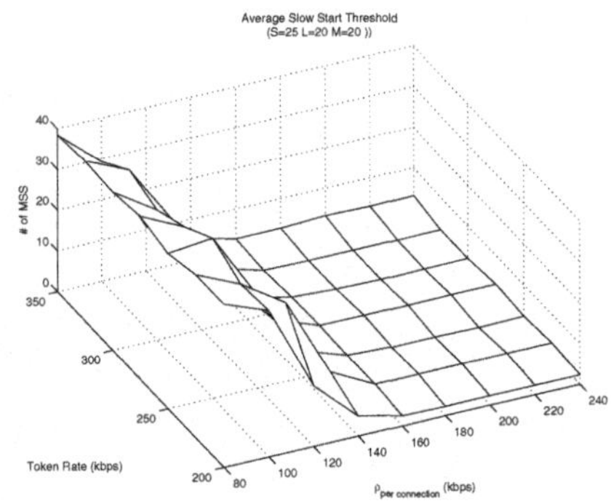

(d) Mean value of the slow start threshold.

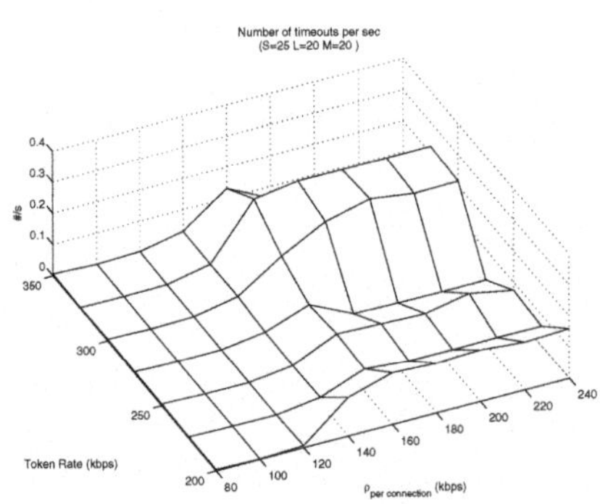

(e) Time-out rate.

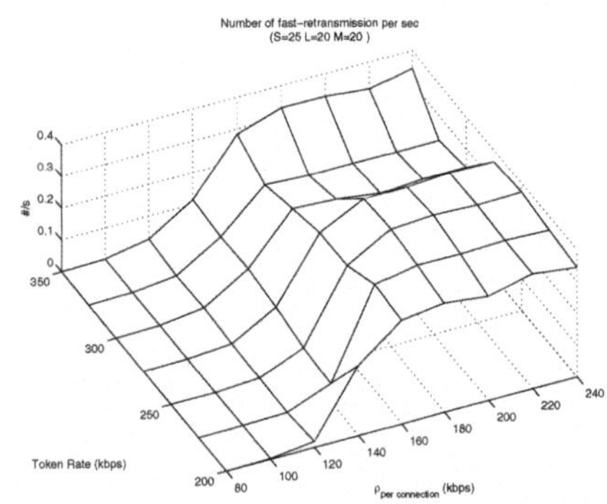

(f) Fast retransmission rate.

FIGURE 6.7
TCP performance as the function of token rate, in the piconet with two slaves.

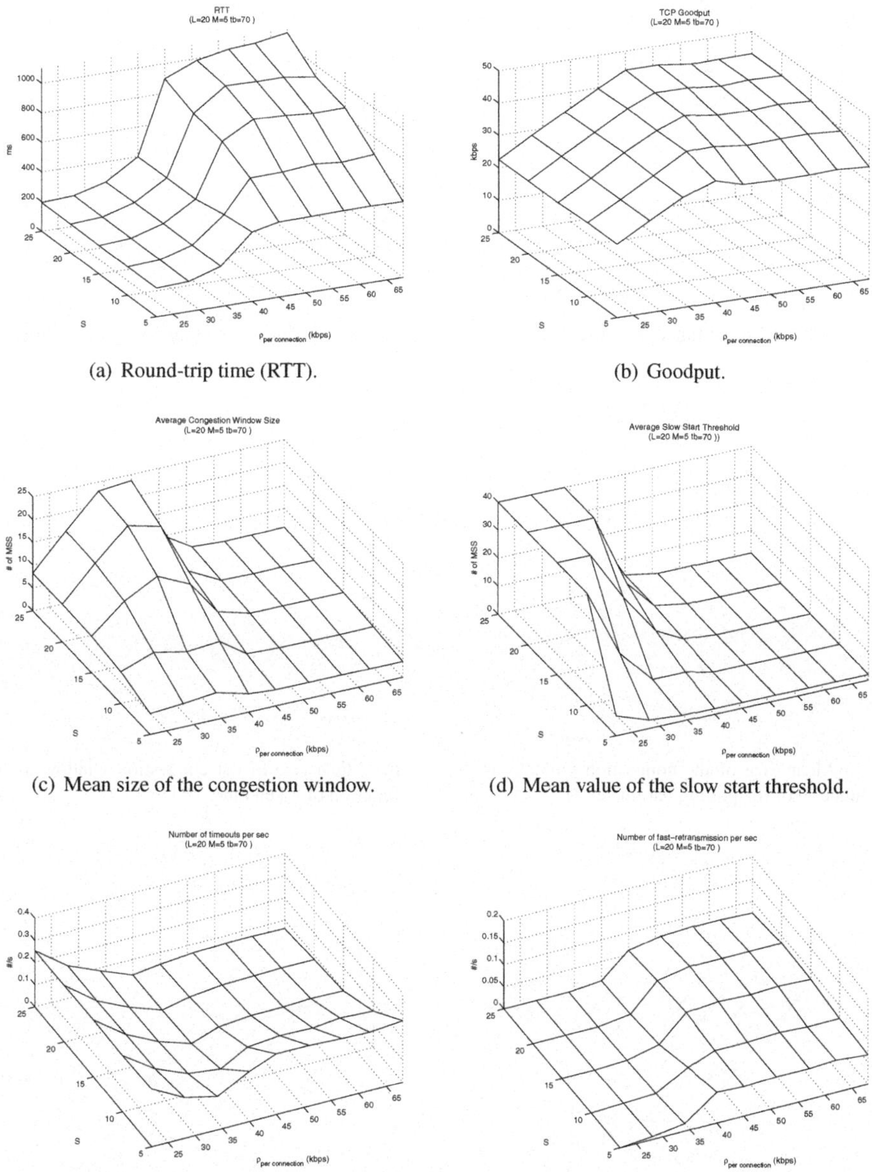

(a) Round-trip time (RTT).

(b) Goodput.

(c) Mean size of the congestion window.

(d) Mean value of the slow start threshold.

(e) Time-out rate.

(f) Fast retransmission rate.

FIGURE 6.8

TCP performance as the function of the buffer size S, in the piconet with seven slaves.

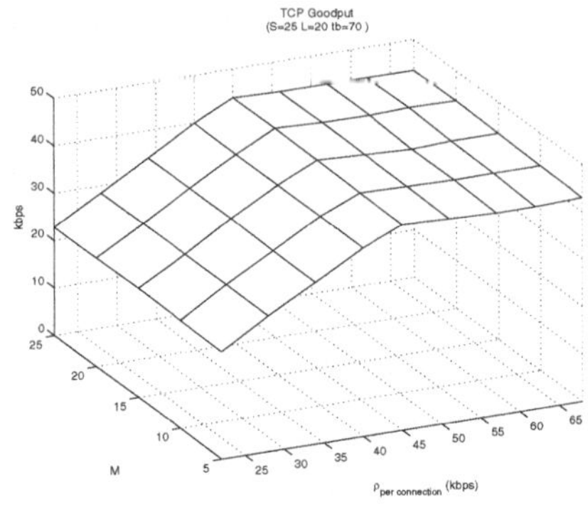

(a) Goodput as the function of the polling parameter M.

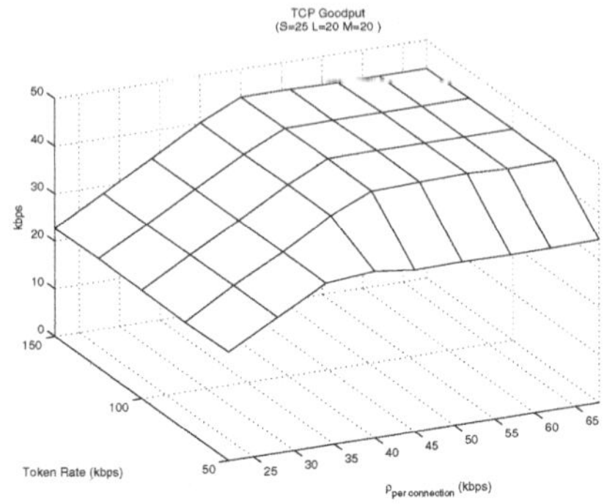

(b) Goodput as the function of token rate.

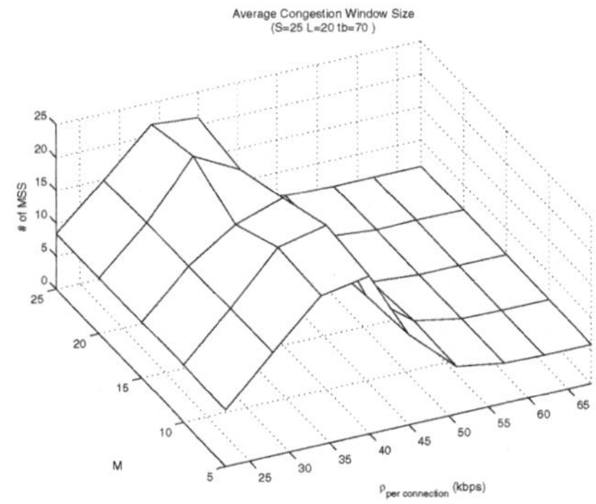

(c) Mean size of the congestion window as the function of the polling parameter M.

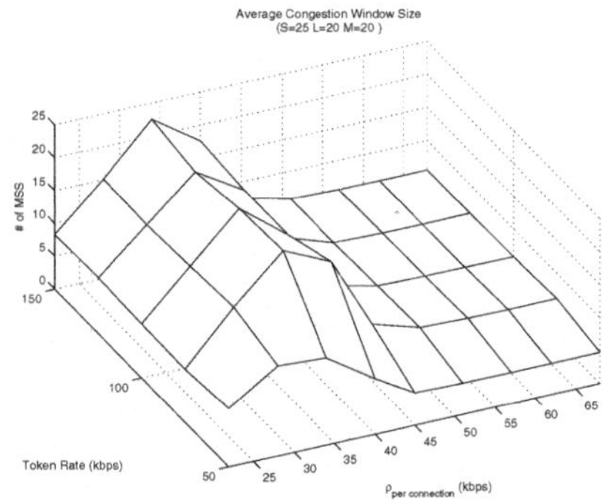

(d) Mean size of the congestion window as the function of token rate.

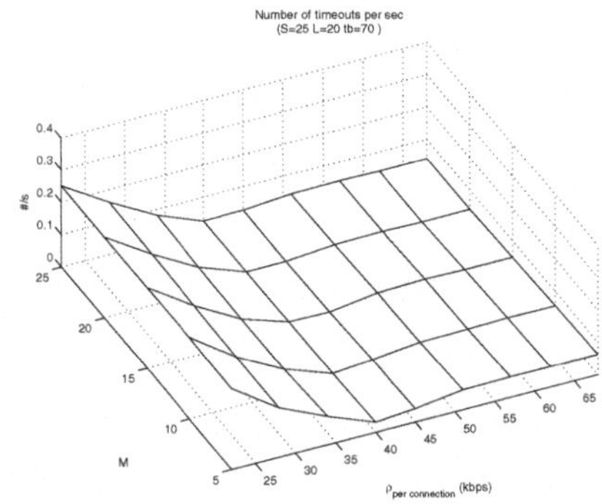

(e) Time-out rate as the function of the polling parameter M.

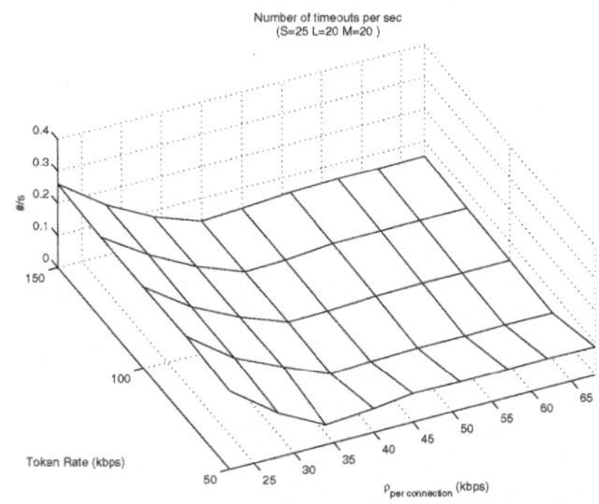

(f) Time-out rate as the function of token-rate.

FIGURE 6.9
TCP performance as functions of token rate and the polling parameter M, in the piconet with seven slaves.

7

Piconets with synchronous traffic

Voice and media communications are repeatedly singled out as important application areas for Bluetooth. The official Bluetooth specification provides a special type of connection – the so-called SCO connection – to carry voice traffic. Yet when such a connection is established, the performance of asynchronous, data traffic in the piconet is drastically reduced. In this chapter we investigate whether an alternative solution would enable the piconet to carry both asynchronous and voice traffic with satisfactory performance. It turns out that a feasible solution, which we will refer to as pseudo-SCO or pSCO for short, may indeed be implemented using only the facilities already provided in the current Bluetooth specification [Bluetooth SIG, 2001*b*], without requiring modifications or extensions of any kind. The performance of asynchronous traffic under the new scheme is much improved over that of the original SCO scheme, while the bandwidth requirements for voice traffic are still easily satisfied.

The chapter begins with the description of the operation of ordinary SCO links and their limitations, in Section 7.1. Then, we present the pSCO scheme in Section 7.2. Section 7.3 analyzes the performance of the new scheme using the theory of $M^{[x]}/G/1$ queues with vacations [Takagi, 1991]. Finally, we present some results that clearly show the benefits that may be obtained through the pSCO scheme.

7.1 Why the built-in SCO links are bad

As noted in Chapter 1, two types of links may exist between the piconet master and its slaves. In an ACL link, the master is free to poll or not to poll the slave at will; the slave may talk back only when addressed by the master, and only immediately after being addressed by the master. This scheme uses packets of DMx and DHx types (where x = 1, 3, or 5) with different length and different information-carrying capacity, listed in Table 7.1. The packet payload is protected with appropriate CRC, so that the receiver can request retransmission in case the content is damaged due to interference and/or noise. Furthermore, some types of packets offer forward error correction (FEC) as well.

The master and an active ACL slave may also establish another type of link: the Synchronous Connection-Oriented (SCO) link. SCO links are designed to support

TABLE 7.1
Packet types for communication over an ACL link.

Type	Slot(s)	Payload (bytes)	FEC
DM1	1	17	2/3
DH1	1	27	none
DM3	3	121	2/3
DH3	3	183	none
DM5	5	224	2/3
DH5	5	339	none

synchronous, constant bit rate (CBR) traffic such as voice [Bluetooth SIG, 2001*b*], and they use special packet types labeled HV1, HV2, and HV3, the characteristics of which are outlined in Table 7.2 (a more detailed description is given in Chapter 1). A fourth packet type (DV) combines 10 bytes of voice data with up to 150 bits of other data; as its behavior is not different from that of the HV type packets, it will not be considered separately. These packets are not equipped with CRC, and retransmission of damaged packets is not supported, in line with the observation that voice communications are much more sensitive to packet delays than to packet loss [Kurose and Ross, 2005].

TABLE 7.2
Packet types for communication over an SCO link.

Type	Slot(s)	Payload (bytes)	Speech duration (ms)	FEC	SCO interval (slots)
HV1	1	10	1.25	1/3 (repetition)	2
HV2	1	20	2.5	2/3 (polynomial)	4
HV3	1	30	3.75	none	6

SCO links are specifically designed to carry voice traffic at the standard, non-compressed rate of 64kbps. An SCO link can optionally use compression, but this does not change the timing scheme, nor does it affect the type of packets used. Issues related to compression are beyond the scope of this text and our analysis will simply use the value of 64kbps as the required bandwidth for voice transmissions.

In order to provide the 64*kbps* bandwidth needed for a voice channel, a strict timing scheme, shown in Fig. 7.1, has to be observed. The SCO link reserves slots in master-slave communication, and ACL traffic is not allowed to use these slots at all. Therefore, the bandwidth available to ACL traffic is substantially reduced, or even eliminated in extreme cases.

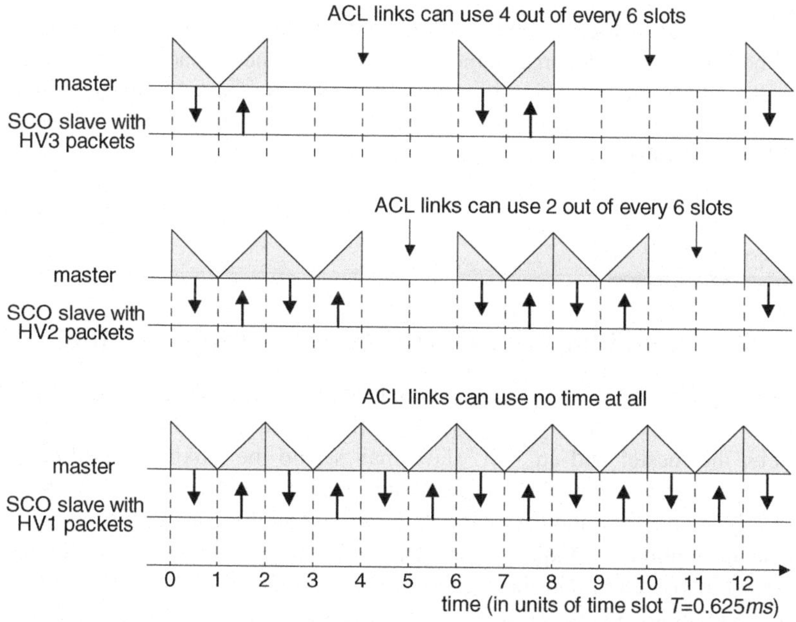

FIGURE 7.1
Timing of SCO communications with different packet types.

Furthermore, the time available to ACL traffic (if any) is partitioned in intervals of four or two time slots apiece, as can be seen from Fig. 7.1. Consequently, five slot packets cannot be used at all, while three slot packets can be used if and only if the other packet in a frame is a single-slot packet. Even this limitation will not suffice in all cases: if the first frame after the SCO link uses two one slot packets, the second frame must use one slot packets too. As these limitations are not part of the original communication protocol, they must somehow be advertised to all the ACL slaves as soon as an SCO link is established. The Bluetooth specification does allow the master to limit the number of slots used by a slave by issuing a LMP_max_slot command [Bluetooth SIG, 2001*b*], but the limitations outlined above are more complex and appropriate extensions would be needed.

Ultimately, the presence of an SCO link in a Bluetooth piconet leads to drastic reductions in maximum achievable data rates and corresponding increase in end-to-end packet delays for asynchronous traffic. The queueing delays for asynchronous traffic will also increase, as the regular service of ACL packets is effectively frozen during SCO frames. As packet delays for asynchronous traffic are mostly determined by queueing delays in different devices [Mišić and Mišić, 2003*b*; Mišić and Mišić, 2003*a*], the increase in delays and the corresponding decrease of data rates will be disproportionately high compared to the simple reduction in available bandwidth.

To summarize, the performance of asynchronous traffic in the presence of an SCO link is far from satisfactory, and it is worth investigating whether a better alternative could be found. Such an arrangement should use higher-capacity ACL packets, yet it should satisfy the strict timing requirements for synchronous traffic. Of course, a solution that uses the mechanisms already provided by the official specification would be preferable.

7.2 pSCO: an improved scheme for synchronous traffic

It turns out that such an arrangement is indeed possible, as the Bluetooth Link Manager provides basic Quality of Service (QoS) capabilities [Bluetooth SIG, 2001*b*]. Namely, the master and an ACL slave may set up the maximum polling interval T_{poll}, i.e., the maximum time between subsequent transmissions. This polling interval is guaranteed in the active mode, except when there are collisions with page, page scan, inquiry, and inquiry scan. This mechanism may be used as the basis for the pSCO improved synchronous transmission scheme.

As noted above, the basic requirements to be satisfied in the pSCO mode are the bandwidth and latency of the synchronous transmission channel. Uncompressed voice transmission requires the bandwidth of 64kbps. Latency primarily depends on the directionality of traffic; for bidirectional traffic (i.e., ordinary telephone-like conversation), the end-to-end (round trip) delays of 100 to 300ms are noticeable but still acceptable [Partridge, 1994]. Of course, for unidirectional traffic, such as a live lecture broadcast, delays are not important. The main portion of those end-to-end delays will be the transmission and queueing delays, which may be controlled through the pSCO scheme. However, the end-to-end delays will also include the time for packetization, compression, and decompression of voice signal (each of which has to be performed twice in a two-way conversation), and possibly some time for buffering in order to compensate for the packet arrival jitter [Kurose and Ross, 2005]. These additional delay times will not be negligible, the more so because the computational power of Bluetooth devices may be limited by the (more important) requirements of low energy consumption. Of course, no hard numbers can be given, but, in general, we would like to keep the latency as low as possible.

The parameters to be adjusted in order to satisfy those requirements are the type of packets to be used and the duration of the polling interval T_{poll}. The obvious task sequence would be as follows.

- First, choose the type of packet. Multi-slot ACL packets have larger payloads, which means they can be sent at longer intervals. Longer polling intervals, in turn, leave more contiguous time available for asynchronous traffic. However, noise and interference conditions might dictate the use of FEC-protected packets (i.e., DM type), which carry somewhat smaller payloads than their DH counterparts.

- Then, calculate the polling interval so as to satisfy the bandwidth requirement. As the polling interval is expressed in Bluetooth time slots T, it should be rounded to the nearest lower integer value.

- In fact, the polling interval may be increased if we allow multiple packets to be exchanged at once. For example, if each exchange takes two packets instead of just one, the polling interval may be doubled. However, polling intervals that are too long might violate the latency requirement.

The corresponding values for different three- and five-slot ACL packet types, assuming single-frame exchanges, are given in Table 7.3. Note that the round trip transmission delay will be twice the duration of the polling interval plus the additional delays described above. For simplicity, we will consider only single-frame pSCO exchanges, although our analysis framework can easily accommodate multi-frame exchanges as well.

TABLE 7.3
Packet types and polling intervals for 64kbps pSCO mode connections. (From J. Mišić, V. B. Mišić, and K. L. Chan, "Talk and let talk: performance of Bluetooth piconets with synchronous traffic," *Ad hoc networks*, **3**(4):451–477, © 2003 Elsevier B. V.)

Type	Slot(s)	Payload (bytes)	Polling interval (slots)	(ms)	Data rate (kbps)
HV3	1	30	6	3.75	64
DM3	3	121	24	15.0	64.5
DH3	3	183	36	22.5	65.1
DM5	5	224	44	27.5	65.2
DH5	5	339	67	41.9	64.8

The packet timing under the pSCO scheme is shown in Fig. 7.2. Although similar in principle to the SCO timing, the use of multi-slot packets for synchronous traffic allows less overall time to be spent on synchronous exchanges on the average, and makes each interval between those exchanges last much longer. Consequently, asynchronous traffic will suffer less disruption than under the original SCO scheme, and its performance will improve, as will be seen in subsequent discussions.

It should be noted that similar schemes have been proposed elsewhere. Chawla, Saran and Singh [2001] have investigated the possibility of bandwidth conservation for voice transmission. To that effect, two schemes have been proposed; in the second one, ACL packets are used to transmit voice and other types of synchronous traffic. However, they provide limited simulation results only, without going into much detail about the performance of asynchronous traffic in this case.

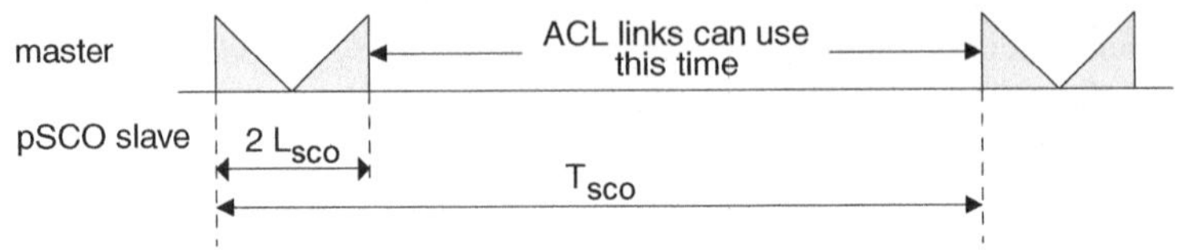

FIGURE 7.2
Timing of pseudo-SCO links.

Also, Famolari and Anjum [2002] have proposed the use of ACL packets for voice transmission, and then analyzed the effect of different traffic priorities, noise, and interference, on voice traffic. Again, the results are obtained by simulations only.

Kapoor, Jyh-Ling, Lee and Gerla [2002] have also proposed the use of ACL packets to carry voice traffic. However, their main focus is the performance of voice traffic, including the resulting delay distribution, rather than asynchronous traffic. A similar proposal by Wu and Todd [2004] utilizes different ACL packets and discusses their performance, in particular with respect to call quality and call blocking in a noisy environment. An earlier paper [Wu, Todd and Shirani, 2003], based on the results of [Xue and Todd, 2001], has shown that user mobility can be exploited to improve the traffic capacity and reduce voice call blocking.

However, none of these papers provide a theoretical analysis of the performance of ACL traffic in the presence of SCO or pSCO links, nor do they analyze the impact of different intra-piconet polling schemes. Therefore, the work described here complements, rather than replicates, the results presented in the aforementioned papers.

Finally, we note that an additional type of link has been introduced in version 1.2 of the Bluetooth specification [Bluetooth SIG, 2003*a*]: the extended SCO, or eSCO link. The eSCO scheme is rather similar to our pSCO scheme, as it uses longer polling intervals and packets that carry larger payloads than the original SCO scheme. As the behavior of the eSCO link is quite similar to that of the pSCO scheme, the analysis presented below applies to eSCO links as well, provided the differences in timing are accounted for.

7.3 Performance of the pSCO scheme

We consider a single Bluetooth piconet with m members. In addition to the piconet master and $m-2$ ACL slaves, there is a single slave device with synchronous (SCO or pSCO) traffic. The master and ACL slaves use downlink and uplink queues, respectively, according to the queueing model of Fig. 2.4. The CBR traffic packets are generated at regular intervals and sent immediately, therefore the pSCO link does not need any queues.

We will assume that all slaves generate ACL traffic with the same arrival rate, and that the master has no ACL traffic of its own. The pSCO link generates packets of length L_{sco} slots on every T_{sco} time slots. Note that the plain SCO connection with HV3 packets, as defined by the Bluetooth specification, may be described by the combination of $L_{sco} = 1$ and $T_{sco} = 6$. Multi-frame exchanges, as well as synchronous transmissions with data rates other than 64kbps, are handled by simply choosing appropriate values for L_{sco} and T_{sco}.

We assume that the piconet master is the member with the index 1, the SCO/pSCO slave has the index 2, while the ACL slaves have indices from 3 to m. We will use S_i^p and S_i^t to denote channel service time for ACL slave i, $i = 3\,..\,m$ without and with the time slots inserted by pSCO connection, respectively. As explained in Chapter 3, this time depends on the chosen polling scheme; we assume that the E-limited polling scheme is used to poll the plain ACL slaves. The Probability Generating Function (PGF) for the channel service time will be denoted as $S_i^p(z)$ and $S_i^t(z)$ and its Laplace-Stieltjes Transform (LST) will be denoted as $S_i^{p*}(s)$ and $S_i^{t*}(s)$ for the cases without and with the slots inserted by the pSCO connection. By the same token, piconet cycle time, i.e., the time needed to visit all the slaves once will be denoted with C^p and C^t. The probability density function (pdf) for the piconet cycle time in these two cases is denoted as $c^p(x)$ and $c^t(x)$. The PGF for the piconet cycle time without and with the pSCO time slots is denoted as $C^p(z)$ and $C^t(z)$ while its LST is denoted as $C^{p*}(s)$ and $C^{t*}(s)$, respectively. As stated in Chapter 3, the vacation time for the ACL slave i is the time while master serves all the other slaves. The vacation time for slave i is denoted as V_i^p and V_i^t, its PGF is $V_i^p(z)$ and $V_i^t(z)$, and its LST is denoted as $V^{p*}(s)$ and $V^{t*}(s)$ for the cases without and with taking into account time slots for pSCO traffic.

Performance analysis of the piconet operating with the mix of ACL and pSCO traffic is similar to the analysis of the piconet with ACL traffic only, discussed earlier in Chapter 3. However, the frame time, vacation time, and piconet cycle time must be modeled without and with taking pSCO packets into account; in order to distinguish between those two cases, we will decorate the corresponding variables with superscripts p and t, respectively. As before, the number of packets in the uplink queue is modeled as a set of imbedded Markov points [Takagi, 1991]. The Markov points correspond to vacation termination times, i.e., the times when the master starts to service the given slave, and service completion times, i.e., the times when the slave finishes one uplink packet transmission.

Thanks to the symmetry of the piconet with respect to the slaves, it suffices to consider the packet exchange between a single slave, say i, and the master. Let $q^i_{k_u,k_d}$ denote the joint probability that a Markov point in the uplink queue of slave i is a vacation termination time, and that there are $k_u = 0, 1, 2\ldots$ packets at the slave i's uplink queue and $k_d = 0, 1, 2\ldots$ packets at the master's i's downlink queue at that time. Also, let $\pi^{i,(\mu)}_{k_u,k_d}$ denote the joint probability that a Markov point is the μ-th frame transmission completion time and that there are k_u packets in the slave's i queue at that time, and k_d packets at the master's queue toward the slave i, where $\mu = 1\,..\,M$ and $k_d, k_u = 0, 1, 2, \ldots$

Let $f_i^t(x)$ and $v_i^t(x)$ stand for the pdfs (probability density functions) of the total frame transmission time and vacation time, respectively, including pSCO packets, at the uplink queue of slave i; their LST transforms will be $(F_i^t)^*(s)$ and $(V_i^t)^*(s)$. We will also make use of the following probabilities:

- the probability of k_u packet arrivals at the slave i's uplink queue during the frame time, denoted with a_{k_u};
- the probability of k_d packet arrivals at the master's downlink queue during the frame time, denoted with a_{k_d};
- the probability of k_u packet arrivals at the slave i's uplink queue during the vacation time, denoted with f_{k_u}; and
- the probability of k_d packet arrivals in the master's downlink during the vacation time, denoted with f_{k_d}.

Those probabilities may be calculated as

$$\begin{aligned}
a_{k_u} &= \sum_{l=0}^{\infty} \frac{1}{k_u!} \frac{d^{k_u}}{dz^{k_u}} (G_b(z))^l \Big|_{z=0} \int_0^{\infty} \frac{(\lambda_{iu}x)^l}{l!} e^{-\lambda_{iu}x} f_i^t(x)dx \\
&= \frac{1}{k_u!} \frac{d^{k_u}}{dz^{k_u}} (F_i^{t*}(\lambda_{iu} - \lambda_{iu}G_b(z)))\Big|_{z=0} \\
a_{k_d} &= \sum_{l=0}^{\infty} \frac{1}{k_d!} \frac{d^{k_d}}{dz^{k_d}} (G_b(z))^l \Big|_{z=0} \int_0^{\infty} \frac{(\lambda_{id}x)^l}{l!} e^{-\lambda_{id}x} f_i^t(x)dx \\
&= \frac{1}{k_d!} \frac{d^{k_d}}{dz^{k_d}} (F_i^{t*}(\lambda_{id} - \lambda_{id}G_b(z)))\Big|_{z=0} \\
f_{k_u} &= \sum_{l=0}^{\infty} \frac{1}{k_u!} \frac{d^{k_u}}{dz^{k_u}} (G_b(z))^l \Big|_{z=0} \int_0^{\infty} \frac{(\lambda_{iu}x)^l}{l!} e^{-\lambda_{iu}x} v_i^t(x)dx \\
&= \frac{1}{k_u!} \frac{d^{k_u}}{dz^{k_u}} V_i^{t*}(\lambda_{iu} - \lambda_{iu}G_b(z))\Big|_{z=0} \\
f_{k_d} &= \sum_{l=0}^{\infty} \frac{1}{k_d!} \frac{d^{k_d}}{dz^{k_d}} (G_b(z))^l \Big|_{z=0} \int_0^{\infty} \frac{(\lambda_{iu}x)^l}{l!} e^{-\lambda_{iu}x} v_i^t(x)dx \\
&= \frac{1}{k_d!} \frac{d^{k_d}}{dz^{k_d}} V_i^{t*}(\lambda_{iu} - \lambda_{iu}G_b(z))\Big|_{z=0}
\end{aligned} \tag{7.1}$$

Similar expressions hold for the number of arrivals in the corresponding downlink queue at the master. The probabilities that the uplink queue of slave i contains k_u packets and that the downlink queue toward slave i contains k_d packets in imbedded Markov points, satisfy the following equations:

$$
\begin{aligned}
\pi^{i,(1)}_{k_u,k_d} &= \sum_{j_u=1}^{k_u+1}\sum_{j_d=1}^{k_d+1} q^i_{j_u,j_d} a_{k_u-j_u+1} a_{k_d-j_d+1} + \sum_{j_d=1}^{k_d+1} q^i_{0,j_d} a_{k_d-j_d+1} a_{k_u} \\
&+ \sum_{j_u=1}^{k_u+1} q^i_{j_u,0} a_{k_u-j_u+1} a_{k_d} \\
\pi^{i,(\mu)}_{k_u,k_d} &= \sum_{j_u=1}^{k_u+1}\sum_{j_d=1}^{k_d+1} \pi^{i,(\mu-1)}_{j_u,j_d} a_{k_u-j_u+1} a_{k_d-j_d+1} + \sum_{j_d=1}^{k_d+1} \pi^{i,(\mu-1)}_{0,j_d} a_{k_d-j_d+1} a_{k_u} \\
&+ \sum_{j_u=1}^{k_u+1} \pi^{i,(\mu-1)}_{j_u,0} a_{k_u-j_u+1} a_{k_d}, \qquad \mu = 2\,..\,M \\
q^i_{k_u,k_d} &= \left(\sum_{\mu=1}^{M-1} \pi^{i,(\mu)}_{0,0} + q^i_{0,0}\right) f_{k_u} f_{k_d} + \sum_{j_u=0}^{k_u}\sum_{j_d=0}^{k_d} \pi^{i,(M)}_{j_u,j_d} f_{k_u-j_u} f_{k_d-j_d}
\end{aligned}
\tag{7.2}
$$

The probability generating functions (PGFs) for the number of packets in the uplink and the corresponding downlink queue at the imbedded Markov points are defined by

$$
\begin{aligned}
\Pi^t_{i,\mu}(z,w) &= \sum_{k_u=0}^{\infty}\sum_{k_d=0}^{\infty} \pi^{i,(\mu)}_{k_u,k_d} z^{k_u} w^{k_d}, \ \mu = 1\,..\,M \\
Q^t_i(z,w) &= \sum_{k_u=0}^{\infty}\sum_{k_d=0}^{\infty} q^i_{k_u,k_d} z^{k_u} w^{k_d}
\end{aligned}
\tag{7.3}
$$

which, after the transformation shown in Section 3.3, p. 37, may be written as

$$
\begin{aligned}
\Pi^t_{i,1}(z,w) &= \frac{F^{t*}_i(\lambda_{iu} - \lambda_{iu} G_b(z)) F^{t*}_i(\lambda_{id} - \lambda_{id} G_b(w))}{zw} \\
&\cdot \Big(Q^t_i(z,w) - (1-w) Q^t_i(z,0) \\
&- \ (1-z) Q^t_i(0,w) + q^i_{0,0}(1-z-w)\Big) \\
\Pi^t_{i,\mu}(z,w) &= \frac{F^{t*}_i(\lambda_{iu} - \lambda_{iu} G_b(z)) F^{t*}_i(\lambda_{id} - \lambda_{id} G_b(w))}{zw} \\
&\cdot \Big(\Pi^t_{i,\mu-1}(z,w) - (1-w)\Pi^t_{i,\mu-1}(z,0) \\
&-(1-z)\Pi^t_{i,\mu-1}(0,w) + \pi^{i,(\mu-1)}_{0,0}(1-z-w)\Big), \\
&\qquad \mu = 2\,..\,M \\
Q^t_i(z,w) &= V^{t*}_i(\lambda_{iu} - \lambda_{iu} G_b(z)) V^{t*}_i(\lambda_{id} - \lambda_{id} G_b(w)) \\
&\cdot \left(\sum_{\mu=1}^{M-1} \pi^{i,(\mu)}_{0,0} + q^i_{0,0} + \Pi_{i,M}(z,w)\right)
\end{aligned}
\tag{7.4}
$$

When $z = 0$ is substituted in the last system, we can find $\Pi_{i,1}(0,w) \ldots \Pi_{i,M}(0,w)$ and $Q_i(0,w)$ as functions of $\pi^{i,(\mu)}_{0,0}$ $\mu = 1 \ldots M$ and $q^i_{0,0}$, and the system (7.4) becomes

$$
\begin{aligned}
\Pi^t_{i,1}(0,w) &= \frac{F_i^{t*}(\lambda_{iu})F_i^{t*}(\lambda_{id}-\lambda_{id}G_b(w))}{w}\cdot\left(Q_i(0,w)-q^i_{0,0}\right)\\
\Pi^t_{i,\mu}(0,w) &= \frac{F_i^{*}(\lambda_{iu})F_i^{t*}(\lambda_{id}-\lambda_{id}G_b(w))}{w}\\
&\quad\cdot\left(\Pi_{i,\mu-1}(0,w)-\pi_{0,0}^{i,(\mu-1)}\right), \qquad \mu = 2\,..\,M\\
Q^t_i(0,w) &= V_i^{t*}(\lambda_{iu})V_i^{t*}(\lambda_{id}-\lambda_{id}G_b(w))\\
&\quad\cdot\left(\sum_{\mu=1}^{M-1}\pi_{0,0}^{i,(\mu)}+q^i_{0,0}+\Pi_{i,M}(0,w)\right)
\end{aligned}
\tag{7.5}
$$

From the system (3.27), we find $\Pi_{i,1}(0,w)\ldots\Pi_{i,M}(0,w)$ and $Q_i(0,w)$. In an analogous fashion, we can find $\Pi_{i,1}(z,0)\ldots\Pi_{i,M}(z,0)$ and $Q_i(z,0)$ as functions of $\pi_{0,0}^{i,(\mu)}$, $\mu = 1\ldots M$, and $q^i_{0,0}$. For clarity, we introduce additional substitutions:

$$
\begin{aligned}
Q^t_0(z,w) &= (1-w)Q_i(z,0)+(1-z)Q^t_i(0,w)-q^i_{0,0}(1-z-w)\\
\Pi^t_{\mu,0}(z,w) &= (1-w)\Pi^t_{i,\mu}(z,0)+(1-z)\Pi^t_{i,\mu}(0,w)\pi_{0,0}^{i,(\mu)}(1-z-w),\\
&\qquad \mu = 1\,..\,M
\end{aligned}
\tag{7.6}
$$

and the system (7.4) becomes

$$
\begin{aligned}
\Pi^t_{i,1}(z,w) &= \frac{1}{zw}F_i^{t*}(\lambda_{iu}-\lambda_{iu}G_b(z))F_i^{t*}(\lambda_{id}-\lambda_{id}G_b(w))\\
&\quad\cdot\left(Q^t_i(z,w)-Q^t_0(z,w)\right)\\
\Pi^t_{i,\mu}(z,w) &= \frac{F_i^{t*}(\lambda_{iu}-\lambda_{iu}G_b(z))F_i^{t*}(\lambda_{id}-\lambda_{id}G_b(w))}{zw}\\
&\quad\cdot\left(\Pi^t_{i,\mu-1}(z,w)-\Pi^t_{\mu-1,0}(z,w)\right) \qquad \mu = 2\,..\,M\\
Q^t_i(z,w) &= V_i^{t*}(\lambda_{iu}-\lambda_{iu}G_b(z))V_i^{t*}(\lambda_{id}-\lambda_{id}G_b(w))\\
&\quad\cdot\left(\sum_{\mu=1}^{M-1}\pi_{0,0}^{i,(\mu)}+q^i_{0,0}+\Pi^t_{i,M}(z,w)\right)
\end{aligned}
\tag{7.7}
$$

Finally, by solving the system (7.7), we are able to obtain the expression for $Q^t_i(z,w)$ in the form

$$
Q^t_i(z,w) = V_i^{t*}(\lambda_{iu}-\lambda_{iu}G_b(z))V_i^{t*}(\lambda_{id}-\lambda_{id}G_b(w))z^M w^M\frac{A}{B} \tag{7.8}
$$

where

$$
\begin{aligned}
A &= \sum_{\mu=1}^{M-1}\pi_{0,0}^{i,(\mu)}+q^i_{0,0}-Q^t_0(z,w)Y^M-\sum_{\mu=1}^{M-1}Y^{M-\mu}\Pi^t_{\mu,0}(z,w)\\
B &= z^M w^M - V_i^{t*}(\lambda_{iu}-\lambda_{iu}G_b(z))V_i^{t*}(\lambda_{id}-\lambda_{id}G_b(w))\\
&\quad\cdot F_i^{t*}(\lambda_{iu}-\lambda_{iu}G_b(z))^M F_i^{t*}(\lambda_{id}-\lambda_{id}G_b(w))^M
\end{aligned}
\tag{7.9}
$$

and

$$Y = \frac{1}{zw} F_i^{t*}(\lambda_{iu} - \lambda_{iu} G_b(z)) F_i^{t*}(\lambda_{id} - \lambda_{id} G_b(w)) \tag{7.10}$$

However, the solution of (7.8) requires two more elements to be calculated: namely, the LSTs of the total frame time and vacation time distributions (with taking pSCO packet into account). In order to do that, we must first find pure channel and vacation times and later extend them with pSCO packets to obtain total duration of these times. Let us find the LST of the pure channel service time first; this is the time from the moment when master polls the slave for the first time, until either an empty frame has been encountered, or a total of M data frames have been exchanged. (As usual, this time is expressed in time slots T.) This service time can take from one up to M frames. The LSTs of the length of k-th data frame without the POLL/NULL packet pair is

$$\begin{aligned}
F^{p*1}(s) &= \frac{\left(Q_i^t(1,0) + Q_i^t(0,1) - 2q_{0,0}^i\right)}{Q_i(1,1)} G_p^*(s) e^{-s} \\
&\quad + \frac{\left(Q_i^t(1,1) - Q_i^t(1,0) - Q_i^t(0,1) + q_{0,0}^i\right)}{Q_i(1,1)} G_p^*(s)^2 \\
F^{p*\mu}(s) &= \frac{\left(\Pi_{i,\mu-1}^t(1,0) + \Pi_{i,\mu-1}^t(0,1) - 2\pi_{0,0}^{i,(\mu-1)}\right)}{\Pi_{i,\mu-1}^t(1,1)} G_p^*(s) e^{-s} \\
&\quad + \frac{\left(\Pi_{i,\mu-1}^t(1,1) - \Pi_{i,\mu-1}^t(1,0) - \Pi_{i,\mu-1}^t(0,1) + \pi_{0,0}^{i,(\mu-1)}\right)}{\Pi_{i,\mu-1}^t(1,1)} G_p^*(s)^2, \\
&\qquad \mu = 2 \mathinner{..} M
\end{aligned} \tag{7.11}$$

Note that probability that Markov point corresponds to the end of vacation for slave i is $Q_i(1,1)$. Therefore, the conditional probability that both the uplink and downlink queue are empty at the end of vacation is $\frac{q_{0,0}^i}{Q_i(1,1)}$. Next, we observe that the probability that master slave transmission will take k data frames is

$$\begin{aligned}
P_{f,0} &= \frac{q_{0,0}^i}{Q_i^t(1,1)} \\
P_{f,1} &= \frac{\left(Q_i^t(1,1) - q_{0,0}^i\right)}{Q_i^t(1,1)} \frac{\pi_{0,0}^{i,(1)}}{\Pi_{i,1}^t(1,1)} \\
P_{f,k} &= \frac{\left(Q_i^t(1,1) - q_{0,0}^i\right)}{Q_i^t(1,1)} \prod_{\mu=1}^{k-1} \frac{\left(\Pi_{i,\mu}^t(1,1) - \pi_{0,0}^{i,(\mu)}\right)}{\Pi_{i,\mu}^t(1,1)} \cdot \frac{\pi_{0,0}^{i,(k)}}{\Pi_{i,k}^t(1,1)}, \\
&\qquad k = 2 \mathinner{..} M-1 \\
P_{f,M} &= 1 - \sum_{k=0}^{M-1} P_{f,k}
\end{aligned} \tag{7.12}$$

Then, the LST for the master-slave pure channel service time is

$$S_i^{p*}(s) = \sum_{k=0}^{M-1} P_{f,k} \prod_{\mu=1}^{k} (F^{p*_\mu}(s))e^{-2s} + P_{f,M} \prod_{\mu=1}^{M} F^{p*_\mu}(s) \tag{7.13}$$

The PGF for the pure vacation time observed by the slave i is

$$V_i^{p*}(s) = \prod_{\substack{j=3 \\ j\neq i}}^{m} S_j^{p*}(s) \tag{7.14}$$

A given frame will contain two data packets when the uplink and downlink queues are not empty, one data and one empty (POLL or NULL) packet when one of the queues is empty, or two empty (POLL and NULL) packets when both queues are empty. Given that the LST of a single-slot packet is e^{-s}, the LST for the pure frame time during the exchange between the master and the slave i is

$$\begin{aligned} F_{is}^{p*}(s) = {} & \left(\sum_{\mu=1}^{M} \pi_{0,0}^{i,(\mu)} + q_{0,0}^{i}\right) e^{-2s} \\ & + \left(Q_i^t(0,1) + Q_i^t(1,0) + PP - 2\left(\sum_{\mu=1}^{M} \pi_{0,0}^{i,(\mu)} + q_{0,0}^{i}\right)\right) G_p^*(s)e^{-2s} \\ & + \left(1 - Q_i^t(0,1) - Q_i^t(1,0) - PP + \sum_{\mu=1}^{M} \pi_{0,0}^{i,(\mu)} + q_{0,0}^{i}\right) (G_p^*(s))^2 \end{aligned} \tag{7.15}$$

where

$$PP = \sum_{\mu=1}^{M} \left(\Pi_{i,\mu}^t(0,1) + \Pi_{i,\mu}^t(1,0)\right) \tag{7.16}$$

Let us denote the portion of time available for asynchronous traffic with $\alpha = 1-2L_{sco}/T_{sco}$. Let us assume that the LST for the frame duration can be represented as: $F_{is}^{p*}(s) = \sum_{m=0}^{\infty} p_m e^{-ms}$. Then the LST for the total frame time has the form

$$F_{is}^{t*}(s) = \sum_{m=0}^{\infty} p_m e^{-ms-\frac{2mL_{sco}}{T_{sco}-2L_{sco}}} = F_{is}^{p*}\left(\frac{s}{\alpha}\right) \tag{7.17}$$

By the same token, the total vacation time for the slave i is

$$V_i^{t*}(s) = V_i^{t*}\left(\frac{s}{\alpha}\right) \tag{7.18}$$

Expressions (7.18) and (7.17) should be substituted in (7.8) which then depends on the packet arrival process for the given slave, as well as on the values $\pi_{0,0}^{i,(\mu)}$ and

$q^i_{0,0}$. The latter two can be found from the marginal PGF $Q_i(z, 1)$ (alternatively, $Q_i(1, w)$ could be used instead), by making use of the fact that $Q_i(z, 1)$ must be analytic function for all $|z| \in (0, 1)$. Therefore, its numerator and denominator must have identical roots. The number of roots of the denominator can be determined by Rouché's theorem [Bak and Newman, 1982] and it is equal to M. Obviously, $z_0 = 1$ is one of the roots, while the remaining $M - 1$ of them can be determined using Lagrange's theorem [Whittaker and Watson, 1952]:

$$z_j = \sum_{n=1}^{\infty} \frac{e^{2\pi jn\sqrt{-1}/M}}{n!} \cdot \frac{d^{n-1}}{dz^{n-1}} \left(V_i^{t*}(\lambda_{iu} - \lambda_{iu}G_b(z))F_i^{t*}(\lambda_{iu} - \lambda_{iu}G_b(z))\right)^{n/M}\Big|_{z=0} \tag{7.19}$$

where $j = 1\,..\,M - 1$ denotes the index of the root in question. In practice, it is possible to truncate the sum (7.19) to the first few members only. The solutions thus obtained may contain negligible imaginary part. When the $M - 1$ roots are substituted in the numerator of (7.8), we obtain a total of $M - 1$ equations for the given slave i, with unknowns $q^i_{0,0}$ and $\pi^{i,(k)}_{0,0}$, i.e.,

$$\left(1 - \left(\frac{F_{is}^{t*}(\lambda_{iu} - \lambda_{iu}G_b(z_j))}{z_j}\right)^M\right) q^i_{0,0} + \sum_{k=1}^{M-1} \left(1 - \left(\frac{F_{is}^{t*}(\lambda_{iu} - \lambda_{iu}G_b(z_j))}{z_j}\right)^{M-k}\right) \pi^{i,(k)}_{0,0} = 0 \tag{7.20}$$

where $k = 1\,..\,M - 1$. The missing M-th equation is obtained from the condition $Q(1, 1) + \sum_{\mu=1}^{M} \Pi_{i,\mu}(1, 1) = 1$, and it reads

$$Mq^i_{0,0} + \sum_{k=1}^{M-1}(M - k)\pi^{i,(k)}_{0,0} = \frac{M(1 - \lambda_{iu}\overline{F^t_{is}}\,\overline{B}) - \lambda_{iu}\overline{B}\,\overline{V^t_i}}{1 - \lambda_{iu}\overline{F^t_{is}}\,\overline{B} + \lambda_{iu}\overline{B}\,\overline{V^t_i}} \tag{7.21}$$

Solving the system

As we see, finding the distribution of the number of packets in the uplink or downlink queue upon the return from the vacation, requires that we know the distribution of the vacation time, and vice versa. In order to break the recursion, we have applied iterative approach. Before we explain the process of obtaining solutions, we will introduce the marginal mass probabilities of queue lengths as

$$q^i_{j_u} = \sum_{j_d=0}^{\infty} q^i_{j_u,j_d}, \qquad q^i_{j_d} = \sum_{j_u=0}^{\infty} q^i_{j_u,j_d},$$
$$\pi^{i,(j_u)}_{ju} = \sum_{j_d=0}^{\infty} \pi^{i,(j_u)}_{ju,jd}, \quad \text{and} \quad \pi^{i,(\mu)}_{jd} = \sum_{j_u=0}^{\infty} \pi^{i,(\mu)}_{ju,jd}. \tag{7.22}$$

Then, the solutions may be obtained with the following approach:

Iteration 0 1. Consider the marginal probability distributions for uplink queues (but the same procedure holds for the downlink queue as well).

2. Assume that $Q_i^{t0}(1,1) = 0.9$ and that $q^i_{j_u} = \pi^{i,(j_u)}_{ju}$, $j_u = 1\,..\,M-1$.

3. Calculate the distribution of service time and vacation time for each slave.

4. Solve the system which consists of $M-1$ equations of the type (3.39) and one of the type (3.40) for each queue (i.e., the overall system consists of $2M(m-1)$ equations).

5. Calculate the new values of $Q_i^t(z,1)$, and find $Q_i^t(1,1)$. Also find $q^i_{j_u}$ for $j_u = 0\,..\,M-1$.

Iterations 1 .. k 1. Calculate the new distributions of the service and vacation times, using $Q_i^t(1,1)$ and values $q^i_{j_u}$ and $j = 0\,..\,M-1$ from the previous iteration.

2. Solve again the system of equations that consists of $2(m-1)$ instances of the system (3.39) and (3.40).

3. With the solutions calculate $Q_i^t(z,1)$ and so on.

We have found that two to three iterations suffice for good accuracy.

Access delay

Since the PGF for the burst length distribution is geometric, $G_b(z) = \dfrac{z}{\overline{B} + z - z\overline{B}}$, we will introduce the substitution $s = \lambda_{iu} - \lambda_{iu} z/(\overline{B} + z - z\overline{B})$ in the expression (7.8). By using the decomposition principle [Takagi, 1991], the LST for the packet access delay at the slave uplink queue becomes

$$W^*_{ai}(s) = \frac{s(1-\lambda_{iu}\overline{F^t_{is}}\,\overline{B})}{s - \lambda_{iu} + \lambda_{iu} G_b(F^{t*}_{is}(s))} \cdot \frac{1 - G_b(F^{t*}_{is}(s))}{\overline{B}(1 - F^{t*}_{is}(s))} \cdot \frac{1 - V_i^{t*}(s)}{s\overline{V_i^t}} \cdot \frac{Q_i^t\left(1 - \dfrac{s}{\lambda_{iu}\overline{B} - s\overline{B} + s}, 1\right)}{Q_i^t(1,1)V_i^{t*}(s)} \tag{7.23}$$

Different terms in the last expression correspond to different events. The first term corresponds to the time needed to serve the first packet in the burst in the $M^{[x]}/G/1$ system. The second term corresponds to the time needed to serve the target packet in the burst. The third term corresponds to the time needed to serve packets which arrive during the vacation but before the target burst. Finally, the fourth term corresponds to time needed to serve packets which were already in the uplink queue when the vacation was started. Then, the mean access delay is obtained as $\overline{W_{ai}} = -W_{ai}^{*'}(0)$, which amounts to

$$\begin{aligned}\overline{W_{ai}} = & \frac{\lambda_{iu}\overline{B}\overline{((F_{is}^{t})^2)}}{(1-\lambda_{iu}\overline{F_{is}^{t}}\overline{B})} + \frac{\overline{B^{(2)}}\overline{2}}{2\overline{B}(1-\lambda_{iu}\overline{F_{is}^{t}}\overline{B})} \\ & + \frac{\overline{(V_i^t)^2}}{2\overline{V_i^t}} - \overline{V_i^t} + \frac{Q_i^{t'}(1,1)}{\lambda_{iu}\overline{B}Q_i^t(1,1)}\end{aligned} \tag{7.24}$$

where $\overline{(V_i^t)^2} = V_i^{t*''}(0)$.

From the last expression, the following observations can be made:

1. Under constant offered load $\rho = \lambda\overline{F_{is}^{t}}\,\overline{B}$ and fixed M, mean access delay will increase when the mean burst size $\overline{B}$ increases. This increase is due to the increased second factorial moment of the burst size and an increase in the number of packets in the uplink queue when the vacation is finished, as can be seen from the second and last term in (7.24), respectively.

2. Under constant offered load and fixed average burst size $\overline{B}$, mean access delay decreases when the value of M increases. This decrease is due to the decreased number of packets in the uplink queue at the end of vacation, as can be seen from the fourth and fifth term in (7.24).

These observations are confirmed through the diagram that shows the dependency of mean access delay on mean burst size $\overline{B}$ and parameter M, shown in Fig. 7.3. It may be interesting to note that the dependency shown in it is similar in form to the one from Fig. 3.10(a), except that the delays are lower, as could have been expected.

Downlink and end-to-end delay

Under E-limited service, the burstiness of the traffic that arrives at the downlink queues will differ from that in the uplink queue, because of possible interleaving of bursts from different sources. We will model this effect by modifying the parameter of the geometric distribution for the uplink queue. Let the uplink burst size be determined by the PGF $G_b(z) = zp_b/(1-z+zp_b)$, where p_b is the parameter of the geometric distribution which determines the mean burst size, $\overline{B} = 1/p_b$. In the downlink queue, bursts from different sources may get interleaved, and the mean burst size becomes $\overline{S_{iu}}/\overline{L}$, where $\overline{S_{iu}}$ is the mean service period of all uplink queues which have traffic toward the slave i. The probability that the source slave i is the only one to transmit in the current cycle, and the transmission is targeted toward slave j (where $i, j = 3..m$ and $i \neq j$), is

$$P_c = \sum_{\substack{i=3 \\ i\neq j}}^{m} \prod_{\substack{k=2 \\ k\neq i,j}}^{m} \frac{q_0^k}{Q_k(1,1)} \cdot \frac{(1-q_0^i)}{Q_i(1,1)} \tag{7.25}$$

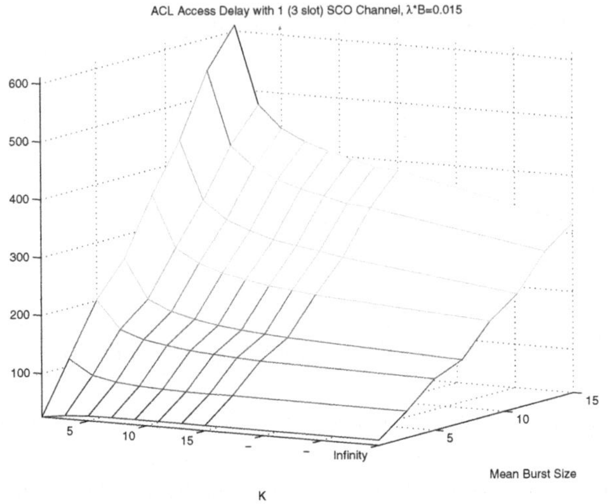

FIGURE 7.3
Mean access delay for asynchronous traffic in the presence of pSCO connections as the function of M and $\overline{B}$, when DH3 packets are used. (From J. Mišić, V. B. Mišić, and K. L. Chan, "Talk and let talk: performance of Bluetooth piconets with synchronous traffic," *Ad hoc networks*, **3**(4):451–477, © 2003 Elsevier B. V.)

The probability that slave i is the only one which transmits to the slave j among two or three active slaves in the same cycle is

$$P_d = \left(1 - \frac{1}{m-4} - \frac{1}{(m-5)^2}\right)^{\left\lceil \frac{\overline{B}}{M} \right\rceil} \tag{7.26}$$

Then, the parameter of the downlink burst length distribution is given by

$$p_{b,j,d} = \frac{1}{\overline{B}} P_c + (1 - P_c) \cdot \left(\frac{1}{\overline{B}} P_d + \frac{\overline{L}}{\overline{S_u}} (1 - P_d) \right) \tag{7.27}$$

where $\overline{S_u} = \sum_{\substack{i=2 \\ i \neq j}}^{m} \overline{S_{iu}} \lambda_{iu} \Big/ \sum_{\substack{i=2 \\ i \neq j}}^{m} \lambda_{iu}$. The adjusted downlink packet arrival rate is

$$\lambda_{jd} = \sum_{\substack{i=2 \\ i \neq j}}^{m} \frac{\lambda_{iu}}{m-2} \overline{B} p_{b,j,d} \tag{7.28}$$

where λ_{iu} denotes the original uplink packet burst arrival rate. The PGF for the distribution of downlink burst size becomes

$$G_{b,j,d}(z) = \frac{z p_{b,j,d}}{1 - z + z p_{b,j,d}} \tag{7.29}$$

and the average burst size is $\overline{B_{j,d}} = 1/p_{b,j,d}$. Note that the packet burst is going to be almost intact in the downlink, provided that the polling parameter M is equal to, or larger than the mean burst size $\overline{B}$.

In order to obtain LST for the downlink delay, we will introduce the substitution $u = \lambda_{id} - \lambda_{id} w/(\overline{B_{i,d}} + w - w\overline{B_{i,d}})$ in (7.8). The LST for the delay in the downlink queue corresponding to slave i is

$$W_{di}^{*}(u) = \frac{u(1 - \lambda_{id}\overline{F_{is}^{t}}\,\overline{B_{i,d}})}{u - \lambda_{id} + \lambda_{id} G_{b,i,d}(F_{is}^{t*}(u))} \cdot \frac{1 - G_{b,i,d}(F_{is}^{t*}(u))}{\overline{B_{i,d}}(1 - F_{is}^{t*}(u))} \cdot \frac{1 - V_i^{t*}(u)}{u\overline{V_i}} \cdot \frac{Q_i^t\left(1, 1 - \dfrac{u}{\lambda_{i,d}\overline{B_{i,d}} - u\overline{B_{i,d}} + u}\right)}{Q_i^t(1,1)V_i^{t*}(u)} \tag{7.30}$$

and the mean downlink delay is

$$\overline{W_{di}} = \frac{\lambda_{id}\overline{B_{i,d}}\,\overline{F_{is}^{t}}}{(1 - \lambda_{id}\overline{F_{is}^{t}}\,\overline{B_{i,d}})} + \frac{\overline{B_{i,d}^{(2)}}\,\overline{F_{is}^{t}}}{2\overline{B_{i,d}}(1 - 2\lambda_{id}\overline{F_{is}^{t}}\,\overline{B_{i,d}})} + \frac{\overline{(V_i^t)^2}}{2\overline{V_i^t}} - \overline{V_i^t} + \frac{Q_i^{t'}(1,1)}{\lambda_{id}\overline{B_{i,d}}Q_i^t(1,1)} \tag{7.31}$$

The LST for end-to-end delay is equal to $W_{ije}^{*}(s,u) = W_{ai}^{*}(s)W_{dj}^{*}(d)$. Mean end-to-end delay is $\overline{W_{ije}} = \overline{W_{ai}} + \overline{W_{dj}}$.

As can be seen, the behavior of the downlink delay when the parameter M changes is different from that of the access delay. This is mainly due to burst interleaving at moderate and high burst arrival rates. The interleaving effectively limits the mean burst size to a value close to M. As a consequence, the mean downlink delay will increase with the polling parameter M until it reaches the mean burst size; further increase of delays is much slower. The other factor that affects the delay is the periodic interruption of ACL traffic by the pSCO traffic. The probability that an ACL packet exchange is interrupted by the pSCO traffic is higher at higher values of M, and the mean downlink delay increases accordingly.

In order to examine the behavior of the overall end-to-end packet delay, we have found the minima of its derivative with respect to M:

$$\frac{\partial\overline{W_{ai}}}{\partial M} + \frac{\partial\overline{W_{dj}}}{\partial M} = 0 \tag{7.32}$$

The optimal value of M depends on the polling interval of the pSCO connection and on the type of packets used, but the single most important factor is the mean burst size. The resulting dependency of optimal M as a function of $\overline{B}$ is shown in Fig. 7.4. As can be seen, the optimal value of M is a nearly linear function of $\overline{B}$, similar to the corresponding diagram in Fig. 3.11, obtained in the piconet with ACL traffic only. The gradient of this dependency is slightly below 1 (i.e., the optimal M is slightly smaller than $\overline{B}$). The optimal value of M will be somewhat lower when DH-type packets are used, due to the longer polling interval.

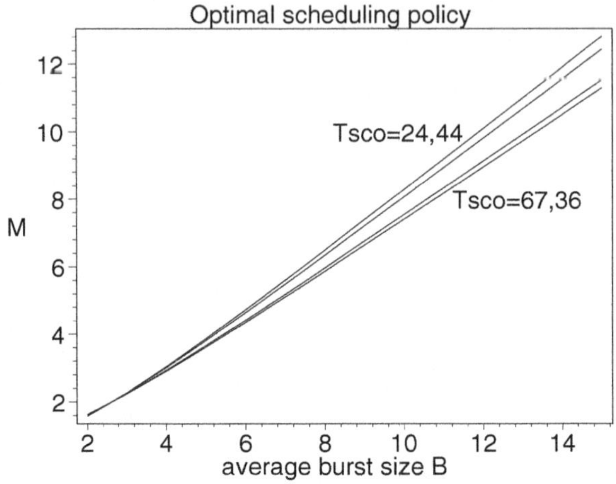

FIGURE 7.4
Optimal value of M as a function of the mean burst size $\overline{B}$. (From J. Mišić, V. B. Mišić, and K. L. Chan, "Talk and let talk: performance of Bluetooth piconets with synchronous traffic," *Ad hoc networks*, **3**(4):451–477, © 2003 Elsevier B. V.)

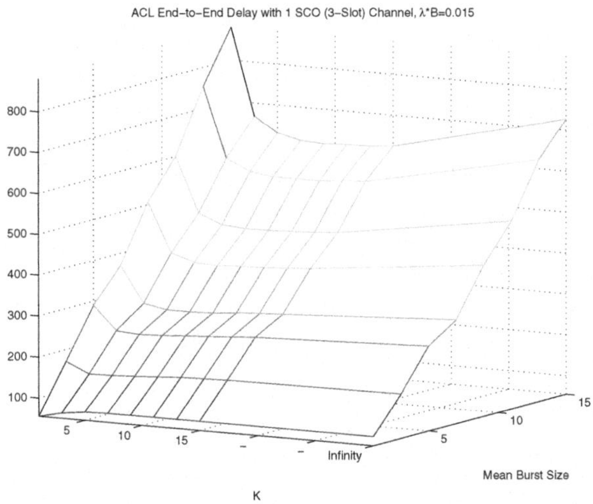

FIGURE 7.5
End-to-end delay for ACL traffic in the presence of pSCO connections with DH3 packets, as the function of M and $\overline{B}$. (From J. Mišić, V. B. Mišić, and K. L. Chan, "Talk and let talk: performance of Bluetooth piconets with synchronous traffic," *Ad hoc networks*, **3**(4):451–477, © 2003 Elsevier B. V.)

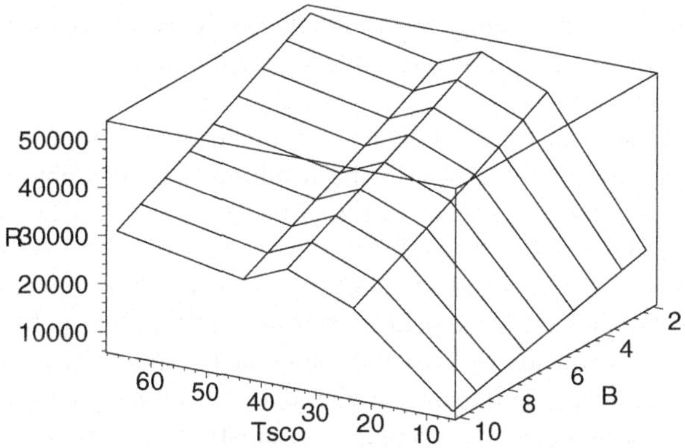

(a) Maximum achievable data rates (analytical solutions).

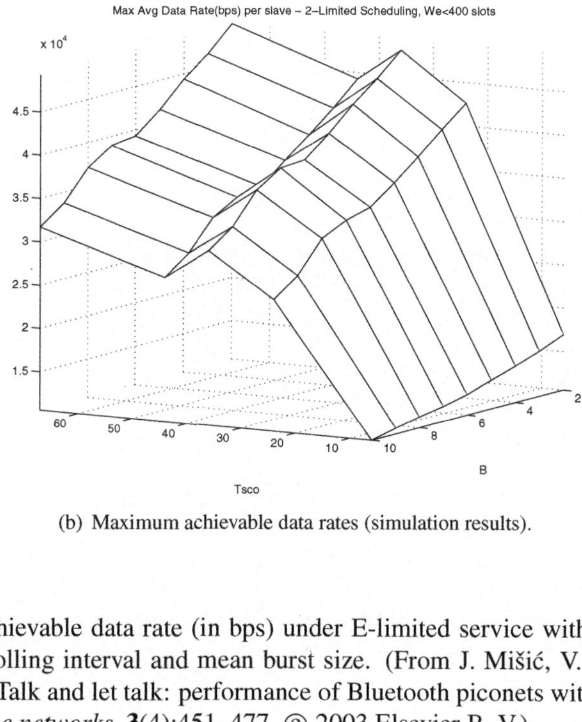

(b) Maximum achievable data rates (simulation results).

FIGURE 7.6
Maximum achievable data rate (in bps) under E-limited service with $M = 2$, as a function of polling interval and mean burst size. (From J. Mišić, V. B. Mišić, and K. L. Chan, "Talk and let talk: performance of Bluetooth piconets with synchronous traffic," *Ad hoc networks*, **3**(4):451–477, © 2003 Elsevier B. V.)

The net result of such behavior is that the dependency of mean end-to-end delay on the value of M will exhibit a minimum for a given mean burst size $\overline{B}$; the existence of such minima is confirmed through simulation, as shown in Fig. 7.5. The shape of this dependency is very similar to that of the end-to-end delay obtained under pure ACL traffic, Fig. 3.10(b). It seems safe to conclude that the E-limited service scheduling policy with the value of M close to the mean burst size $\overline{B}$ offers excellent overall performance while being fair to all slaves. This observation could serve as the foundation for the design of a suitable segmentation and reassembly policy for Bluetooth piconets.

We have also found the maximum data rates that result in end-to-end packet delays for asynchronous traffic below $400T$; the resulting diagrams obtained analytically and through simulation are shown in Fig. 7.6, for $M = 2$. The far right side of the diagrams corresponds to the original SCO (where $T_{SCO} = 6T$), while the left part corresponds to different pSCO packet types. The use of longer packets (which can carry larger payloads) for synchronous traffic allows the polling interval to be longer, which in turn leads to higher throughput for the ACL traffic. The pSCO scheme clearly achieves significantly higher data rates for asynchronous traffic than the original SCO scheme, whilst being able to satisfy the timing requirements for the synchronous traffic.

8

Adaptive polling and predefined delay bounds

Voice and multimedia communications are repeatedly singled out as important application areas for Bluetooth, yet few of the polling schemes proposed so far provide support for any predefined delay bounds. This observation has motivated us to consider a lightweight adaptive polling scheme derived from the E-limited scheme considered in previous chapters.

In this scheme, the number of time slots allocated to each slave depends on its data traffic, and is dynamically determined in each piconet cycle. The number of wasted (POLL and NULL) packets is thus minimized, especially in piconets where slaves have highly asymmetric traffic. Optionally, the scheme introduces cycle length control, which makes it suitable for piconets with synchronous, multimedia traffic.

This chapter is organized as follows. In Section 8.1, we outline the adaptive bandwidth allocation scheme and develop the suitable Markov chain model. We introduce hard cycle control in Section 8.2, where we describe the ACLS scheme and its algorithms in detail and analyze their performance. We compare the performance of the new scheme to that of other polling schemes in Section 8.3 and discuss possible modifications to improve its performance in Section 8.4.

8.1 Adaptive bandwidth allocation

The basic concept of E-limited polling, analyzed in detail in Chapter 3, can be extended to provide dynamic bandwidth allocation according to the behavior in previous piconet cycle. To achieve this, different values of the polling parameter M can be allocated to each slave, according to its traffic in the previous piconet cycle. Slaves that have finished their packet exchange, which is detected through an empty (POLL-NULL) frame as is customary in Bluetooth [Bluetooth SIG, 2003*c*]), will get less bandwidth (and, consequently, a lower value M_L of the polling parameter) in the next cycle. Slaves that still have undelivered packets for exchange at the time the master ends its visit will get more bandwidth through a higher value M_H of the polling parameter. This scheme is described with the following pseudocode.

```
procedure adaptive E-limited polling
do forever
    poll slave i with M = current M(i)
    if both uplink and downlink queues are empty
    then set M(i) = M_L
         move on to next slave
    else if M(i) packets are exchanged
    then set M(i) = M_H
         move on to next slave
    end if
    if all slaves have been visited
    then move on to first slave
    end if
end do
```

The queueing model for this scheme is very similar to the model of the E-limited polling scheme given in Chapter 3, except that the value of the polling parameter will not be fixed. Instead, some slaves will be assigned a lower value, say, M_L, of the polling parameter, while the others will have a higher value, say, M_H. Their respective bandwidth allocation will thus be different.

Let us consider just one slave and disregard for a moment the bandwidth fluctuations due to other slaves. We model the system at the imbedded Markov points which correspond to the end of the piconet cycle. Any given slave i can be in the state of low or high bandwidth allocation during any given piconet cycle, depending on their usage of allocated bandwidth during the previous piconet cycle. Let us first recall Equation (3.33) from Chapter 3, which states the probabilities that transmission between master and the slave will take k data frames:

$$
\begin{aligned}
cal\,P_{f,0} &= \frac{q^i_{0,0}}{Q_i(1,1)} \\
\mathcal{P}_{f,1} &= \frac{Q_i(1,1) - q^i_{0,0}}{Q_i(1,1)} \cdot \frac{\pi^{i,(1)}_{0,0}}{\Pi_{i,1}(1,1)} \\
\mathcal{P}_{f,k} &= \frac{(Q_i(1,1) - q^i_{0,0})}{Q_i(1,1)} \prod_{\mu=1}^{k-1} \frac{\Pi_{i,\mu}(1,1) - \pi^{i,(\mu)}_{0,0}}{\Pi_{i,\mu}(1,1)} \cdot \frac{\pi^{i,(k)}_{0,0}}{\Pi_{i,k}(1,1)}, \\
& \qquad k = 2\,..\,M-1 \\
\mathcal{P}_{f,M} &= 1 - \sum_{k=0}^{M-1} \mathcal{P}_{f,k}
\end{aligned}
\tag{8.1}
$$

This result can be extended to any combination of low and high bandwidth allocations for current piconet cycle and slave i can have the value of the polling parameter of either M_L or M_H. Then, the state transition probabilities for slave i become

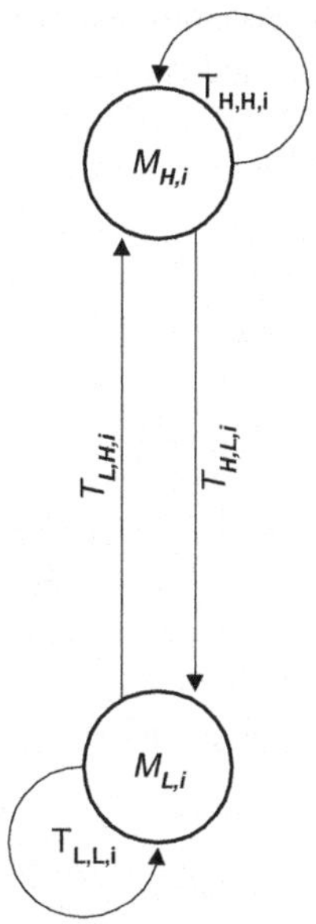

(a) Markov chain for one slave.

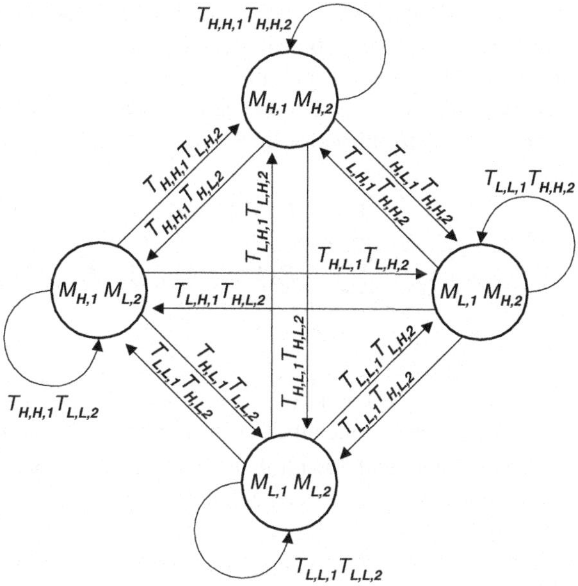

(b) Markov chain for two slaves.

FIGURE 8.1
Markov chains that describe adaptive bandwidth allocation.

$$
\begin{aligned}
T_{L,L,i} &= \text{low state}|\ \text{low state} = \sum_{k=0}^{M_L-1} \mathcal{P}_{f,k} \\
T_{L,H,i} &= \text{high state}|\ \text{low state} = 1 - T_{L,L,i} \\
T_{H,L,i} &= \text{low state}|\ \text{high state} = \sum_{k=0}^{M_H-1} \mathcal{P}_{f,k} \\
T_{H,H,i} &= \text{high state}|\ \text{high state} = 1 - T_{H,L,i}
\end{aligned}
\tag{8.2}
$$

For both states, we can determine the joint queue length distributions for the uplink and downlink queues corresponding to the slave i, $Q_i(z, w)$ and $\Pi_{i,k}(z, w)$, where $k = 1 \mathinner{..} M_{L,i}$ for the low state and $k = 1 \mathinner{..} M_{H,i}$ for the high state.

In the presence of k slaves in the piconet, the number of states of the Markov chain is 2^k. One state of the chain is described by a k-tuple of states corresponding to each slave. Components of the transition probabilities between the states are given by (8.2). For every state of the chain, the slot/frame allocations have to be determined. The Markov chains for the piconets with one and two slaves are shown in Fig. 8.1.

When the actual bandwidth allocations are found for each slave in each state of the Markov chain, the joint uplink-downlink probability distributions of queue length for each slave, in each state, have to be found. Then, the transition probabilities for the Markov chain have to be calculated. Finally, we can set the balance equations for the Markov chain and find all the state probabilities $P(M_{L,1}/M_{H,1} \ldots M_{L,i}/M_{H,i} \ldots M_{L,i}/M_{H,i})$. The state probabilities are important since the final queue length probability distribution for the slave i is determined as

$$\begin{aligned} Q_i^F(z,w) &= \sum_{over\_all\_states} P(M_{L,1}/M_{H,1} \ldots M_{L,i}/M_{H,i} \ldots M_{L,i}/M_{H,i}) Q_i(z,w) \\ \Pi_{i,\mu}^F(z,w) &= \sum_{over\_all\_states} P(M_{L,1}/M_{H,1} \ldots M_{L,i}/M_{H,i} \ldots M_{L,i}/M_{H,i}) \Pi_{i,\mu}(z,w), \\ & \mu = 1 \,..\, \max(M_{H,i}) \end{aligned} \quad (8.3)$$

Note that $Q_i(z,w)$ and $\Pi_{i,\mu}(z,w)$ should be calculated separately for each state of the Markov chain, since they correspond to different bandwidth allocations.

Access delay and downlink delay can be calculated in the form given in Chapter 3 with the use of $Q_i^F(z,w)$ calculated above instead of $Q_i(z,w)$.

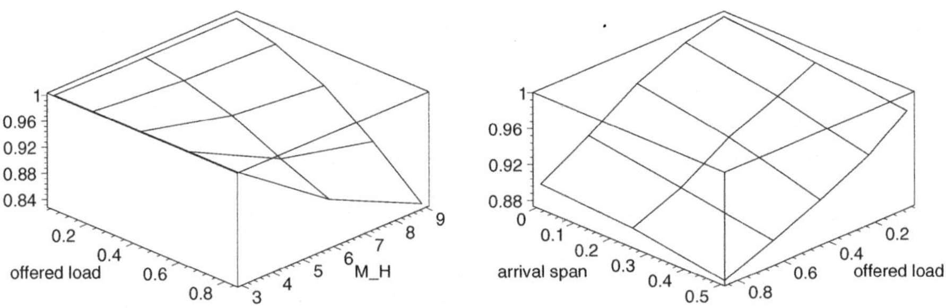

(a) Performance with fixed $M_L = 3$ and varying M_H.

(b) Performance under asymmetric traffic.

FIGURE 8.2
Delay improvement due to adaptability with respect to E-limited polling with $M = 3$.

The improvements obtained through adaptive E-limited polling may be seen in Fig. 8.2. The diagram on the left shows the ratio of end-to-end packet delay obtained under adaptive E-limited polling with fixed $M_L = 3$ and variable M_H, to those obtained under the original E-limited polling with $M = 3$ (or $M_L = M_H = 3$). As the upper limit increases, the delays decrease; this decrease is more pronounced for higher offered loads, where the adaptiveness of the proposed scheme leads to performance improvements over 10%. The diagram on the right shows the delay ratio under asymmetric loads, where $M_L = 3$ and $M_H = 6$. The adaptive scheme performs better than the simple E-limited scheme at higher traffic load, and also when the asymmetry between the traffic of individual slaves becomes more pronounced.

8.2 Adaptive polling with cycle control: the ACLS scheme

The reader may recall that the piconet cycle time has been shown to mostly depend on the total piconet load, i.e., the load of all slaves together, rather than the traffic load of individual slaves. Still, the variance of the piconet cycle time is high and any constraints on the cycle time are difficult to impose. However, the adaptive E-polling scheme described in the previous section can easily be extended to include the limit on cycle control, which is important for multimedia traffic. The new scheme will be referred to as Adaptive Cycle-Limited polling Scheme, or ACLS [Mišić, Mišić and Ko, 2004]; its pseudocode is given below.

```
procedure new_cycle()
int add_on, avg_slots, reservoir, weighting;
// initialize auxiliary variables
avg_slots = C/num_slaves;
reservoir = C;
weighting = 0;
// initialize slots array
for (i=0; i < num_slaves; i++)
{ if (queues[i] < 2)
      slots[i] = MIN_SLOTS;
  else slots[i] = avg_slots;
  // update auxiliary variables
  reservoir -= slots[i];
  weighting += queues[i];
}
if (weighting > 0) // calculate weighted increment
{ add_on = reservoir/(2.0*weighting);
  for (i=0; i < num_slaves; i++)
  // refill according to current traffic
  { slots[i] += add_on*(queues[i]);
    reservoir -= add_on*(queues[i]);
  }
}
return // end of procedure
```

The scheme assumes that the duration of the piconet cycle C (in time slots T) is defined in advance. The time within the cycle is allocated as follows. At the beginning of each cycle, slaves are categorized according to the state of their uplink and downlink queues at the end of service in the previous cycle. Slaves where both uplink and downlink queues were emptied (i.e., all the traffic has been serviced) are assigned to category 0, slaves that have emptied one of the queues but not the other are assigned to category 1, while slaves left with undelivered packets in both uplink and downlink queues are assigned to category 2.

Each slave is allocated at least MIN_SLOTS in each piconet cycle; the remaining time is allocated according to the category of that particular slave. Slaves that have some traffic left from the previous cycle will get more time. Note that all slaves will get some time in each cycle, unlike some other schemes (*cf.* Chapter 2). The actual allocation is performed within the procedure **new_cycle()** which is executed at the end of the current piconet cycle.

The minimum guaranteed time allocation MIN_SLOTS which also defines the lower bound for the piconet cycle time: $C > (m-1)$MIN_SLOTS. A safe value to choose is $10T$, which corresponds to the maximum duration of a single frame. (Alternatively, time may be allocated in units of packets or frames.)

Note, however, that the sum of guaranteed allocations does not add up to C time slots. The remaining part is kept in the reservoir, which is accessible to the slaves on the basis of their current traffic.

Then, the master polls its slaves using the **poll_slave()** procedure, with slave number and current value of the reservoir as parameters. Polling starts from the reference slave, the role of which will be explained below.

```
procedure poll_slave (int i, int reservoir)
int remaining;
int used = 0;
// initialize, start polling
remaining = reservoir + slots[i];
while (remaining >= (used + MIN_SLOTS + 10))
{ used += sent(i).len; // poll slave
  used += rcvd(i).len; // receive response
  if ((sent(i).type == POLL) &&
        (rcvd(i).type == NULL)) // no more data left
      { queues[i] = 0;
        break; // finish polling the current slave
      }
  else if ((sent(i).type == POLL) ||
             (rcvd(i).type == NULL))
           queues[i] = 1; // data left for one direction only
  else queues[i] = 2; // data left for both directions
}
// update global variables
reservoir -= used - slots[i];
return // end of procedure
```

The communication between master and the slave during the single exchange is performed as follows. First, the current value of the reservoir is added to the initial time allocation of the slave giving the current time allocation. The master polls the slave in the usual way. After each packet exchange, the count of remaining slots in decremented accordingly, and the corresponding `queues` information is updated. The exchange ends when the number of `remaining` slots is less than ten – the maximum length of a frame – or when both downlink and uplink queues are empty, i.e., when the master sends a POLL packet and the slave responds with a NULL packet. In the latter case, unused slots are returned to `reservoir`, so as to allow the remaining slaves to use them for their traffic. The master then moves on to the next slave.

The role of the reference slave

The reference slave, as noted above, is the slave from which the piconet polling cycle starts. Any of the slaves could be chosen to be the reference slave, and the master could simply poll the slaves in fixed order, as shown in Fig. 8.3(a). However, this would destroy the fairness of bandwidth allocation to individual slaves. Namely, the first slave to be polled in the cycle may utilize the entire `reservoir`, if necessary; the second can use only what's left after the first slave has finished, and so on. In fact, every slave but the first one might well end up getting only the guaranteed `MIN_SLOTS`. Moreover, the worst case cycle time and, consequently, the maximum polling interval will suffer as well.

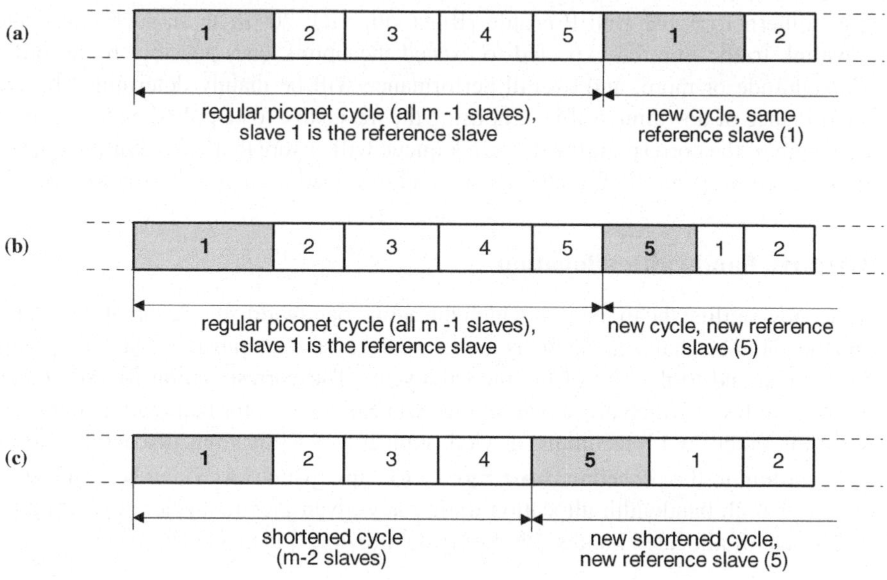

FIGURE 8.3
Pertaining to the choice of reference slave. (From J. Mišić, V. B. Mišić, and E. W. S. Ko, "Fixed cycles and adaptive bandwidth allocation can coexist in Bluetooth," *Canad. J. Elect. Comput. Eng.* **29**(1/2):135–147), © 2004 CJECE.)

Fairness can be restored in different ways. The first solution is a deterministic one: when the cycle ends, the role of the reference slave in the new cycle could be assigned to the last slave in the current one, as shown in Fig. 8.3(b). The last slave will thus get *two* chances to exchange packets with the master, so that any leftover traffic will be taken care of immediately. In the worst case, when there is no data to or from this slave, two frames will be wasted instead of one.

This approach may further be simplified by shortening the polling cycle to $(m-2)$ slaves, as shown in Fig. 8.3 (c). In this case, the **`new_cycle()`** procedure allocates bandwidth to the *current* set of $m-2$ slaves, with the `C` parameter suitably modified; the `queues` value for the missing slave could be taken from the cycle before the last. Each slave will thus get equal attention in a super-cycle of $m-1$ shorter cycles, or $m-2$ normal piconet cycles.

In the probabilistic approach, the fairness is achieved by choosing the next slave in the cycle with the equal probability. For example, first slave in the cycle will be chosen with probability $1/(m-2)$, second slave with probability $1/(m-2)$ and so on. However, this approach is computationally more complex than the deterministic one.

Another probabilistic approach would be to reorder the slave in each piconet cycle according to the LDQF principle (*cf.* Chapter 2), which may be accomplished by modifying the procedure **`update()`**. This modification is particularly well suited to the situation where the piconet master acts as the access point to another network (e.g., Ethernet, as per BNEP profile [Bluetooth SIG, 2001*a*]). In such cases, the downlink traffic may be expected to exceed the uplink one, possibly by an order of magnitude or more, and overall performance will be mainly determined by the performance of downlink traffic. Choosing the next slave to be polled on the basis of the length of the corresponding downlink queue will ensure that the downlink queues are serviced in the most efficient manner, and thus lead to improved performance.

Adaptive bandwidth allocation

Let us now outline the manner in which the actual bandwidth allocation of the ACLS scheme could be analyzed, again by modeling the system at imbedded Markov points that correspond to the end of the piconet cycle. The corresponding Markov chain (two examples of which are shown in Fig. 8.1) can be used for bandwidth allocation modeling in the ACLS technique as well, subject to modifications discussed below.

According to the procedure **`new_cycle()`**, any given slave i can be in the state of low or high bandwidth allocation during any given piconet cycle, depending on their usage of allocated bandwidth during the previous piconet cycle:

- The slaves that did not use all of the allocated slots in the previous cycle will be given low bandwidth of $M_{L,i}$ frames (corresponding to `MIN_SLOTS`).
- The slaves that have used all of the allocated slots will be given high bandwidth of $M_{H,i}$ frames (corresponding to `C/num_slaves` slots).

For both states, we can determine the joint queue length distributions for the uplink and downlink queues corresponding to the slave i, $Q_i(z, w)$ and $\Pi_{i,k}(z, w)$, $k = 1 \,..\, M_{L,i}$ for low state and $k = 1 \,..\, M_{H,i}$ for the high state. (The index i has been previously omitted for simplicity.) The transition probabilities between the states are determined according to the corresponding value of `queues[i]`:

$$\begin{aligned} T_{L,L,i} &= \mathcal{P}mbox(\texttt{queues[i]} = 0 \text{ or } 1)|\text{ low state} \\ &= 1 - \mathcal{P}_{f,M_{L,i}} \\ T_{L,H,i} &= \mathcal{P}_{f,M_{L,i}} \\ T_{H,L,i} &= \mathcal{P}(\texttt{queues[i]} = 0 \text{ or} 1)|\text{ high state} \\ &= 1 - \mathcal{P}_{f,M_{H,i}} \\ T_{H,H,i} &= \mathcal{P}(\texttt{queues[i]} = 2)|\text{ high state} \\ &= \mathcal{P}_{f,M_{H,i}} \end{aligned} \tag{8.4}$$

In the presence of k slaves in the piconet, the number of states of the Markov chain is 2^k. One state of the chain is described by the k-tuple of states corresponding to each slave. Components of the transition probabilities between the states are given by (8.4). For every state of the chain, the slot/frame allocations have to be determined.

We note that actual bandwidth allocations for low and high bandwidth states for one slave are not equal across the states of the Markov chain. For example, for states $(M_{H,1}, M_{L,2})$ and $(M_{H,1}, M_{H,2})$, the values for $M_{H,1}$ may, in general, be different. To illustrate that point, assume that we have two slaves, and the cycle is defined as 36 slots (which corresponds to 6 frames, or 12 packets with the mean length $\overline{L} = 3$). Also note that the rotation of the reference slave equally distributes the remaining bandwidth (reservoir) among the slaves. Now, $M_{L,2}$ is 12 slots (for both `queues[2]` $= 0$ or 1), and $M_{H,1} = 24$. However, for the state $M_{H,1}, M_{H,2}$, both slaves will get 18 slots.

When the actual bandwidth allocations are found for each slave in each state of the Markov chain, the PGFs $Q_i(z, w)$ and $\Pi_{i,\mu}(z, w)$ for each slave, in each state, have to be found. Then, the transition probabilities for the Markov chain have to be calculated. Finally, we can set the balance equations for the Markov chain and find all the state probabilities $P(M_{L,1}/M_{H,1} \ldots M_{L,i}/M_{H,i} \ldots M_{L,i}/M_{H,i})$. The state probabilities are important since the final queue length probability distribution for the slave i is determined as

$$\begin{aligned} Q_i^F(z, w) &= \sum_{over\_all\_states} \mathcal{P}(M_{L,1}/M_{H,1} \ldots M_{L,m-1}/M_{H,m-1}) Q_i(z, w) \\ \Pi_{i,\mu}^F(z, w) &= \sum_{over\_all\_states} \mathcal{P}(M_{L,1}/M_{H,1} \ldots M_{L,m-1}/M_{H,m-1}) \Pi_{i,\mu}(z, w), \\ &\qquad \mu = 1 \,..\, \max(M_{H,i}) \end{aligned} \tag{8.5}$$

where summation is performed over all possible states. Note that $Q_i(z, w)$ and $\Pi_{i,m}(z, w)$ should be calculated separately for each state of the Markov chain, since they correspond to different bandwidth allocations.

Queue stability and fairness

The piconet operating under the ACLS algorithm is stable if all of its queues are stable, i.e., if the average number of packet slots C that have arrived during the piconet cycle (i.e., the total number of slots that belong to all the packets arrived) will be serviced during that cycle:

$$2\overline{B}C\overline{L}\sum_{i=2}^{m}\lambda_{iu} \leq C \tag{8.6}$$

In case of slaves with identical load, the stability condition becomes

$$2(m-1)\lambda\overline{B}\,\overline{L} \leq 1 \tag{8.7}$$

As before, we consider the piconet polling scheme to be fair if the average number of slots devoted to service a slave per piconet cycle is the same for each slave, given that they have identical load. This can be proved by considering Equations (8.4) and Figs. 8.1(a) and 8.1(b). When the arrival rates at all slaves are equal,

$$\begin{aligned} T_{L,L,i} &= T_{L,L,j}, & i,j &\in 2..m-2, i \neq j \\ T_{L,H,i} &= T_{L,H,j}, & i,j &\in 2..m-2, i \neq j \\ T_{H,L,i} &= T_{H,L,j}, & i,j &\in 2..m-2, i \neq j \\ T_{H,H,i} &= T_{H,H,j}, & i,j &\in 2..m-2, i \neq j \end{aligned} \tag{8.8}$$

In that case, the probability of all slaves being in the low bandwidth allocation state p_L or in the high bandwidth allocation state $1 - p_l$, respectively, will be the same. Actual values of these two probabilities will depend on the offered load. This further means that each slave will have the same average allocated bandwidth, i.e., the same average number of allocated slots per piconet cycle.

For slaves with symmetric load, the probabilities that k slaves out of $m-1$ are in low bandwidth allocation state following the binomial distributions are

$$\mathcal{P}(M_{L,i}, M_{H,i})_k = \binom{m-1}{k} p_L^k (1-p_L)^{(m-1-k)} \tag{8.9}$$

This simplification greatly reduces complexity of calculating the PGFs from (8.5).

Access and downlink delay can be determined using the results for E-limited scheme in Chapter 3 with the use of $Q_i^F(z, w)$.

8.3 ACLS performance

In order to evaluate the performance of the ACLS scheme, we have considered three different scenarios. Scenario 1 considers the piconet with seven ACL slaves with symmetric traffic, while scenario 2 considers that same piconet but with asymmetric slave traffic (i.e., variable packet arrival rates). (Note that those two scenarios are fully supported through the analytical model.) Finally, scenario 3 considers the piconet with six ACL slaves with symmetric traffic and a slave with synchronous (CBR) traffic.

In all three scenarios, we have measured the end-to-end delay (from the moment the packet enters an uplink queue at the source slave, to the moment it is received by the destination slave) and cycle time (the time to visit all the slaves in the piconet starting with the current reference slave). Initially, we have applied rotation of the reference slave, as described in Section 8.2.

For comparison purposes, simulations were performed for the same scenarios, but with the piconet master polling the slaves using simple E-limited service with the polling parameter M fixed to the mean burst size $\overline{B}$.

Scenario 1: symmetric traffic

Measured values of mean end-to-end delay under ACLS as functions of the mean burst size $\overline{B}$, are shown in Fig. 8.4. The burst arrival rate was made variable so as to keep the aggregate packet arrival rate constant at $\overline{B}\lambda = 0.015$, which gives the total offered load of $\rho = 0.63$.

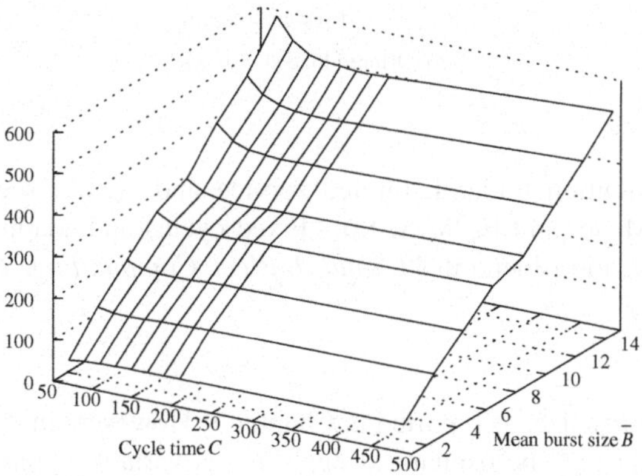

FIGURE 8.4
Mean end-to-end packet delay as a function of cycle time C and mean burst size, scenario 1. (From J. Mišić, V. B. Mišić, and E. W. S. Ko, "Fixed cycles and adaptive bandwidth allocation can coexist in Bluetooth," *Canad. J. Elect. Comput. Eng.* **29**(1/2):135–147), © 2004 CJECE.)

As could be expected, ACLS offers similar delay characteristics to E-limited service: delays increase with increased traffic burstiness, and there is a broad minimum

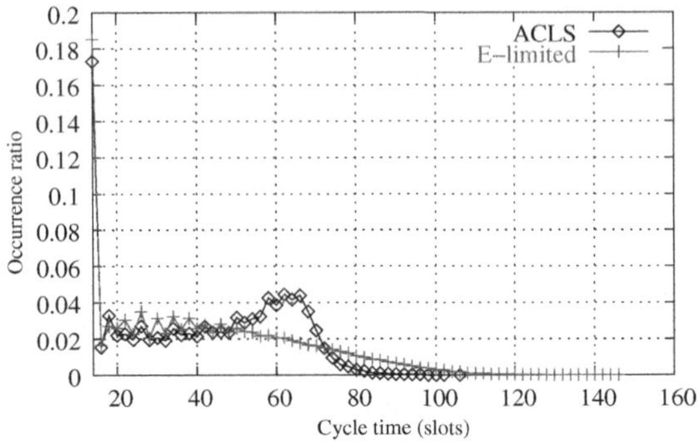

(a) Offered load $\rho = 0.63$.

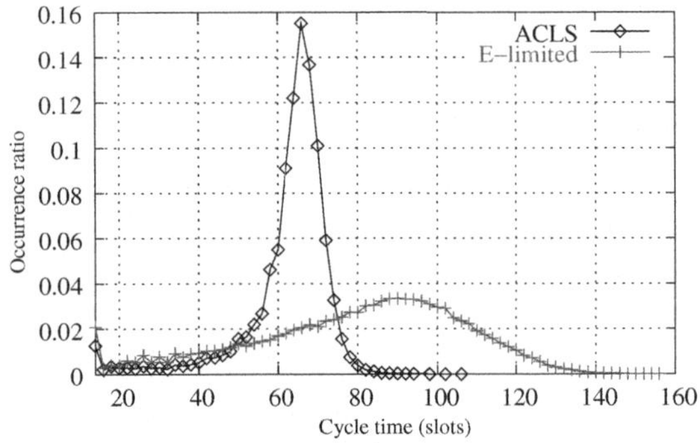

(b) Offered load $\rho = 0.84$.

FIGURE 8.5

Cycle time distribution under E-limited service and ACLS, scenario 1. (From J. Mišić, V. B. Mišić, and E. W. S. Ko, "Fixed cycles and adaptive bandwidth allocation can coexist in Bluetooth," *Canad. J. Elect. Comput. Eng.* **29**(1/2):135–147), © 2004 CJECE.)

for given mean burst size, as reported in Chapter 3. However, under ACLS the number of frames that may be exchanged per slave is variable – unlike the E-limited service where it cannot exceed M.

The distribution of cycle times at different offered loads is shown in Fig. 8.5, with the results for E-limited polling shown for reference. In both diagrams, the mean burst size was fixed at $\overline{B} = 3$, and $M = \overline{B}$ for E-limited service, while the ACLS cycle time has been set to $\mathsf{C} = 74T$.

At lower load, $\rho = 0.63$, there is a noticeable peak at 14 slots for both schemes, which is caused by the significant number of cycles without any data packets. The mean delays for ACLS appear clustered around $60T$, whilst those for E-limited do not exhibit any discernible peak.

Much more interesting, though, are the results obtained at higher load, $\rho = 0.84$.

Under E-limited service, the cycle times are spread fairly wide, with mean value of $\overline{C} = 79.6T$ and variance $Var(C) = 26.9T$. There is a fairly mild peak at about $90T$, which can be explained by the fact that the E-limited scheme can serve any number of frames between 1 and M in a cycle regardless of the queues' status in the previous cycle. Under ACLS, the cycle time distribution is noticeably clustered around the mean value of $\overline{C} = 62.4T$ with variance of $Var(C) = 11.3T$. (Note that the preset cycle time was $74T$.) Also, the maximum cycle time (which obviously does not occur very frequently) is only $108T$ under ACLS, while piconet cycles under E-limited polling can last for as long as $140T$. Note that the purpose of the procedure **`new_cycle`** in ACLS is to distribute the slots among the slaves such that their sum is close to C as much as possible.

Therefore, it seems safe to conclude that ACLS is much better equipped to deal with QoS requirements than the E-limited service, even though their delay characteristics appear comparable.

Scenario 2: asymmetric traffic

The settings in this scenario are similar to those of the previous one, except that different slaves will have different packet burst arrival rates. For simplicity, we have assumed that the burst arrival rates are uniformly distributed between $(1 - \delta)\lambda_{mean}$ and $(1 + \delta)\lambda_{mean}$. Fig. 8.6 shows mean end-to-end packet delay as a function of the average packet burst arrival diagrams λ_{mean} and the variability range δ. For all slaves, mean burst size was kept constant at $\overline{B} = 6$, while the ACLS scheme used $\mathtt{C} = 137$.

As can be seen, the ACLS scheme is able to handle asymmetry extremely well, especially at higher loads and higher variability δ. This may be attributed to the fact that ACLS allocates slots in a flexible manner, according to traffic in both the previous *and* the current cycle.

Scenario 3: symmetric ACL traffic with some CBR traffic

In this scenario, the piconet contains six ACL slaves with symmetric traffic load and one ACL slave with synchronous (CBR) traffic. CBR traffic is a synchronous byte stream generated at a rate of 64kbps, with the piconet master as the destination. CBR traffic is queued for delivery in a designated accumulator, which is emptied using DH5 and DH3 packets; if a DH3 or DH5 packet cannot carry its full payload, the remaining bytes are left in the accumulator until the next cycle. Note that the number of packets will be variable, because the cycle length may vary. The CBR slave is always polled at the beginning of the cycle. Mean burst size for asynchronous traffic was $\overline{B} = 3$, and burst arrival rate was $\lambda = 0.0067$; the total offered load of the piconet, taking CBR traffic into account as well, was $\rho = 0.72$, and the ACLS cycle time was set to $\mathtt{C} = 74$.

In this setup, we have measured mean end-to-end delay for ACL traffic, shown in Fig. 8.7(a), and the delay distribution of CBR traffic under the same conditions, shown in Fig. 8.7(b). (The delay for CBR traffic is defined as the waiting time for

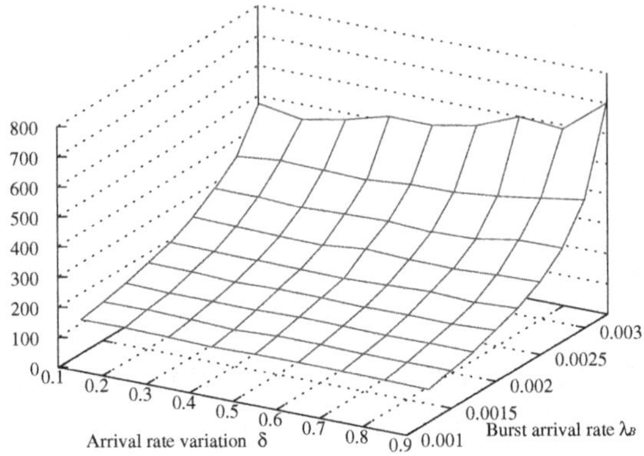

FIGURE 8.6
Mean end-to-end packet delay in a piconet with asymmetric traffic, scenario 2. (From J. Mišić, V. B. Mišić, and E. W. S. Ko, "Fixed cycles and adaptive bandwidth allocation can coexist in Bluetooth," *Canad. J. Elect. Comput. Eng.* **29**(1/2):135–147), © 2004 CJECE.)

the first byte in the accumulator to be packed and transmitted.) As before, results obtained using E-limited polling with $M = 3$ are included as a reference. In terms of delay, the ACLS scheme outperforms the E-limited one, but the difference between the two schemes is small. In terms of delay distribution, however, the diagrams show the major difference between E-limited service and ACLS. Under ACLS, there is a local maximum at $72T$, very close to the preset cycle time; the mean delay is $65.3T$, with a variance of $22.8T$. Under E-limited service, the mean delay is lower at $59.4T$, but the delay variance is much higher at $38.2T$. Therefore, we may conclude that ACLS is much better than E-limited service at maintaining the preset cycle time, while achieving much smaller delay variance for CBR traffic.

8.4 Improving the performance of ACLS

A promising enhancement for ACLS is to apply slave reordering as per LDQF policy (*cf.* Chapter 2). In this Section, we have repeated our simulations for all scenarios described in the previous Section.

Mean end-to-end delays under Scenario 1 (symmetric ACL traffic only) are shown in Fig. 8.8; all other settings were kept the same as in the previous experiment, the results of which were shown in Fig. 8.4. In this case, the use of LDQF results in an improvement of up to ten percent.

The LDQF scheme improves the delay performance in the asymmetric traffic case (scenario 2) as well, as can be seen from the diagram in Fig. 8.9. (This diagram should be compared to the one in Fig. 8.6.)

Finally, we have measured the mean end-to-end delay for ACL traffic in the scenario 3 (six ACL slaves with symmetric traffic and one CBR slave); the corresponding diagram is shown in Fig. 8.10. Furthermore, mean ACL delays are noticeably lower than under the original ACLS, as is the case with the delay variation. Both improvements may be attributed to the simple measure of reordering the slaves according to the LDQF policy.

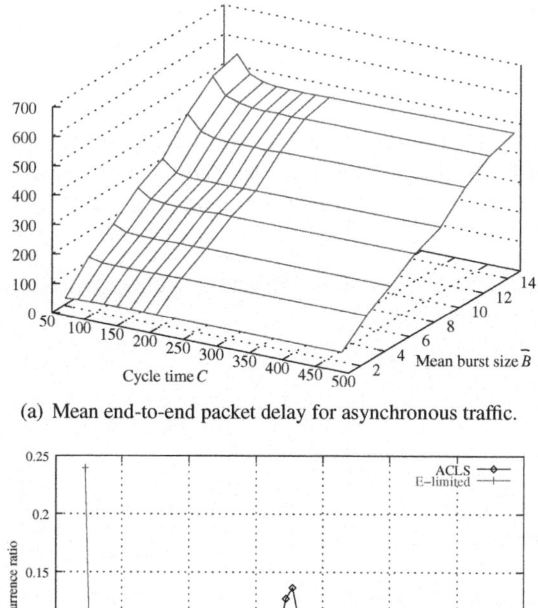

(a) Mean end-to-end packet delay for asynchronous traffic.

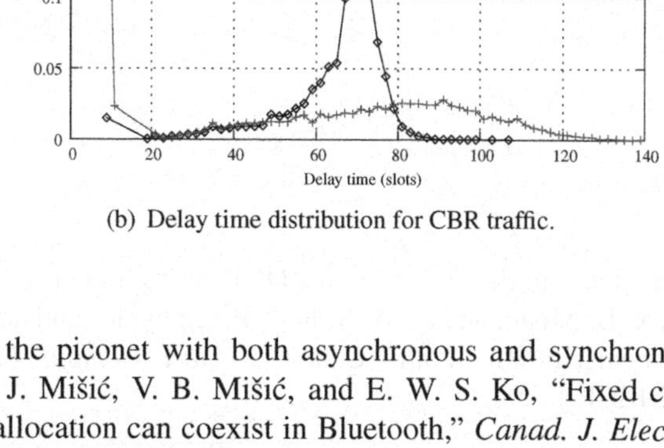

(b) Delay time distribution for CBR traffic.

FIGURE 8.7

Performance of the piconet with both asynchronous and synchronous traffic, scenario 3. (From J. Mišić, V. B. Mišić, and E. W. S. Ko, "Fixed cycles and adaptive bandwidth allocation can coexist in Bluetooth," *Canad. J. Elect. Comput. Eng.* **29**(1/2):135–147), © 2004 CJECE.)

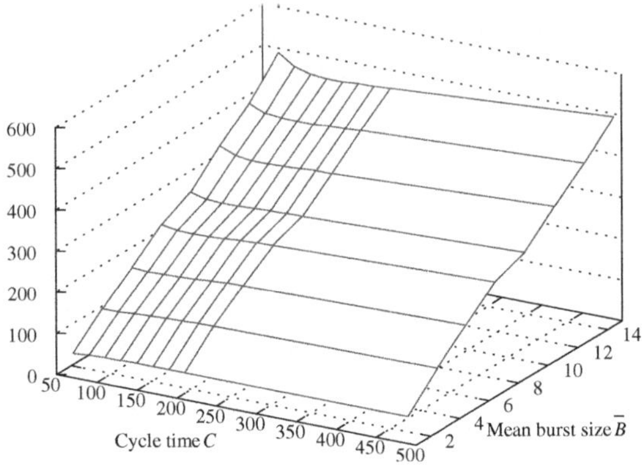

FIGURE 8.8
Mean end-to-end packet delay of ACLS with LDQF, scenario 1. (From J. Mišić, V. B. Mišić, and E. W. S. Ko, "Fixed cycles and adaptive bandwidth allocation can coexist in Bluetooth," *Canad. J. Elect. Comput. Eng.* **29**(1/2):135–147), © 2004 CJECE.)

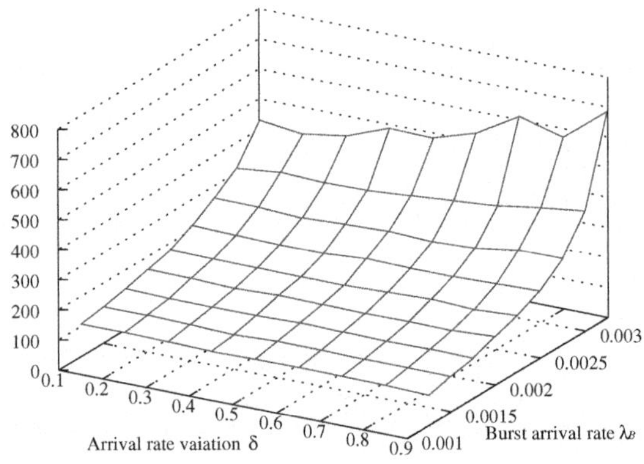

FIGURE 8.9
Mean end-to-end delay under ACLS with LDQF for asymmetric traffic, scenario 2. (From J. Mišić, V. B. Mišić, and E. W. S. Ko, "Fixed cycles and adaptive bandwidth allocation can coexist in Bluetooth," *Canad. J. Elect. Comput. Eng.* **29**(1/2):135–147), © 2004 CJECE.)

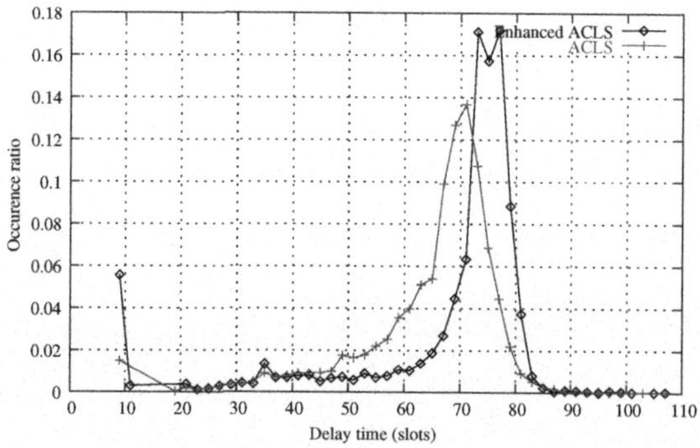

FIGURE 8.10

Delay time distribution for CBR traffic under ACLS with and without LDQF, scenario 3. (From J. Mišić, V. B. Mišić, and E. W. S. Ko, "Fixed cycles and adaptive bandwidth allocation can coexist in Bluetooth," *Canad. J. Elect. Comput. Eng.* **29**(1/2):135–147), © 2004 CJECE.)

9

Bluetooth scatternet formation in ad hoc wireless networks

Ivan Stojmenovic and Nejib Zaguia
University of Ottawa, Ontario, Canada

Bluetooth standard allows the creation of piconets, with one node serving as its master and up to seven nodes serving as slaves. Additional slaves must be parked, with significant overhead involved for parking and unparking them. Although the standard allows for the creation of a collection of connected piconets, called scatternet, it does not give any particular protocol for it. In a unit disk graph, two nodes can communicate with each other if and only if the distance between them is at most R, where R is transmission radius, which is equal for all nodes. Given set of Bluetooth nodes which are positioned so that their unit disk graph is connected, the *Bluetooth scatternet formation* (BSF) problem is to select piconets, and master and slave roles in each piconet, so that the obtained scatternet is connected, has some desirable properties and good performance with respect to some metrics. This chapter surveys the solutions proposed so far in literature for the BSF problem.

9.1 Introduction

Related scheduling problem

When two Bluetooth devices establish communication, one of them assumes the role of a *master* node while the other is a *slave* node. Two nodes in a scatternet can communicate by finding a route between them, where each hop is a master-slave pair of nodes from the same piconet. A node may serve as the master in, at most, one piconet, and as a slave in an unlimited number of other piconets. However, while it serves as slave in other piconet, its own piconet (if it has master role in any of them) will be idle. Obviously there are problems of *scheduling* transmissions and the time division for each master and slave node so that the overall operation is synchronized and delay minimized. The scatternet characteristics will have a direct impact on performance of scheduling protocols. Miklós, Rácz, Valkó and Johansson [2000] concluded that piconet switching poses a significant overhead and has a major impact on system performance. It is therefore important for the overall scatternet performance not only that the scatternet topology is carefully constructed, but also that the piconet switching is scheduled as efficiently as possible. This chapter is concerned only with the Bluetooth scatternet formation (BSF) problem; the scheduling problem is non-trivial and is discussed in subsequent chapters of the this book.

Preliminary taxonomy of BSF protocols

We will now describe a classification taxonomy for the most known Bluetooth scatternet formation (BSF) algorithms. The main criteria used to evaluate these protocols will be on how they achieved the main mission, of providing a *connected* and a *degree limited* scatternet topology. Starting from a connected unit disk communication network and assuming that each node is aware of all its neighbors within a communication range, the algorithms will be classified into those that *guarantee connectivity* and those that do not. Obviously, there is tiny probability that two nodes will never find each other; therefore, connectivity cannot be guaranteed in that sense even for a network consisting of two nearby nodes. Thus the assumption is natural, and connectivity is judged subject to established neighbors' knowledge.

The next classification is based on observing the *degree limitation*. The protocols will be divided into those that *guarantee degree limitation* for each created piconet (that is, always no more than seven slaves for each master) and those that *do not guarantee degree limitation.*

Existing protocols may also be divided into those that work properly and are designed for the *single-hop scenarios* only, where each device is within communication range of any other device in the network. In this case, and using the graph terminology, the unit disk communication graph is a complete graph. More general protocols can be applied for single-hop scenarios as well, but they are designed to work properly for arbitrary type of unit disk graphs, that is, for *multi-hop scenarios*, where some nodes are not within transmission range of each other, but are connected via other nodes in multi-hop fashion. This chapter will primarily classify the existing protocols into these two categories, and describe them in separate sections.

The protocols can also be divided into those that require that each node learns about *all its neighbors* for proper functioning, and those that decide about scatternet links after learning about *some of its neighbors*. More classification criteria will be listed in the sequel.

Device discovery

A closely related problem is *neighbor discovery* (or *device discovery*), that is, how two nodes find each other and establish communication. Almost all proposed solutions assume that Bluetooth technology is used for both neighbor discovery and data communication in the created scatternets. Most BSF protocols use the following device discovery scheme, which is described in [Salonidis, Bhagwat, Tassiulas and LaMaire, 2001]. Device discovery is performed by each node randomly entering into an *inquiry* or *inquiry scan mode* (with equal probabilities) and randomly selecting the time to stay in that mode; this is repeated until a time-out expires (the time-out should be carefully selected to enable one hop information with high probability, but within reasonable time). Inquiry nodes select a repeated pattern of 32 frequencies (out of a total of 79 available frequencies) and send signal on selected frequency in a given spot. Inquiry scan nodes also select a frequency at random in each spot and listen to the transmission at the selected frequency. The discovery (and establishment of master-slave relationship) occurs when both sender and receiver nodes are at the same frequency. The time-out for overall device discovery protocol (for finding sufficient number of neighbors) is experimentally determined to have the best value of about 8 seconds. A different 'recipe' is proposed in [Ferraguto, Mambrini, Panconesi and Petrioli, to appear] where each device executes the device discovery protocol until it is connected with c neighbors (c is between 5 and 7). If the visibility graph (normally it is the unit disk graph) is connected, then the resulting graph is experimentally shown to be connected with high probability.

Basagni, Bruno and Petrioli [2003*b*] have studied the reasons for the Bluetooth based device discovery being so inefficient. The overall connectivity is established fairly quickly, but the full awareness of all neighbors is slow. By means of thorough ns2-based performance evaluations, the authors identified at least two major features of the Bluetooth specifications that are responsible for this lack of performance. First, the overly long backoff intervals adds a considerable time to every handshake and, second, the impossibility of the node in inquiry mode to identify itself in the ID packet results in handshakes between nodes that have already discovered each other.

Joung and Huang [2004] proposed to modify device discovery protocol by using node's ID to decide a pseudo-random sequence, so that scan or inquiry state is decided deterministically. When two nodes discover each other, the smaller ID node replaces its ID with the ID of the other node. They discussed the time-out duration to guarantee the connectivity, which depends on diameter, since the node with largest ID will propagate its ID to all the other nodes in the network.

Bluetooth and Wi-Fi are two widely adopted technologies for wireless communication between two nodes. They are competing and complementary at the same time, since Bluetooth is more suitable for short communications and provides direct communication between any two nodes, while Wi-Fi requires access points and generally is applied on somewhat longer distances. Many equipments use both Bluetooth and Wi-Fi as two independent applications, which may cause interference when both are running. Some latest products allow them to be synchronized based on time slicing technique. Further steps would be to provide software that will further synchronize the two technologies and allow, for instance, that two nodes discover each other using Wi-Fi technology and then communicate by Bluetooth chips. This device discovery protocol has been proposed in [Li, Stojmenovic and Wang, 2004]. One of route discovery schemes for on-demand BSF construction in [Liu, Lee and Saadawi, 2003] also uses single channel communication for broadcasting the destination search packet. The use of Wi-Fi or another single channel medium access technology for device discovery is a predicable approach since Bluetooth based neighbor discovery is proven to be slow, especially when each device is required to discover all its neighbors before a good scatternet can be created. Therefore *neighbor discovery* phase of scatternet formation can be classified as being Bluetooth- or Wi-Fi-based.

Communication and time requirements

A formation algorithm should not depend on a central component for otherwise it will contradict the character of ad hoc networks. When Bluetooth is applied to a hybrid ad hoc network, that is, a network attached to a fixed infrastructure (e.g. Internet) the point of attachment may run a *centralized algorithm* and distribute master-slave decisions to each node. A single node in ad hoc network may also collect all the information and run a centralized algorithm. A better approach is to design a formation algorithm in a decentralized and *distributed* manner. Distributed algorithms can be further classified into globalized and localized protocols. In a *localized protocol*, each node makes formation decisions solely based on the information from its neighbors (possibly also 2-hop neighbors). Localized protocols are further divided into *local* and *quasi-local* based on maintenance cost. The algorithms should support mobile devices, and devices entering and leaving the network. When nodes move, appear or disappear from the network, the *scatternet maintenance* (or self healing) protocol is required. If the changes made by a single node have impact only on piconets associated with 1-hop and 2-hop neighbors then the protocol is local. Otherwise (e.g., when local change sometimes trigger global updates, that is, cause a 'chain' effect), the protocol is quasi-local. An implicit solution is to apply localized schemes that provide local maintenance, as in [Li et al., 2004].

Further division can be made according to *message complexity*, *time complexity* and *memory requirements*. In general, the protocols along these lines can be classified as being *constant* and *non-constant size* with respect to the number of nodes in the network. The message complexity is particularly important, since it is directly related to the energy needs of the algorithm. The QoS requirements of the application may also be considered. The topology could be set in such a way that the QoS requirements of the user applications in the network are met. The scatternet formation problem appears very difficult even without QoS considerations. Existing work on creating scatternets that provide QoS guarantees appears either centralized or has considerable communication overhead. Therefore, QoS provision largely remains for future studies.

Scatternet design criteria

Fig. 9.1 illustrates a scatternet with four piconets. Master nodes are labeled $M1-M4$ and the slave nodes have indices that correspond to the piconet numbers (for example $S14$ is a slave of piconets 1 and 4). Common slave $S23$ serves as a bridge to connect piconets 2 and 3. In order to connect piconets 1 and 2, a (new) piconet 4 is created. Node $M4-S2$ is its master, and plays also the role of a bridge node for two piconets, serving as slave node in piconet 2. Bridges participate in piconets on time division basis. It is therefore anticipated that scatternet formation protocols should have (at least) the following goals in mind:

- Minimization of the number of piconets and, therefore, the number of master nodes;

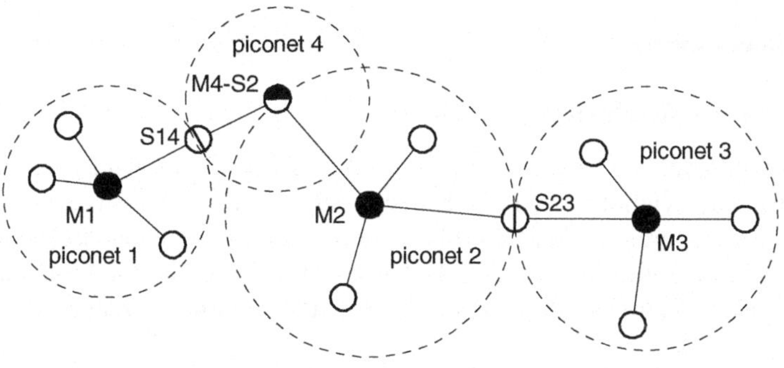

FIGURE 9.1
A scatternet consisting of four piconets.

- Minimization of the number of slave roles for each node (a rough division is into protocols with *constant* and *non-constant maximum number of slave roles* for each node);
- Minimization of the number of master-slave bridge nodes; a rough division is into those that *do not have master-slave bridges* (bipartite graphs) and those that *allow master-slave nodes*).

There are various metrics that can be used for evaluating scatternets. For example, Melodia and Cuomo [2004*a*] favoured scatternets with maximum capacity, scatternets with maximum residual capacity or minimum average load, and metrics associated with path lengths (average path length, average path capacity). Capacity related metrics may require a priori knowledge of traffic demands. Persson, Manivannan and Singhal [2004] listed the following criteria for constructing scatternets: complete scatternet connectivity, maximized aggregate bandwidth, minimized average routing path length, maximized average node availability, minimized bridge switching overhead, communication group clustering, self healing, multi-hop node participation, and on-demand scatternet formation. This survey concentrates on traffic independent measures when evaluating the performance of scatternets. Hodge and Whitaker [2004*a*] listed the following such measurements: number of piconets, average number of slaves per piconet, average number of roles per device, average number of bridges per piconet, average number of bridges between piconets, number of master-slave bridges, average shortest-path length, bottleneck and average path latency.

9.2 BSF in single-hop networks

In a single-hop ad hoc network (or a complete unit disk graph), all wireless devices are in the radio vicinity of each other, e.g., electronic devices in a laboratory, or laptops in a conference room. In this section, we only focus on designing scatternet formation algorithms for single-hop networks. Note that the initial single-hop network, after creating scatternet, is converted into a multi-hop scatternet.

Centralized BSF protocols for complete graphs

Traffic and capacity based scatternets

Miorandi, Trainito and Zanella [2003] investigated the relationship between the network capacity and topology for Bluetooth scatternets. They started by considering the intrinsic capacity limits of a scatternet structure, and show that limiting capacity may be achieved for very local traffic and under specific conditions on the scatternet structure. Then, they provided a description of the performance achievable with two basic scatternet configurations, namely, star and closedloop topologies, and then showed the role played by interpiconet interference in the choice of efficient configurations. Finally, they presented some efficient topologies, based on Platonic solids structures. A centralized BSF solution for single-hop networks, where the traffic between any pair of nodes is known a priori, is described by Miorandi and Zanella [2002].

Super-master election for central decisions

Bhagwat and Rao [2001] described an efficient technique for enumerating all feasible Bluetooth scatternet topologies as well as several constrained subsets of topologies. These results are useful in the design of optimization algorithms for Bluetooth networks.

Salonidis, Bhagwat, Tassiulas and LaMaire [2001] proposed a BSF topology construction algorithm which first collects neighborhood information using an inquiry procedure, where senders search for receivers on randomly chosen frequencies, and the detected receivers reply after random backoff delay. In the process one leader for each connected component is elected. Leader then collects the information about the whole network, decides the roles for each node, and distributes back the roles. Since this is a centralized approach, it is not scalable and not localized. Moreover, [Salonidis, Bhagwat, Tassiulas and LaMaire, 2001] did not elaborate how to assign the roles and they also assumed that the network could have up to 36 nodes only.

Huang, Chen, Sivakumar, Kashima and Sezaki [2004], argue that Bluetooth characteristics preclude the formation of very big network, and propose a centralized scatternet formation scheme that is optimized for conference scenarios. Only one node, the super-master, will do inquiry continuously, while all the other nodes will be continuously in the inquiry scan mode. The super-master makes decisions about the topology on the basis of a load metric instead of hop count metric, and communicates those decisions to the other nodes. It needs to know the number of nodes in the network (and enter it at the beginning of the protocol) to create scatternet.

Tree structure is not a desirable scatternet topology since there is exactly one path between any two nodes, and failure of node on the path will disconnect the network. Also, root node is likely to be a bottleneck.

Ring topology

Lin, Tseng and Chang [2003] described a *ring topology* for scatternet formation in single-hop networks. A ring is created by the master nodes and the slave-slave bridges. Each master can have further slaves outside the ring. The protocol uses park mode, and is centralized. Therefore, it is not scalable and the park mode introduces long message delays. Foo and Chua [2002] presented a similar ring structure consisting of master-slave bridges only, which simplifies the routing and offers larger fraction of bandwidth to each device. Ring structure for Bluetooth has the simplicity and the easy creation as advantage, but it suffers from a large diameter (i.e., the maximum number of hops between any two devices) and a large number of piconets.

Distributed BSF protocols for complete graphs

Tree scatternet structure

Law, Mehta and Siu [2001] described an algorithm that creates connected degree bounded scatternet in single-hop networks. The final structure is a tree-like scatternet, which limits efficiency and robustness.

Sun, Chang and Lai [2002] described a self-routing topology for single-hop Bluetooth networks. Nodes are organized and maintained in a search tree structure, with Bluetooth IDs as keys (these keys are also used for routing). It relies on a sophisticated scatternet merge procedure with significant communication overhead for creation and maintenance. The procedure generates maximally filled piconets [Sun et al., 2002].

Tan, Miu, Guttag and Balakrishnan [2001] proposed a method for multi-hop networks and which is restricted to single-hop scenarios and where every node is active in at most two piconets. This method is similar to the one proposed in [Zàruba, Basagni and Chlamtac, 2001] (described below) for multi-hop networks.

Mimicking known topologies

A single-hop Bluetooth scatternet formation scheme based on 1-factors is described by Baatz, Bieschke, Frank, Martini, Scholz and Kühl [2002]. However, piconets are not degree limited in that scheme.

Barrière, Fraigniaud, Narayanan and Opatrny [2003] described a connected degree limited and distributed scatternet formation solution based on projective geometry for single-hop networks. They assume that only slave nodes can act as bridges. They described procedures for adding and deleting nodes from the networks and claimed that it uses $O(\log^4 n \log^4 \log n)$ messages and $O(\log^2 n \log^2 \log n)$ time in local computation, where n is the number of nodes in the network. The degree of the scatternet can be fixed to any $q + 1$, where q is a power of a prime number. However, in their method, every node needs to hold the information of the projective plane and the master node who has the 'token' needs to know the information of the projective scatternet (which label should be used for the new coming master and which existing nodes need to be connected to it). Barrière et al. [2003] did not discuss in detail how to compute the labels for the new master and its slaves, and what will happen when the number of nodes reaches the number of nodes of a complete projective scatternets. Also, notice that the method has large overhead for construction and maintenance.

Daptardar [2004] proposed to use cube structure in two and three dimensions for creating scatternets in single-hop scenarios. The author [Daptardar, 2004] argues that the structure provides higher connectivity, lower diameter, less node contention, multiple paths between any two nodes, in-built routing, easy inter-piconet scheduling, and the ability to reconfigure for dynamic environments.

Song, Li, Wang and Wang [2005] adopted the well-known *de Bruijn graph* structure to form the backbone of Bluetooth scatternet, called *dBBlue*, such that every master node has at most seven slaves, every slave node is in at most two piconets, and no node assumes both of the master and slave roles. Their dBBlue structure also enjoys a nice routing property, i.e., the diameter of the graph is $O(\log n)$ and there exists a path with at most $O(\log n)$ hops for every pair of nodes without any routing table. Moreover, the congestion of every node is at most $O((\log n)/n)$, assuming that a unit of total traffic demand is equally distributed among all pair of nodes. In the same paper, Song et al. [2005] discuss in detail a vigorous method to locally update the dBBlue structure when a node joins or leaves the network, using at most $O(\log n)$ communications. In most cases, the cost of updating the scatternet is actually $O(1)$ since a node can join or leave without affecting the remaining scatternet. The number of nodes affected when a node joins or leaves the network is always bounded from above by a constant. To facilitate self-routing and easy updating, the authors design a scalable MAC assigning mechanism for piconet, which guarantees the packet delivery during scatternet updating. The dBBlue scatternet can be constructed incrementally when the nodes join the network one by one. The proposed method, therefore, has a number of desirable characteristics.

Minimal spanning tree based scatternets

Wang, Stojmenovic and Li [2004] addressed the problem of scatternet formation for single-hop Bluetooth based personal area and ad hoc networks, with minimal communication overhead. The scatternet formation schemes by Li et al. [2004] (described below) are position based and were applied for multi-hop networks. These schemes are localized and can construct degree limited and connected piconets, without parking any node. They also limit to 7 the number of slave roles in one piconet. The creation and maintenance require small overhead in addition to maintaining location information for one-hop neighbors. In [Wang et al., 2004], the authors apply this method to single-hop networks, by showing that position information is then not needed. Each node can simply select a virtual position, and communicate it to all neighbors in the neighbor discovery phase. Nodes then act according to the scheme using such virtual positions instead of real ones [Li et al., 2004]. In addition, Wang et al. [2004] used Delaunay triangulation instead of partial Delaunay triangulation proposed in [Li et al., 2004], since each node has all the information needed. Likewise, Wang et al. [2004] applied minimum spanning tree (MST) as the planar topology in their schemes. The experiments confirm good functionality of the created Bluetooth networks in addition to their fast creation and straightforward maintenance [Wang et al., 2004]. If MST is used as the scatternet topology, some long edges can be added to provide shorter routes, following the suggestions given by Stojmenovic [2004*b*].

Loop scatternet structure

Zhang, Hou and Sha [2003] proposed a BSF scheme for single-hop scenario. The new loop scatternet structure [Zhang et al., 2003] preserves connectivity and maximum node degree, and minimizes number of piconets. Additionally, it incurs a much smaller network diameter and much smaller maximum node contention. The main idea in [Zhang et al., 2003] is to create smaller scatternet structures, and to make some changes in master-slave relations whenever two such scatternets merge into one. The goal is to create a loop rather, than a tree-like structure. The final structure has a form of a loop, with a number of additional slave nodes attached to loop masters. In the first phase, piconets are created, with at most six slaves each. In the second phase, a slave from each piconet is explicitly selected and shared with another piconet to reduce the diameter. The protocol creates only slave-slave bridges. The authors did clearly show, theoretically and experimentally, the advantages of their method.

On-demand scatternet formation and maintenance

Sivakumar, Chen and Huang [2004] proposed a framework for continuously optimizing the network topology in order to produce the best suitable one for current data streams. The optimization process also takes care of the network maintenance to accommodate the node mobility. A high level description of a protocol is designed for single-hop networks to follow the framework (however, a precise protocol description is missing).

9.3 BSF in multi-hop networks

Centralized algorithms

Ajmone Marsan, Chiasserini, Nucci, Carello and De Giovanni [2002] described a centralized solution for finding a Bluetooth topology (for multi-hop case) that provides full network connectivity, fulfills the traffic requirements and the constraints posed by the system specification, and minimizes the traffic load of the most congested node in the network. The solution is based on a linear optimization formulation, and the formulation leads to an NP-complete problem suited only for small and stationary networks.

Sreenivas and Ali [2004] described a centralized BSF protocol based on genetic algorithm, which selects random groups of nodes as an initial population. Each group corresponds to a combination of masters, slaves, and bridge nodes and is represented by a string. The goal is to minimize the number of piconets created. The protocol is described on half page without sufficient details on the population coding (master-slave relations are apparently not coded according to given description), crossover, mutation, and fitness function used. Genetic algorithm approach to scatternet formation is also suggested in [Hodge and Whitaker, 2004*a*].

Mehta and El Zarki [2004] outlined an approach, centered on the Bluetooth technology, to support a sensor network composed of fixed wireless sensors for health monitoring of highways, bridges, and other civil infrastructures. They present a topology formation scheme that not only takes into account the traffic generated by different sensors but also the associated link strengths and buffer capacities. The algorithm makes no particular assumptions as to the placement of nodes, nor the assumption that all nodes need to be in radio proximity of each other. The output is a tree-shaped scatternet rooted at the sensor hub (data logger), that is balanced in terms of traffic carried on each of the links. The solution is centralized (data logger collects network information and makes all decisions), and is based on a combinatorial optimization formulation followed by simulated annealing based solution.

Yun, Kim, Kim and Ma [2002] presented an approach that forms master-slave mesh topology. Their Bluestars approach models the discovery neighborhood as an inquiry graph I, with $2^{|I|}$ topology subsets available. The solution requires a costly determination of the optimal topology subset (presumably in a centralized fashion) before the scatternet is formed.

Ramachandran, Kapoor, Sarkar and Aggarwal [2000] proposed a BSF algorithm based on growing a tree from the root, where master node is not always directly connected to its slave node. They presented a deterministic as well as randomized algorithm. Both approaches involve a leader election of a super-master, which subsequently forms the actual topology in a centralized manner.

A preliminary account on how to deal with changing network topology has been presented by Chiasserini, Ajmone Marsan, Baralis and Garza [2003]. They presented an optimized approach for scatternet formation that attempts to minimize the traffic load. The authors assume that traffic patterns and routes are known a priori and formalize the topology formation as a min-max problem, which finds bottleneck node and minimizes the traffic load at that node. They also discuss a distributed approach.

Growing tree based distributed BSF

Zàruba et al. [2001] proposed two protocols for forming connected scatternet. In both cases, the resulting topology is termed a bluetree. The number of roles each node can assume is limited to two or three. The first protocol is initiated by a single node, called the blueroot, which will be the root of the bluetree. A rooted spanning tree is built as follows. The root will be assigned the role of a master node. Every one hop neighbor of the root will be its slave. The children of the root will now be assigned an additional master role and all their neighbors that are not assigned any roles yet will become slaves of these newly created masters. This procedure is repeated recursively until all nodes are assigned.

Each node is a slave for only one master, the one that paged it first. Each internal node of the tree is a master on one piconet, and a slave of another master (its parent in the initial tree). In order to limit the number of slaves, Zàruba et al. [2001] observe that if a node in the unit disk graph has more than five neighbors, then at least two of them must be connected. This observation is used to reconfigure the tree so that each master node has no more than five slaves. If a master node has more than five slaves, it selects its two slaves $s1$ and $s2$ that are connected and instructs $s2$ to be the master of $s1$, and then disconnects $s2$ from itself. Such branch reorganization is carried throughout the network.

Dong and Wu [2003] proposed three modifications to the Bluetree algorithm described by Zàruba et al. [2001]. The modifications aim to minimize the overheads introduced by Bluetooth's piconet and multi-hop scatternet. Their modified algorithms use the neighbor's neighbor set and/or neighbor's location to construct the Bluetree to efficiently balance two conflicting goals between the number of piconets and the average shortest path ratio. The modifications are to select as bridge the slave nodes that has maximal degree, to select closest slave and instruct it to become master, and to apply the Yao structure similar to the one by Li et al. [2004].

Pagani, Rossi and Tebaldi [2004] proposed an improvement to the Bluetree algorithm of Zàruba et al. [2001], in order to address the issues involved in practical implementation, and to describe mechanisms to support mobility, joining and leaving the network by proposing an on-demand BSF algorithm. The main improvement is to start the scatternet formation when needed, by an initiator node that becomes the tree root, while other nodes progressively join the tree so that the overall structure is optimized with respect to the latency in data forwarding and on-demand formation of scatternet.

Huang, Yang, Bai and Huang [2003] modified the Bluetree scatternet formation scheme of Zàruba et al. [2001] by limiting the number of children slaves of each node to 5, and using two more slave roles for up to two siblings. Thus additional links between nodes on the same level are added in the tree. This creates a structure with less critical links for connectivity, but the maintenance is still expensive as in the Bluetree construction scheme [Zàruba et al., 2001]. Also the scheme by Huang et al. [2003] is apparently working properly only for single-hop networks, since otherwise links between nodes on the same level may not exist.

In the second protocol of Zàruba et al. [2001], several roots are initially selected. As in the first protocol, each of them creates its own scatternet. In the second phase, subtree scatternets are connected into one scatternet spanning the entire network. Each node is active in up to three piconets.

Bhatnagar and Kesidis [2002] proposed to run first a leader election process to first find a root for the whole network, and then to use the proximity information of this election protocol to create a tree scatternet by merging subtrees.

Pamuk and Karasan [2003] described a Bluetooth scatternet formation scheme which creates a tree, with nodes selected as masters using a measure based on device characteristics (that is a measure combining battery type, battery level, and traffic generation rate). Nodes that have better device characteristics are preferred in master-slave decisions and, therefore, end at levels closer to the root (which is the node with the best device characteristics). In several existing tree based approaches, master nodes take the lead in selecting the slaves. This has contributed to a counterexample for connectivity of the final scatternet (see Fig. 9.2). In [Pamuk and Karasan, 2003], however, each slave node takes the lead and selects one other node as its master. Each node, however, accepts up to 6 requests from potential slaves. During the first phase of neighbor discovery process, the selected master is the discovered neighbor that accepted the slave and has best device characteristics among such neighbors (once accepted master can be replaced in the process if a better one is discovered later). This phase is run for a certain predefined time. In the next phase, only roots of created trees participate in completing the scatternet. In case of single-hop network, roots enter the second phase which runs in the same way as the first phase (with fewer nodes, since only the roots will participate). This phase runs until a single root remains. In case of multi-hop networks, each root needs to label all its descendants (slaves in the corresponding tree). The labels are then exchanged with all neighbors discovered in the first phase, in order to identify all bridges (nodes with endpoints having different labels, that is, roots). To connect the two trees, master-slave relations on the path from the node with label of lower priority (which also becomes the slave of other node) toward its root are reversed. Scatternet is then connected, since each node may receive at most one additional slave role in the second phase, keeping the maximum to 7. We observe that the initial maximum can be lower than 6, allowing the addition of more links and the creation of a mesh rather than a tree, by using extra slave connections.

Guerin, Sarkar and Vergetis [2002] proposed depth first search (DFS), breadth first search (BFS), and MST-based scatternet formation schemes for unit disk graphs in two and three dimensions. They construct a tree where all nodes at each level (the tree they construct is seen as a bipartite graphs) are either masters or slaves. Their construction does not guarantee the maximum degree bound unless the structure itself provides the bound. For example, MST in two dimensions has a maximum degree of five, but in three dimensions, some nodes can have degrees up to 13. The schemes are not localized.

The communication overhead in growing tree based BSF algorithms [Zàruba et al., 2001; Ramachandran et al., 2000; Pamuk and Karasan, 2003; Guerin, Sarkar and Vergetis, 2002; Bhatnagar and Kesidis, 2002; Pagani et al., 2004; Dong and Wu, 2003; Huang et al., 2003] is significant, especially when the appropriate maintenance procedures are designed and added to the protocol.

Context, on-demand and QoS based distributed BFS

Context based BFS

Siegemund [2002] discusses context based scatternet formation for sensor networks. Sensor information is included in the formation process, and nodes with the same context are included in the same piconet. For instance, nodes with the same temperature and noise are assumed to be in the same context and are arranged in the same piconet to enhance their mutual communication.

Gonzales-Valenzuela, Vuong and Leung [2004] proposed BlueScouts, a mobile agent-based on-demand scatternet formation protocol. Their protocol runs in a fully asynchronous fashion, with device discovery being decoupled from actual topology formation. Agents are spread through the existing links in a controlled fashion and recursively signal back the state of the last computation's outcome, leading up to the further replication of the mobile process or its termination. They conduct a coordinated spatial depth-first search over a logical backbone (excluding leaf nodes) in an attempt to reconfigure the role of a new device. The proposed 'programmable' approach introduces unmatched flexibility by allowing context-aware topology formation.

Scatternet queueing models developed by Kapoor, Sanadidi and Gerla [2003] and Mišić and Mišić [2003*a*] were used to compare the delay and throughput characteristics of various topologies. It was found that the best topology is application dependent.

On-demand scatternet formation

Liu et al. [2003] proposed to combine the scatternet formation with on-demand routing, thus eliminating unnecessary link and route maintenances. Conventional Bluetooth broadcast consists of discovering all neighbors, paging them one by one, and sending to each node a route request. Neighbors first enter the piconet, then, after the piconet is created, a route request packet is released to all neighbors. The neighbors can continue spreading the destination search. The authors proposed two variants. In one variant, route request is released only after the piconet is fully created. In the other variant, route request is released to each neighbor immediately after paging it, without waiting for piconet to be fully created.

In the same paper, Liu et al. [2003] also introduced an extended connectionless broadcast scheme where master node and its slaves use the same channel for communication; master node send inquiry messages, while neighbors scan the channel periodically to catch possible inquiry message. It achieves significantly shortened route discovery delay. Liu et al. [2003] also propose to synchronize the piconets along each scatternet route to remove piconet switch overhead and obtain even better channel utilization. They also present a route-based scatternet scheduling scheme to enable fair and efficient packet transmissions over scatternet routes. The proposed method provides high network utilization and extremely stable throughput, being especially useful in the transmission of large batches of packets and real time data in wireless environment [Liu et al., 2003]. Zhen, Park and Kim [2003] proposed an approach called 'blue-star' similar to one by Liu et al. [2003]. This approach lowers the formation delay compared to [Liu et al., 2003] but also needs to perform several consecutive inquiry operations. Other variants of protocol are proposed in [Choi and Choi, 2002; Ghosh, Kumar, Wang and Qiao, 2003] where nodes concurrently form the scatternet (based on a flooding scheme) and route the data traffic.

Kawamoto, Wong and Leung [2003] proposed a two-phase scatternet formation protocol to support dynamic topology changes while maintaining a high aggregate throughput. In the first phase, a control scatternet is constructed, which is not degree limited (more precisely, the number of bridge nodes is limited to 6, while pure slave nodes are parked) to support topology changes and route determination. The second phase creates a separate on-demand scatternet whenever a node wants to initiate data communication with another node, similar to the approach found in [Liu et al., 2003]. The on-demand scatternet is torn down when the data transmissions are finished. Since all the time slots are dedicated to a single communication session, a high aggregate throughput is achieved at the expense of a slightly higher connection setup delay.

QoS based scatternet formation

Augel and Knoor [2004] presented a survey of available BSF solutions and then proposed to consider QoS criteria for constructing scatternets. They argue that allowing nodes with larger degrees reduces diameter but has a bad influence on throughput since the piconet capacity has to be shared among more devices. This is correct only if each node in a piconet receives an equal amount of time for sending the data. However, some scheduling schemes are based on the actual traffic amount; therefore, the time allocated to each slave does not need to be the same. Augel and Knoor [2004] proposed that nodes with a high degree stop paging and instruct neighbor with a low degree to start paging instead. Each device may try to influence topology depending on QoS requirements. Their article describes BSF design guidelines for QoS applications, and does not describe any particular BSF protocol. Threshold based schemes advocated by Augel and Knoor [2004] may fail to construct scatternet and/or provide QoS although both may be possible by alternative schemes. One bad link or a node on a route does not immediately fail the QoS criteria on a longer route.

Pabuwal, Jain and Jain [2003] proposed to switch between several different BSF algorithms depending on the application requirements. However, no information about the possible criteria on which this switching can be based is given, which is needed to have a flexible control of the topology. Instead of switching between various formations algorithms, it might be better to have one algorithm which controls

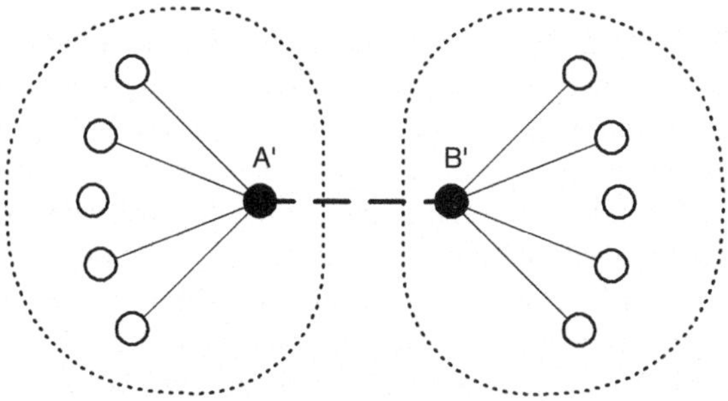

FIGURE 9.2
Creation of disconnected scatternets in several algorithms.

the topology in an application-oriented manner using some specific related parameters.

Melodia and Cuomo [2004*a*], Melodia and Cuomo [2004*b*] and Cuomo, Melodia and Akyildiz [2004] provided an integrated approach to address scatternet formation, scatternet maintenance, and the quality of service support for small and moderate size personal area networks. They first propose a self healing algorithm producing multi-hop scatternets (called SHAPER) which produces tree-shaped scatternets. The initially created tree is converted into a logical tree. The initial physical links in the tree are maintained as much as possible by inserting other links for each network change. When a logical link cannot be supported by a chain of physical links, the overall scatternet is reconfigured. Such reconfiguration guarantees the connectivity of the tree. A procedure that produces a meshed topology (based on combinatorial optimization formulation with initial centralized solution) applying a distributed scatternet optimization algorithm (DSOA) on the network built by SHAPER is then defined. The main issue in the proposed protocols [Melodia and Cuomo, 2004*a*; Melodia and Cuomo, 2004*b*; Cuomo et al., 2004] is their (lack of) scalability. Therefore, further research is needed for the challenging problem of providing QoS in Bluetooth based multi-hop networks.

Clustering based BSF

Basagni and Petrioli [2002] and Petrioli, Basagni and Chlamtac [2003] described a multihop scatternet formation scheme called BlueStar, based on the clustering scheme of Lin and Gerla [1997], taking into account several Bluetooth issues which do not pertain to clustering. Clusterhead (master role) decisions are based on node weights, rather than node IDs, as in [Lin and Gerla, 1997], that express their suitability to become masters. It follows a variant of the clustering method described in [Basagni, 1999]. All clusterhead nodes are declared master nodes in a piconet, with all nodes belonging to their clusters as their slaves. In order to assure connectivity some of the slaves become masters of additional piconets (following, e.g., [Alzoubi,

Wan and Frieder, 2002]). However, piconets may have more than seven slaves. This may result in performance degradation, as slaves need to be parked and unparked in order for them to communicate with their master. The topology discovery phase is performed before clustering in order to provide each node with information about all its neighbors. Each device executes the device discovery protocol for about eight seconds. Then, if the visibility graph (e.g., a unit disk graph) is connected, then the resulting network will be connected with high probability. Device discovery is performed according to procedure in [Salonidis, Bhagwat, Tassiulas and LaMaire, 2001] which we outlined above. A performance evaluation of the clustering-based scatternet formation scheme [Basagni and Petrioli, 2002], [Petrioli et al., 2003] is given in [Basagni, Bruno and Petrioli, 2002*b*].

Basagni et al. [2003*b*] described the results of an ns2-based comparative performance evaluation among the three major solutions for forming multihop scatternets [Li et al., 2004], [Petrioli et al., 2003], and [Zàruba et al., 2001]. They found that device discovery is the most time-consuming operation, independently of the particular protocol to which it is applied. The comparative performance evaluation showed that, due to the simplicity of its operations, BlueStars [Petrioli et al., 2003] is by far the fastest protocol for scatternet formation. However, BlueStars produces scatternets with an unbounded, possibly large number of slaves per piconet, which imposes the use of potentially inefficient Bluetooth operations.

Ferraguto et al. [to appear] proposed Blue Pleiades for device discovery and BSF in multi-hop networks. As soon as a node has discovered c neighbors, it proceeds to the next phases of piconet formation and interconnection. Nodes with less than c neighbors will, upon time-out expiration, exit the device discovery phase [Ferraguto et al., to appear]. Their extensive simulations show that $c = 6, 7$ are excellent choices that guarantee the connectivity of the topology with high probability. The authors [Ferraguto et al., to appear] combine their new device discovery protocol with the BlueStar protocol [Basagni and Petrioli, 2002; Petrioli et al., 2003] to obtain a simple, fast, and effective scatternet formation protocol that overcomes the degree limitation problem of BlueStar. This protocol is termed Blue Pleiades [Ferraguto et al., to appear]. In short, clustering based BSF [Basagni and Petrioli, 2002; Petrioli et al., 2003] is applied on the topology created after the degree limited neighbor discovery [Ferraguto et al., to appear].

A greedy centralized multihop algorithm, where a hypothetical central entity knows the complete topology, has been proposed in [Balai, Kapoor, Nanavati and Ramachandran, 2001]. Distributed algorithms have also been proposed in [Balai et al., 2001], which assume a 2-hop neighborhood information. This is achievable in Bluetooth since the identities of the neighboring nodes are known at the end of the device discovery procedure. The nodes are made to exchange this neighborhood information with each of its neighbors so that they have a 2-hop information and a partial view of the underlying topology. The algorithm [Balai et al., 2001] applies a variant of a clustering algorithm where the number of nodes in each cluster is limited to seven, in accordance to Bluetooth restriction. A node with a highest degree among all its undecided neighbors will become the master node and choose up to seven slaves among neighboring nodes, giving priority to lower degree nodes. However, there are examples where the scatternet is disconnected. This may occur when two clusterheads were originally connected, but formed clusters and erased their link without leaving an alternate connection between their piconets. For example, as illustrated in Fig. 9.2, assume that the graph contains two connected nodes A and B, each with its own seven more neighbors. Thus, A and B have degrees eight, and will become masters of two piconets, containing their own seven neighbors as slaves. However, the graph will then be disconnected since the link between A and B is not part of a scatternet.

Wang, Thomas and Haas [2002] and Guerin, Kim and Sarkar [2002] essentially proposed variants of clustering-based scatternet formation schemes, where the clustering process does not use any ID to decide clusterheads, that is, master nodes. Instead, the decisions are made at random. Already existing master nodes have the priority in attracting more slaves, up to the limit. Initial connections are made by nodes entering a scan or an inquiry scan phases at random. Whenever a node is assigned master or slave role, or is unable to join any piconet or attract any neighbor as its slave in order to create its own piconet, then bridge piconets are added to connect the scatternet. However, the process does not always lead to a connected structure. The counterexample is the same that applies to the approach of Balai et al. [2001] (*cf.* Fig. 9.2). On the positive side, Wang et al. [2002] proposes two excellent measures for the performance of scatternets: average shortest-path length and maximum traffic flow.

Connected out-degree limited scatternets

Petrioli and Basagni [2002] and Petrioli, Basagni and Chlamtac [2004] described a scheme, called BlueMesh, that guarantees connectivity and limits the number of slaves in each piconet. Their neat scheme does not require position information, but instead the local information is extended to a two-hop information, with a two round device discovery phase for obtaining necessary information. It is a modified clustering process, where selection of slaves is performed in such a way that, if a master has more than seven neighbors, it chooses up to seven slaves among them so that it can reach all the others via the chosen ones. Such coverage is always possible with up to five slaves [Zàruba et al., 2001]. The scatternet formation proceeds in iterations. Nodes that participate in a given iteration perform the modified clustering process. Initially all nodes are undecided. In each iteration, init-nodes (nodes having the largest weight among their immediate undecided neighbors) create piconets, by choosing at most seven neighbors as slaves, and deleting remaining edges. The iteration stops when all nodes are decided. All created masters, together with slaves that are not selected for links with slaves from other piconets, withdraw from the next iteration. Simulations by the authors show that the created scatternets have a low average number of roles per node (about two), with an average path length increase between 20% and 80%. The number of iterations grows slowly with the number of nodes (it is about 4.5 for networks with 200 nodes). The method may show weaknesses on some other metrics, especially about the worst-case number of slave roles a node can assume. For instance, in case of dense networks (e.g., complete graph),

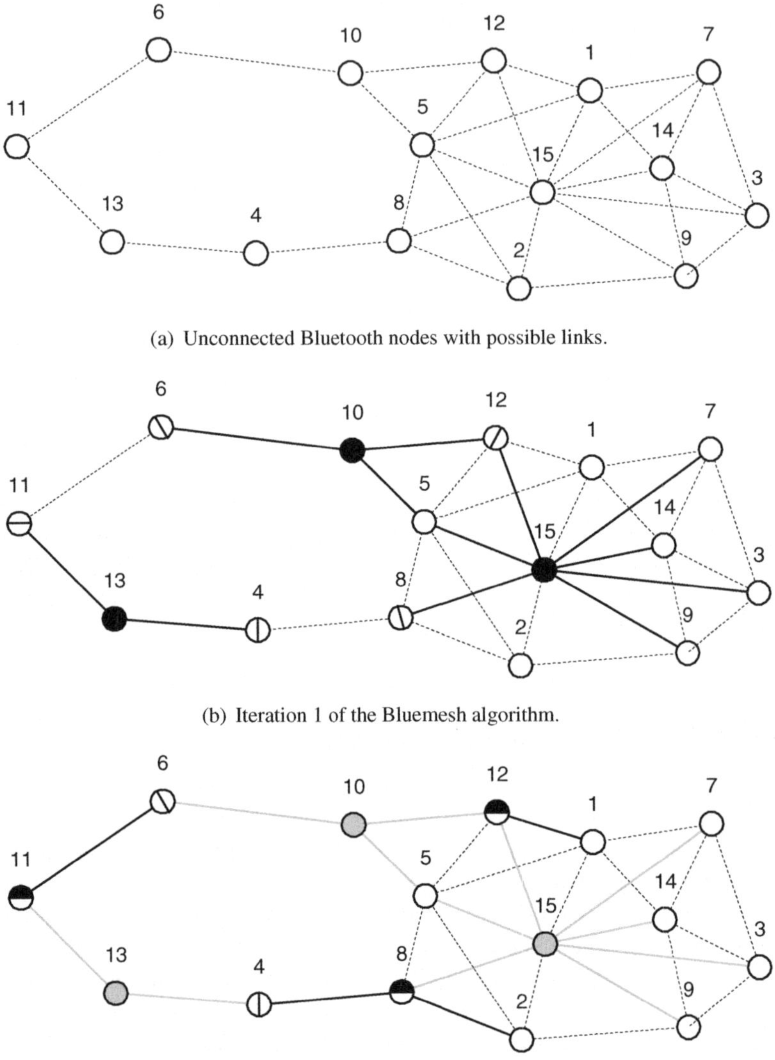

(a) Unconnected Bluetooth nodes with possible links.

(b) Iteration 1 of the Bluemesh algorithm.

(c) Iteration 2 of the Bluemesh algorithm.

FIGURE 9.3
The BlueMesh scatternet formation algorithm.

the second largest node in a neighborhood may end up serving as slave to all the masters in the same neighborhood. Nevertheless, among all methods that do not use position information, the method [Petrioli and Basagni, 2002; Petrioli et al., 2004] appears to be currently the best available method for multi-hop networks. An attempt to improve it further is given in the next subsection (and in [Stojmenovic, 2004*a*]).

Scatternet formation according to the BlueMesh algorithm [Petrioli and Basagni, 2002; Petrioli et al., 2004] is illustrated in Fig. 9.3. The initial node structure with possible links is shown in Fig. 9.3(a). In the first iteration, clustering scheme will select node 15 as master node, with its seven neighbors as slaves (all but nodes 1 and 2), following as selection criteria and whenever needed the largest ID. Node 13 also creates a piconet, having the largest ID among its neighbors. Finally, node 10 also creates a piconet, after its neighbor 12 announces to be 'defeated' by node 15 in the process. Master nodes 15 and 10 select node 12 as gateway, master nodes 15 and 13 select nodes 4 and 8 to connect them, while piconets mastered by 10 and 13 select nodes 11 and 6 for connection.

The results of the first iteration are shown in Fig. 9.3(b), where the nodes 10, 13, and 15 (shown in black) are masters of created piconets, while the nodes 3, 5, 7, 9, and 14 are slaves that are not needed to connect piconets; the nodes 4, 6, 8, 11, and 12 are possible gateways (i.e., bridges). Therefore, nodes 1, 2, 4, 6, 11, and 12 are selected for the second iteration.

In the second iteration, shown in Fig. 9.3(c), three more piconets, mastered by 'black' nodes 8, 11, 12, are created to connect the overall structure. The piconets and links created in the first iteration are shown in gray for clarity.

Maximal independent set based BSF

An attempt to simplify the BlueMesh procedure was made in [Stojmenovic, 2004*a*]. It essentially interprets the slave selection as the *maximal independent set* problem, and reduces the process to two iterations. In the first iteration, every node creates a piconet with itself as a master node. In the second iterations, following a clustering based approach, each node estimates whether or not its piconet is needed for the overall connectivity. If not, it deletes its piconet.

The maximal independent set $MIS(X)$ of a set of nodes X is a set of nodes Y from X such that no two nodes from Y are connected ('independent set'), and Y is not a proper subset of another set with the same property ('maximal'). In the first iteration of MIS based scheme [Stojmenovic, 2004*a*], each node selects the MIS of its neighboring nodes as the set of its slaves. Each node A has $key(A)$ which can be defined in a variety of ways, known to all its neighboring nodes. Let X be a set of neighbors of a given node S. To find $MIS(S)$, node S chooses a node A from X with maximal $key(A)$. Note that here the algorithm can use minimal $key(A)$ instead, which has impact on the performance of the second iteration. Node A is declared a slave of S, and is eliminated from X, together with all its neighbors. This is repeated until X becomes empty. If the number of selected slaves is less than seven then, as in BlueMesh, additional slaves can be selected at random up to the limit.

In the second iteration, the network is not the original unit graph, but the scatternet structure where each node has its own piconet. A clustering based confirmation/elimination scheme is performed. The decisions are made by the masters in S that have higher ID than any of nodes in any of neighboring piconets (this includes the slaves of S and masters of piconets where S is a slave). Such node S verifies whether or not the piconet structure would remain locally connected if its piconet is

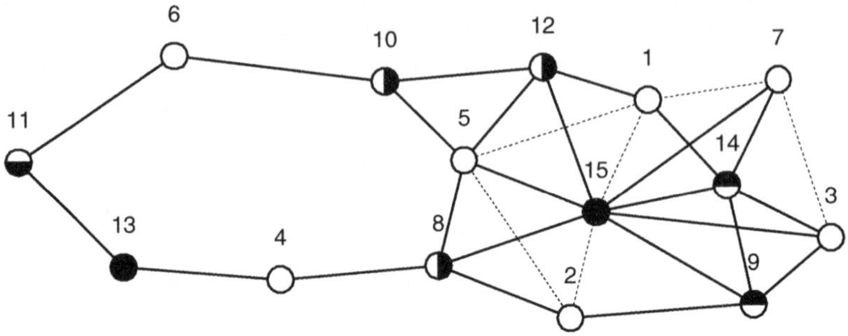

FIGURE 9.4
The second iteration of MIS based BSF.

to be destroyed. If it still remains connected, its piconet is not needed. The decision can be communicated to all piconets where S is participating, which enables other nodes to make their own decision.

For example, in Fig. 9.3(b), the first iteration will create all piconets with all the indicated edges except the two dashed ones. In Fig. 9.4, nodes 15 and 13 first decide to keep their piconets; as they are masters but not slaves, they are shown in black. Node 14 then decides to keep its piconet because of node 1. Node 12 preserves its piconet because of node 10. Node 11 preserves its piconet to connect node 6. Node 10 also preserves the piconet to connect to piconet 11 via 6. Node 9 keeps its piconet because of node 2, while node 8 preserves its piconet in order to remain linked to piconet 13 via node 4. All these nodes are masters as well as slaves, which is why they are half black. The remaining nodes do not need to preserve their piconets and remain ordinary slaves.

While the number of slaves of each master is limited and the scatternet is connected, the number of slaves for each node is not limited, and in some cases, e.g., in a complete graph, one node can be selected as slave to all other nodes. This is the same problem encountered with the BlueMesh. The only advantage of the new algorithm is to reduce the number of iterations to two and therefore obtain a faster BSF.

Position based connected and degree limited BSF

Stojmenovic [2002] and Li et al. [2004] proposed (and made available in June 2001) the first BSF schemes that construct degree limited and connected piconets in multi-hop networks, without parking any node. Moreover, these methods also provide the limit on the number of slave roles for each node, and (if desirable) planarity which is important for performance of routing scheme that guarantees delivery [Bose, Morin, Stojmenovic and Urrutia, 2001]. The BlueMesh scheme [Petrioli and Basagni, 2002; Petrioli et al., 2004] achieved the same main objectives (connectivity and limitation of the piconet size), but did limit the number of slave roles and did not construct a planar graph. However, [Stojmenovic, 2002; Li et al., 2004] achieved their objectives by using a stronger assumption, position information, and some geometrical structures.

Position based BSF schemes [Stojmenovic, 2002; Li et al., 2004] require that nodes first discover their all neighboring devices before the scatternet is established. This phase (learning the underlying unit disk graph) can be implemented by using either a Bluetooth based discovery [Salonidis, Bhagwat, Tassiulas and LaMaire, 2001] or using a single channel medium access such as IEEE 802.11 (another name for Wi-Fi), as discussed above. The protocols then continue by *limiting the degree* of each node, while preserving the connectivity, and by *assigning master* and slave roles to each node. There are several approaches to accomplish this. The basic differences are in the order of these two tasks, in the order of making decisions within each task, and in the way master-slave roles are assigned. Deciding master-slave relations can be done during the degree limitation process, or *after* the degree limitation process is finished as a separate phase. The degree limitation (and/or master-slave decisions) can be done *simultaneously* or *iteratively*. In the *simultaneous approach*, all nodes, independently and at the same time, decide about the links to preserve or the master-slave roles, using a scheme that will assure that the final choices are symmetric (commonly selected links are preserved; master-slave roles decisions are in agreement). Alternatively, nodes can make degree limitation or master-slave decisions at different times, and decisions already made by neighbors have impact on decisions to be made by a given node *(iterative approach)*.

One approach is, for instance, *after-simultaneous-iterative*, where nodes make degree limitations simultaneously (and, in some cases e.g., when *LMST* or *Yao* structure is applied, gather decisions from neighbors to decide which links remain in the final structure), followed by an iterative procedure for deciding master and slave relations (e.g., clustering based procedure such as BlueStar [Basagni and Petrioli, 2002; Petrioli et al., 2003]). This variant of the originally proposed procedure [Li et al., 2004] was proposed by Basagni et al. [2003*b*]. If degree limitation step requires message exchange, then it may be faster if master-slave roles are also assigned at the same time. Such *during-simultaneous-iterative* and *during-iterative-iterative* procedures are the ones proposed in [Li et al., 2004], and they are also based on clustering. The same article [Li et al., 2004] also elaborates on *during-iterative-iterative approach*. If the clusterhead decisions are based on node degrees (number of neighbors in the original unit graph) as the primary key in comparing (that is, nodes with more neighbors have more chance to become master nodes), then another round of information exchange (similar to the device discovery round) is needed following the first neighbor discovery. In [Stojmenovic, 2002], the *during-simultaneous-simultaneous* and *during-iterative-simultaneous* approaches are described. The major difference is in the master-slave decision process. It was proposed to use *(dominating set membership, node degree, node identifier)* as the key for comparing two nodes, and assign master role to higher key node on any link. This means that nodes that belong to a dominating set (see [Stojmenovic and Wu, 2004] for the definition and survey of existing schemes) have priority in becoming the master node on a link. If this primary key is the same for both nodes, or the key is not used at all, then node degree can

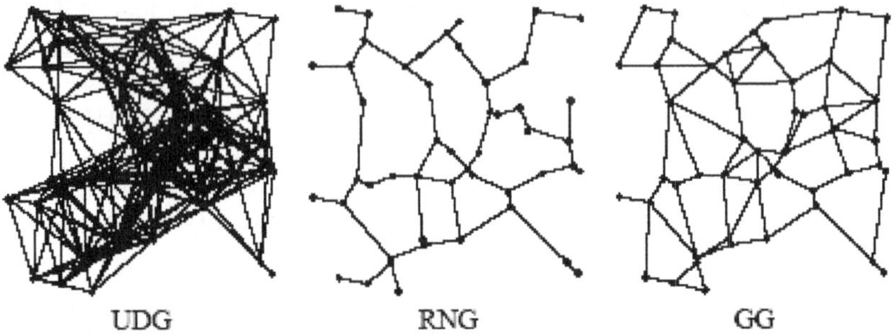

FIGURE 9.5
Unit disk graph (UDG), RNG and GG of a set of nodes.

be used. If node degrees are the same, or not used at all in comparison, then node identifiers can be used for making the role assignment. Note that *after-simultaneous-simultaneous*, *after-iterative-simultaneous*, and *after-iterative-iterative* BSF protocols can also be considered.

Geometric structures for degree limitation

We will now describe several geometric structures that can be used to achieve the degree limitation in scatternets. The basic solution is to apply minimum spanning tree (MST) or a structure that contains it. MST is a subgraph of a given unit disk graph which contains all the nodes, is connected, and whose sum of edge lengths is minimized. The average number of neighbors (the average degree) of each node of a MST is ≈ 1.99, while the maximum number is 6. However, MST is not a localized structure, since its computation requires a global network knowledge at each node. We therefore need to use other structures.

A localized MST ($LMST$) based topology control algorithm was proposed by Li, Hou and Sha [2003]. Each node u first collects positions of its one-hop neighbors $N1(u)$. The node u then computes the minimum spanning tree $MST(N1(u))$ of $N1(u)$. The node u keeps a directed edge uv in $LMST$ if and only if the edge uv is also an edge in $MST(N1(v))$. If each node already has a 2-hop neighboring information, then the construction does not involve any message exchange between neighboring nodes; otherwise each node contacts the neighbors along its $LMST$ link candidates, in order to verify the status at other nodes. The average number of neighbors (average degree) of nodes is ≈ 2.04, while the maximum is still limited by 6. MST and $LMST$ are both planar graphs (a graph is planar is no two edges of it intersect except possibly at common endpoints).

Relative neighborhood graph (RNG) is introduced by Toussaint [1980], and can be defined, in the simplest form, as follows. An edge uv is included in RNG if and only if it is not the longest edge in any triangle uvw. Fig. 9.5 shows an example of an RNG of a UDG (unit disk graph).

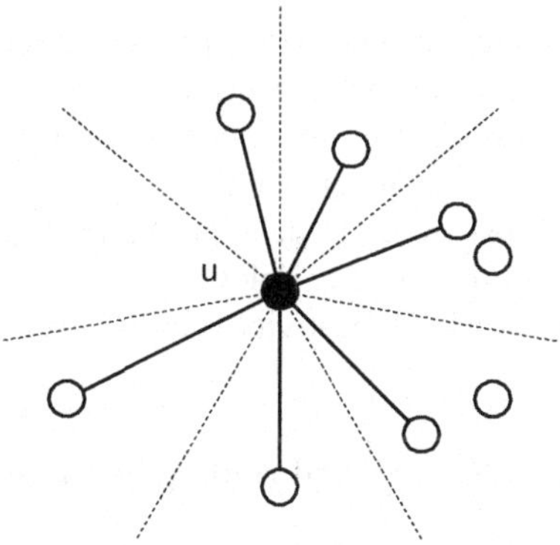

FIGURE 9.6
Yao graph degree limitation for $p = 7$.

Using this definition, some edges may have very large degrees in several particular scenarios. To obtain a degree limited structure, the record w(AB)=(—AB—, min(id(A),id(B)), max(id(A),id(B))) can be used instead for edge comparisons, since no two edges have the same record. We refer to this structure in the sequel, assuming a random node placement and a very low chance of any two edges being of the same length. The degree of each node in $LMST$ and RNG ($LMST$ is a subset of RNG) is limited to 6 (for nodes located in a plane). The average degree of a node in RNG is ≈ 2.4. Note that the construction of $LMST$ and RNG does not require that the exact positions of nodes and their neighbors to be known; in fact, only the corresponding mutual distances are required. In both cases, each node requires the knowledge of its distances to the neighbors, and the distances between any pair of neighbors.

The Gabriel graph (GG) is proposed in [Gabriel and Sokal, 1969], and is defined as follows. GG contains an edge uv if and only if the disk with diameter uv contains no other node inside it. This criterion can be tested in two ways. For an edge uv to be included in GG, each common neighbor w of nodes u and v should be located at a distance of at least $|uv|/2$ from the midpoint of uv. Alternatively, one can verify the angles from neighbors to uv. If for a common neighbor w of u and v, $\angle uwv > \pi/2$ then uv is not in GG. It should be observed, as in the case of $LMST$ and RNG, that the construction of GG requires only the knowledge of the location of a node and those of its neighbors. Fig. 9.5 shows an example of a GG. The average degree of a node of GG is ≈ 3.8. However, there is no worst case limit for the node degree. Therefore, to assure degree limitation, another geometric structure needs to be applied, at least at nodes whose degree exceeds the Bluetooth limit of seven. Note that each node can decide, for each of its links, whether or not it belongs to RNG or GG without communicating with its neighbors. To construct the $LMST$, it needs to exchange the decisions made with the neighbors, since only links selected by both endpoints are in the final structure.

The Yao_p graph is proposed by Yao [1982] to construct MST efficiently in high dimensions. Any p equally separated rays originated at each node u define p cones. In each cone, u then chooses the closest node v within the transmission range, if there is any, and then selects a directed link uv (see Fig. 9.6 for an illustration). Links which are not selected by u are deleted. Since Yao_p contains an MST as a subgraph (for $p = 6$), and that is after deleting all links which are not selected by its both endpoints, the network connectivity is still preserved. Note that Yao_p is not necessarily planar.

Degree limited structures for BSF and routing in scatternets

There are a number of options to obtain degree limited scatternets. One option is to apply Yao_7 to UDG constructed after the device discovery phase (note that, in this option, it is not necessary to complete this phase; it suffices to merely achieve the overall connectivity). The drawback of this option is that the obtained scatternet is not planar. The planar structure may be desirable in order to provide the routing with a guaranteed delivery, since the best existing protocol that achieves that [Bose et al., 2001] requires planar graph in the recovery mode.

The next option is to use $LMST$ or RNG, which are planar and guaranteed to be degree limited (therefore, it is not necessary then to apply the Yao graph construct). The drawback of using them is that these structures are quite sparse, and therefore the greedy routing (forwarding message to a neighbor that is closest to the destination), will frequently fail, leading to long routes with protocol that guarantees delivery [Bose et al., 2001].

Further option is to apply GG, followed by Yao_7, applied only on nodes whose degree exceeds 7. The number of such nodes is small, if any, but they may exist. This structure is planar, localized, and the densest known that is defined with so little local knowledge and zero messages (besides device discovery to learn neighbors). Note that Li et al. [2004] proposed a partial Delaunay triangulation (PDT) as an alternative locally defined structure which is more dense; however, subsequent measures show that the difference in density is only about 1% in favor of PDT over GG, thus we are not covering it in this chapter.

$LMST$, RNG, and GG have average degrees $\approx$ 2.04, $\approx$ 2.5, and $\approx$ 3.8, respectively. They all (and the Yao structure as well) tend to select short edges for the scatternet structure. Stojmenovic [2004*b*] observed that, to improve the routing performance of scatternet, some additional edges may be carefully selected; a note about it is also made in [Li et al., 2004]. The selection depends on the criterion being applied in measuring routing performance. If the criterion is to minimize the hop count then one can add several 'long' edges to the scatternet. It is desirable to spread the added edges in several directions, in order to complement the existing short edges, which can be done by applying the same angular range division used in Yao construct and selecting long edges in sectors where no short edge exists among edges of $LMST$, RNG, or GG. This assures a balanced edge structure in all directions. Addition of randomly selected long edges can also be considered (especially if distances rather than position information were used to define the structure).

On the other hand, if power consumption was used as criterion, the additional edges should have a length close to the ideal one, following the discussion made in [Stojmenovic and Xu, 2001].

9.4 Conclusions

There are a number of Bluetooth scatternet formation protocols already proposed in the literature. It can be observed that very few of them satisfy most of the desirable characteristics. There are relatively few actual implementations and comparisons. For example, Basagni et al. [2003*b*] and Basagni, Bruno, Mambrini and Petrioli [2004] compared BlueTree [Zàruba et al., 2001], BlueStar [Petrioli et al., 2003], BlueNet [Wang et al., 2002] and the position based approach [Li et al., 2004]. They concluded that the device discovery is the most time consuming operation. Their final conclusion is that forming scatternets is still a formidable task, because of the device discovery and the extra complexity imposed by the Bluetooth technology on the implementation of distributed algorithms. Similar conclusions were made by Vergetis, Guerin, Sarkar and Rank [2005*a*]. The reader can find alternative surveys on BSF in Basagni, Bruno and Petrioli [2004], Persson et al. [2004], and Whitaker, Hodge and Chlamtac [2005].

We anticipate that more Bluetooth scatternet formation schemes will be developed in the near future, and that some modifications to Bluetooth specifications could be made to find solutions which satisfy a number of desirable properties and make it suitable for commercial applications in the multi-hop scenarios. An interesting and a major open problem in the area is to design a BSF algorithms that will guarantee connectivity and degree limitation (for both the master and the slave roles) and without using the position information.

10

Bridge topologies and scheduling

Data communications within a Bluetooth scatternet require shared devices or bridges that join the individual piconets. The bridges share their time between the piconets they belong to, switching from one to another in a cyclical sequence. Data packets with destinations in other piconets are queued by the piconet master, and delivered to the bridge during its residence in the piconet. The manner in which the bridge operates in the piconets, the actual duration of bridge residence in each piconet, and the manner in which the switch-over times are arranged, are all determined by the bridge scheduling algorithm.

In this chapter, we will first describe possible bridge topologies and their operation, together with some notes on the termination of bridge exchanges and synchronization delays incurred in bridge operation. We will then describe different approaches to bridge scheduling in Section 10.2, and present some considerations regarding their practical implementation in Section 10.3. Finally, Section 10.4 will present the queueing model and the basic assumptions used in subsequent theoretical analyses.

10.1 Bridge topologies

Let us now describe the possible bridge topologies and their operation in more detail, and discuss some assumptions that enable us to derive analytical measures of scatternet performance.

As mentioned above, the bridge is a Bluetooth device that shares its time between two or more piconets. The bridge may act as the master in at most one of the piconets it belongs to and a slave in all others. This restriction is due to the fact that all communications in a piconet use the frequency hopping sequence (channel) defined by the piconet master. Two piconets with the same master would have to communicate using the same hopping sequence, hence they would in fact be the same piconet. The bridge which is the master in one piconet and the slave in others is referred to as the Master/Slave (MS) bridge; the bridge which is the slave in all the piconets it belongs to is referred to as the Slave/Slave (SS) bridge. Note that complex scatternets may contain bridges of both kinds.

A generic topology of the scatternet with two piconets and an MS bridge is shown

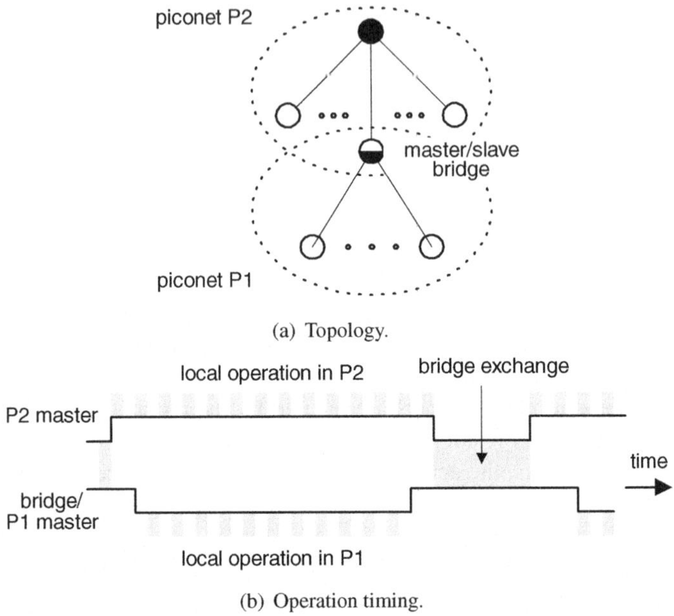

FIGURE 10.1
The operation of the MS bridge.

in Fig. 10.1(a). Scatternet operation may be divided in two phases, local operation and bridge exchange, which occur in alternation. During local operation, the bridge acts as the master of piconet P_1: it routes the packets to and from its slaves, and queues the packets for destinations in the other piconet. At the same time and in the same fashion, the master of P_2 services the slaves of its own.

The time when the bridge leaves the piconet P_1 and joins the piconet P_2 as a slave, will be referred to as *rendezvous*. Some time afterwards, the master of P_2 will contact (i.e., poll) the bridge in order to initiate the packet exchange. Once the contact between the two is established, the actual packet exchange may begin. During the exchange, the packets queued in one piconet are sent to the other piconet, where they are again queued for subsequent delivery.

After the exchange is terminated, the master of P_2 returns to servicing its own slaves, while the bridge takes up its role as the master of P_1 and returns to servicing the slaves of its own. Note that all communications in the piconet P_1 are frozen during the period of bridge exchange, since the piconet is left without its master.

The operation of the scatternet with an MS bridge is schematically depicted in Fig. 10.1(b). The time between two successive bridge visits to the piconet P_2 (i.e., two successive rendezvous) will be denoted as the *bridge cycle*. This cycle should be distinguished from the *piconet cycle*, the time it takes the master to visit all slaves exactly once.

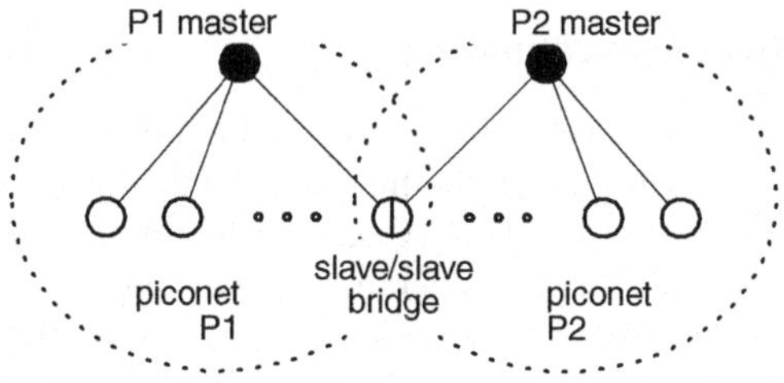

(a) Topology.

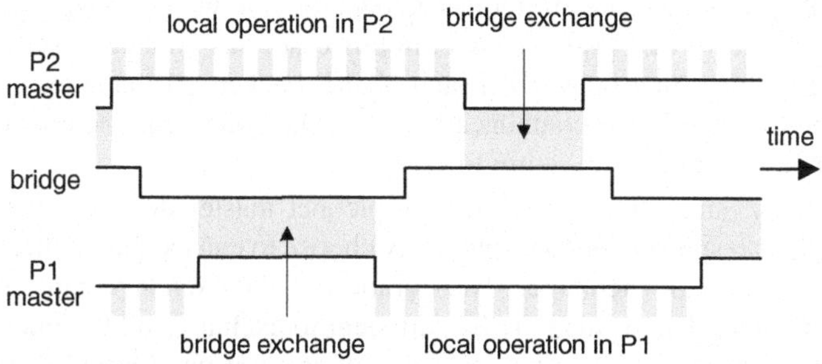

(b) Operation timing.

FIGURE 10.2
The operation of the SS bridge.

The operation of the SS bridge is schematically shown in Fig. 10.2(a). In this case, the bridge again alternates between the two piconets, but it acts as the slave in both of them. When the bridge is present in a piconet, the piconet master can initiate the packet exchange. After finishing the exchange, the master returns to local operation, while the bridge is free to switch to the other piconet. Local operation of the piconet lasts until the bridge joins that piconet again. Note that this scenario, shown schematically in Fig. 10.2(b), is totally symmetric with respect to piconets P_1 and P_2. As before, each piconet has alternating phases of local operation and bridge exchange; however, these phases do not occur at the same time in both piconets.

We stress that the Bluetooth specification poses no limits on the number of bridges in a single piconet, nor does it limit the number of piconets that a single bridge may choose to join. (On the other hand, neither does it offer much help as to the operation of the scatternet and the scheduling of the bridges.) Therefore, local operation within each piconet will include communication with ordinary slaves, but also with other bridges, if present. However, the total number of active piconet members, be they slaves or bridges, is limited to seven at any given time [Bluetooth SIG, 2003*b*].

During the actual exchange, the piconet master may exclusively poll the bridge, or it may poll the bridge as well as other slaves in its piconet. This decision is entirely up to the piconet master, due to the fact that it controls and initiates all communications in the piconet.

Termination of bridge exchanges

There are several possible criteria that can be used to terminate the bridge exchange. The simplest one is probably when the master polls the bridge in exhaustive mode, i.e., for as long as there are packets to be sent to the bridge or received from it. We will sometimes refer to this mode of operation as the *complete exchange*. When both the master and the bridge queues are emptied, a POLL packet from the piconet master will cause the bridge to reply with a NULL packet. This is the signal for both the bridge and the master that the exchange is over; the reader should remember that a POLL packet means that the master has no data to send, while a NULL packet means that the polled slave (the bridge device, in this case) has no data to send [Bluetooth SIG, 2003*b*]. (Note that both POLL and NULL packets still take exactly one time slot T, even though neither contains any data.) The bridge can then detach from the piconet while the master can return to service its slaves.

Alternatively, since the bridge and the piconet master do know the number of packets in their respective queues, they may choose to can exchange this information before the actual data exchange. The duration of the exchange can then be set to any desired value. This value may be sufficient to exchange all the queued packets, or it may be determined according to some other criterion. In the latter case, some packets may be left in one or the other queue, waiting until the next rendezvous and next exchange. This approach will be referred to as *limited exchange*, since it will effectively limit the number of packets exchanged, or their total duration (in Bluetooth clock time slots of $T = 625\mu s$). Once the specified number of packets are exchanged, the bridge exchange may stop without any special notification.

The third reason why the exchange may have to be terminated is that the queue capacity is exhausted. Bluetooth devices are small, mobile, battery-powered devices where energy consumption is to be minimized, and this leads to the use of small chips with limited capabilities. This implies that the buffers implementing the queues on those devices will likely have small capacity, of the order of several kilobytes only. Long exchanges can easily lead to overflow of such buffers; remember that the master has to service other slaves and, possibly, bridges in its piconet, while the bridge may link two or more piconets. The problem of finite buffers may be solved by simple queue blocking and, consequently, rejection of packets, or by redesigning the master-bridge negotiation protocol to incorporate the ability to terminate the exchange if the buffer capacity is reduced below some threshold. The latter approach has been adopted by Tan and Guttag [2002], where appropriate provisions are built in the protocol itself. The performance of the former approach is analyzed in detail in Chapter 14, using the queueing theoretic approach similar to the one used to analyze the performance of a single piconet with finite buffers, in Chapter 4.

It should be noted that the number of packets waiting to be sent from the piconet master to the bridge may differ from the corresponding number to be sent in the opposite direction. If all the queued data packets are to be exchanged, the excess data packets must be paired with empty packets in the corresponding frames. Since the bridge acts as the slave during the exchange, POLL or NULL packets may used for that purpose; this process will be referred to as *padding*. As a consequence, the

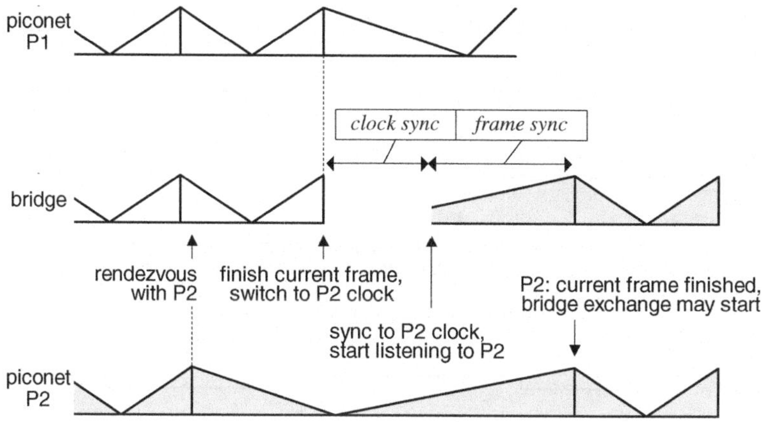

FIGURE 10.3
Additional synchronization intervals when switching from one piconet to another.

total duration of the exchange may exceed the sum of durations of exchanged data packets.

Synchronization delays

Each switch from one piconet to another incurs some synchronization intervals or delays in which data communications cannot occur. These delays are related to the operation of the Bluetooth baseband and physical layers.

First, all transmissions in Bluetooth must be synchronized, in frequency and phase, with the time slot clock of the master device. When the bridge joins the piconet, it has to adjust the frequency and phase of its own clock to the frequency and phase of the clock of the piconet master. The mean value of this synchronization delay depends on the phase difference of corresponding clocks; it may take any value from 0 to $2T$ [Baatz, Frank, Kühl, Martini and Scholz, 2001], but its mean value may easily be calculated to be equal to the Bluetooth time slot T.

Second, each frame must start with the downlink transmission, and this transmission must take place on an even-numbered slot. When the bridge has synchronized its clock with that of the master of piconet P_2, there may be a frame transmission already in progress. The actual frame exchange between the piconet master and the bridge cannot begin until the current frame ends. Since Bluetooth packets may be one, three, or five packets long, this may take as much as $8T$. The mean value of this synchronization delay depends on the mean packet length.

In case of the SS bridge, both synchronization delays occur every time the bridge switches from one piconet to another. In the case of the MS bridge, the two synchronization delays occur only when the bridge joins the piconet in which it acts as a slave. At alternate switches, i.e., when the bridge returns to the piconet in which it

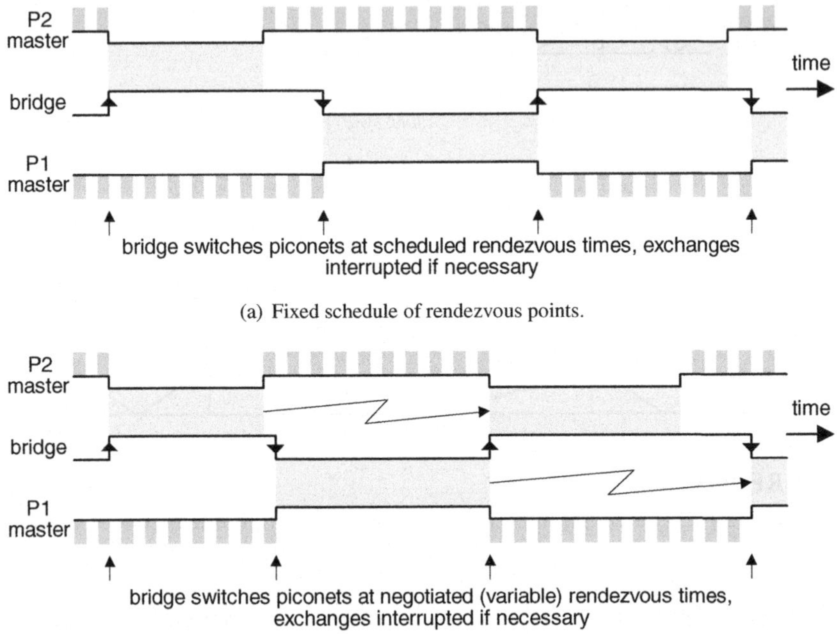

(a) Fixed schedule of rendezvous points.

(b) Variable schedule of rendezvous points.

FIGURE 10.4
Rendezvous-based bridge scheduling in the scatternet with two piconets linked through a SS bridge.

acts as the master, only one of the delays occurs but not other. Namely, upon returning to its own piconet, the bridge has to revert to its own clock. Switching the clock does incur the clock synchronization delay. However, the frame synchronization delay does not exist – the transmissions are initiated by the master/bridge anyway, and it can start polling its slaves as soon as it synchronizes to its own clock.

Fortunately, both of these delays (shown in Fig. 10.3) are small compared to other delays in the scatternet, and may safely be ignored without any noticeable effect on results.

As in the case of the single piconet, the performance of data traffic in Bluetooth scatternets is critically dependent on the manner in which the bridge shares its time between the piconets, and the manner in which the piconet masters communicate with the bridge during its residence in their respective piconets. Unfortunately, neither of these is specified in the current version of the Bluetooth specification [Bluetooth SIG, 2001*b*]. A detailed overview of alternative approaches to bridge scheduling and their characteristics follows in the next Section.

10.2 Approaches to bridge scheduling

Rendezvous-based bridge scheduling

The Bluetooth specification does not prescribe any particular approach to bridge scheduling. It should come as no surprise, then, that a number of proposals have been made regarding bridge scheduling, many of which are based on the concept of rendezvous points.

Rendezvous points, as explained above, are time instants at which the bridge should be present in the piconet in order to exchange data with its master [Johansson, Kapoor, Kazantzidis and Gerla, 2002]. Rendezvous-based scheduling assumes that there exists a sequence or schedule of rendezvous points, which both participants in the exchange are aware of. The rationale is that both the piconet master and the bridge should join the scheduled exchanges simultaneously, so as to minimize idle waiting. Namely, the exchange cannot begin unless both participants are ready; if one of them attempts to initiate the exchange before the other one is ready, it will have to wait without being able to send or receive any data. Such waiting will waste time and, ultimately, bandwidth. This approach is schematically depicted in Fig. 10.4(a).

Given the topology of the scatternet and the traffic requirements of individual devices, some authors have tried to construct a globally optimal schedule of rendezvous points. Such a schedule can be advantageous when traffic patterns are known in advance and vary little over time. Johansson, Körner and Tassiulas [2001] have investigated this approach and found that the construction of an optimal scatternet-wide schedule is a formidable task – it has been shown to be NP-complete, even under very favorable conditions. Schedule maintenance is another problem with this approach, as its complexity is comparable to that of the initial construction, and the distribution of the updated schedule to all devices will inevitably lead to additional communication delays.

A different approach, described by Johansson, Alriksson and Jönsson [2001], is based on a locally maintained rendezvous schedule. In this case, the bridge device (denoted as participant in multiple piconets, or PMP) divides its time into windows of pseudorandom length. Each window is then assigned to one of the piconets they visit, and the masters of those piconets are in turn informed about those windows with little overhead. The windows are then used to coordinate packet exchanges between the masters and the bridges, and the use of pseudorandom scheduling avoids problems of starvation. However, the scheme does not address scalability, and it is based on a novel operational mode which has not been adopted as part of the official Bluetooth standard.

A similar scheme, labeled PCSS, has been proposed by Rácz, Miklós, Kubinszky and Valkó [2001]. In PCSS, every node randomly chooses a communication checkpoint (i.e., rendezvous) that is computed on the basis of the masters clock and the slaves device address. In order to adapt to various traffic conditions, PCSS measures the link utilization and adjusts the checking period accordingly. In this manner, coordination among nodes is achieved with very little overhead. However, the interval between successive rendezvous is increased or decreased by a fixed multiple, which makes adjustments rather coarse grained. Furthermore, there is no coordination between links, and the increase in node density will increase the frequency of scheduling conflicts, leading to missed rendezvous and the corresponding reduction in efficiency.

The aforementioned approaches do not account for widely varying traffic patterns and frequent topology changes that are to be expected in Bluetooth networks. Both of these are highly likely to occur in an ad hoc network formed by devices of widely varying capabilities operated by mobile users. In such cases, schedule conflicts can easily occur, forcing devices to miss the scheduled rendezvous, and the overall efficiency of the scatternet will suffer. The problem is even more pronounced in complex scatternets where a single piconet can have more than one bridge and/or a bridge can participate in more than two piconets.

It should be noted that rendezvous-based scheduling cannot ensure maximum efficiency, even if we disregard the problems of constructing a global schedule without conflicts. First, some mismatch regarding the beginning of the exchange must be tolerated, due to the synchronization delays described above. Second, the actual duration of the master-bridge packet exchanges may not always match the interval between two successive rendezvous points, because of traffic randomness. If the exchange ends too early, the bridge may be left with nothing to do until the next pending appointment. (The piconet master is not affected, as it can always poll its local slaves during such intervals.) On the other hand, an exchange that lasts too long may interfere with the next pending appointment of either piconet master or the bridge. A pending rendezvous cannot be deferred, since it is considered to be at higher priority than the current exchange. In such cases, the exchange will be interrupted and the remaining packets will have to wait until the next visit of that particular bridge to that particular piconet.

Adaptive rendezvous-based scheduling

In order to alleviate these problems, a number of authors have suggested the use of a schedule that adapts to varying traffic conditions. In this case, rendezvous points are determined one by one, according to local traffic conditions, through negotiation between the piconet master(s) and the bridge(s). This approach is schematically depicted in Fig. 10.4(b).

Baatz, Frank, Kühl, Martini and Scholz [2002] proposed a loose pre-calculated schedule which utilizes a local credit-based scheme to adjust the level of service offered to each device to its actual traffic. The scheme is inspired by the leaky bucket filter [Bertsekas and Gallager, 1991] and the Deficit Round Robin fair scheduler [Shreedhar and Varghese, 1996], but it uses the SNIFF mode which may cause scalability problems.

Zhang and Cao [2002] described a scatternet-wide adaptive approach in which piconet masters and bridges keep their own appointment schedules, referred to as switch-tables. The switch-tables are updated according to the traffic load observed at each node; nodes with both asynchronous and synchronous traffic are accounted for. The switch-tables are coordinated between nodes so as to ensure conflict-free operation; however, the speed of convergence is not high, and the communication overhead is comparatively costly. The scheme is deliberately limited to piconets that have at most two bridges, as well as to bridges of the SS type only, which further limits its applicability in larger scatternets.

Mišić and Mišić [2003*c*] have described a simple adaptive algorithm in which the rendezvous times are calculated according to current and future traffic to and from the bridge; the algorithm, denoted as LAMS, targets MS bridges only. A similar approach has been taken by the LCS algorithm of Tan and Guttag [2002], in which each device keeps a list of its outstanding appointments, and each appointment includes the start time (i.e., rendezvous) and the finish time of the future meeting. There is only one outstanding appointment for each link the device participates in (i.e., one per piconet in the case of bridge, and one per bridge in the case of piconet master). The details of the next appointment are negotiated at the end of the previous exchange, taking into account the current and (estimated) future traffic load of both participating devices as well as of their other links. The piconet master actually computes and proposes several possible rendezvous times and durations, from which the bridge selects one that suits it the best. In this manner, the schedule is constructed and maintained locally, on a per-piconet and per-bridge basis, and it is guaranteed to be conflict-free.

A simple adaptive approach for the SS bridge connecting two piconets, known as LASS, has been proposed in Mišić and Mišić [2003*c*]; this algorithm strives to minimize the end-to-end packet delays by adjusting the time for next rendezvous according to the estimated traffic in the next exchange. This algorithm, however, is not scalable to larger scatternets.

Scheduling without rendezvous: the walk-in approach

The main problem with rendezvous-based bridge scheduling, the construction and maintenance of the schedule of rendezvous points, may be avoided if the bridge (or bridges) could operate without such a schedule. It turns out that such an approach, which will be referred to as walk-in bridge scheduling, is indeed feasible [Mišić, Mišić and Chan, 2005*b*]. Under walk-in scheduling, the bridges may switch between piconets at will, without any prior arrangement. The piconet masters will poll their slaves as determined by the chosen intra-piconet polling scheme, which includes the bridge as well as other slaves. The master will, therefore, poll the bridge in each piconet cycle, and the exchange will start only if the bridge is found to be present. The operation of the scatternet under this scheduling scheme is schematically depicted in Fig. 10.5; again, for simplicity we show only two piconets connected through an SS bridge.

The main advantage of the walk-in scheme lies in the absence of rendezvous points, which means that any given piconet can accommodate several bridge devices simultaneously, and any given bridge may visit several piconets in sequence. The walk-in bridge scheduling can thus be applied with ease to scatternets of arbitrary

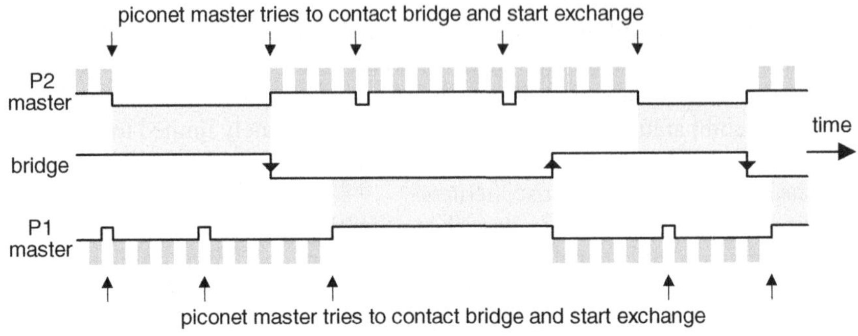

(a) Complete exchange mode.

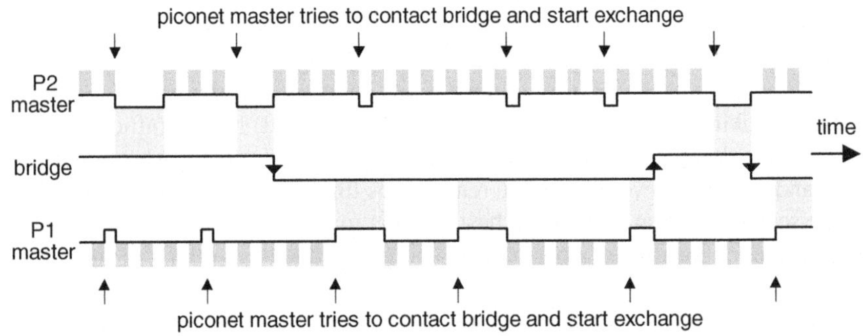

(b) Limited exchange mode.

FIGURE 10.5
Walk-in bridge scheduling in the scatternet with two piconets linked through a SS bridge.

size, and there is no performance penalty due to the construction and subsequent maintenance of the schedule of rendezvous points. Neither of these features can be achieved with rendezvous-based scheduling.

There are certain inefficiencies as well. First, slots will be wasted whenever the master attempts to contact the bridge which is absent; however, the overhead of an unsuccessful poll is just $2T$ [Bluetooth SIG, 2003*b*], which is small compared to other delays in the scatternet.

Second, the bridge is polled like an ordinary slave, which means it will not be polled immediately after joining the piconet, but later on. The exact delay will depend on the chosen intra-piconet polling scheme; it will take, one half of the piconet cycle time. Note that rendezvous-based scheduling is not totally efficient either – the bridge may wait idle *after* an exchange ends, rather than *before* it starts, as in the case of walk-in scheduling; and a few slots must be wasted in negotiating the next rendezvous. Therefore, the inefficiencies are not significant when compared to the rendezvous-based scheduling.

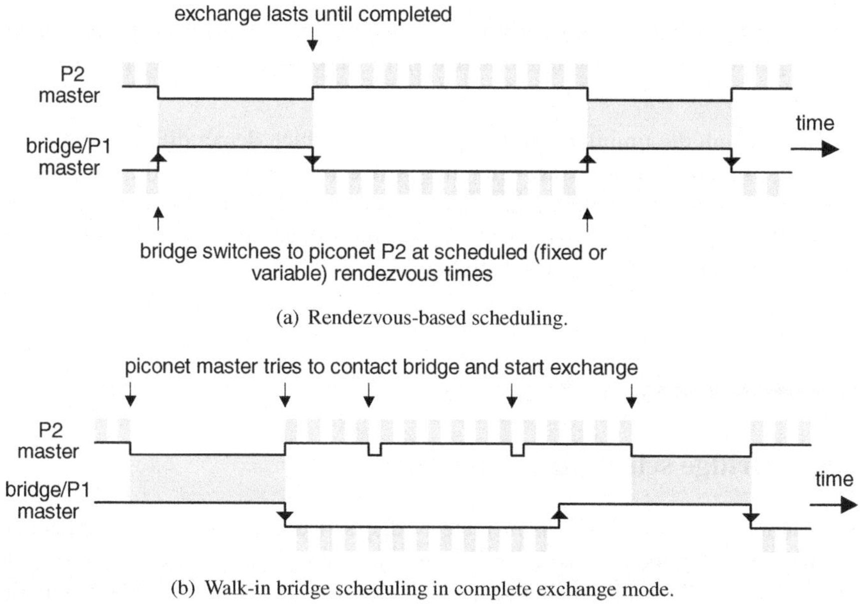

FIGURE 10.6
Bridge scheduling in the scatternet with two piconets linked through an MS bridge.

The walk-in approach solves the question of how should the exchange start, but does not specify how or when it should end. The simplest solution is to let the master and the bridge exchange all the queued packets, since there are no pending rendezvous to honor. This mode, which is schematically depicted in Fig. 10.5(a), is known as the complete exchange mode. Note that the complete exchange may take one or more piconet cycles, depending on the traffic profile and delay requirements. It should be noted that, in this case, the walk-in scheduling behaves in the same manner as the adaptive schemes described above, with the added bonus that there are no rendezvous points to fix, no schedule of such points to construct and maintain, and no administrative overhead associated with those activities.

The non-local traffic will actually benefit from the walk-in scheduling, while bridge exchanges that last too long may harm the local traffic. If cases where this is undesirable and we want to favor local traffic over non-local one, the bridge may be operated in another mode, known as the limited exchange mode. In this case, the duration of the bridge residence in any given piconet is limited, which means the bridge will leave the piconet even though there may still be packets to exchange. This arrangement is shown in Fig. 10.5(b). The limit on the duration of bridge exchanges may also be necessary to prevent buffer overflows, as explained above.

We note that the limit on bridge residence time may be expressed in absolute time units, such as seconds or Bluetooth clock time slots T. Alternatively, the limit may be expressed as the maximum duration of the exchange T_{xm}. In either case, the limit on the duration of the exchange may be coupled with the maximum number of exchanges to be executed (or, alternatively, the maximum number of piconet cycles) within a single bridge visit to the piconet.

Even though the timing diagrams of Fig. 10.5, which depict different approaches to bridge scheduling, are drawn for the bridge that operates in the SS mode, all three bridge scheduling modes and their variants can easily be applied in scatternets with MS bridges as well. Fig. 10.6 depicts the operation of the MS bridge under rendezvous-based and walk-in scheduling.

10.3 Bridge scheduling in practice

It remains to see in which manner the bridge scheduling may be implemented in practice. Parking the bridge device to allow it to visit their piconets is too long and inefficient, and the bridge will lose its device address in the process. For compatibility reasons, implementing a proprietary mechanism is out of the question. Therefore, we can choose between the two modes provided by the specification: SNIFF and HOLD (both of these were described in more detail in Chapter 1).

Indeed some proposals have used the SNIFF mode for bridge scheduling. However, the SNIFF mode is negotiated to last for an indefinite number of SNIFF periods, and it has to be explicitly terminated before the new one is negotiated. Furthermore, the SNIFF mode has a predefined window in which the master-bridge exchange must start, otherwise, both participants have to wait an extra SNIFF interval. Both constraints limit the usability of the SNIFF mode to rendezvous-based scheduling with a fixed schedule. The use of the SNIFF mode for dynamic approaches such as adaptive or walk-in scheduling is likely to incur high overhead due to the need to set up and terminate the SNIFF mode too frequently.

A much better solution is to use the Bluetooth HOLD mode [Bluetooth SIG, 2003*b*], which allows the master and the slave to negotiate their next rendezvous at the end of the current one; all three bridge scheduling modes can be implemented in this manner.

Under walk-in scheduling, the bridge may come and go without an associated schedule, and even the use of HOLD mode is not necessary, strictly speaking. At the same time, the bridge should take care to visit each of its piconets often enough. Namely, if the bridge is absent from the piconet for a prolonged period, there is a risk that the supervision timeout will be reached. (The concept of supervision timeout is explained in Section 1.4.) In this case, the master might be forced to conclude that the bridge device has left the piconet for good. Furthermore, the link to the bridge may have an associated poll interval value set for the purpose of QoS support. Both constraints affect the choice of the time interval between bridge visits.

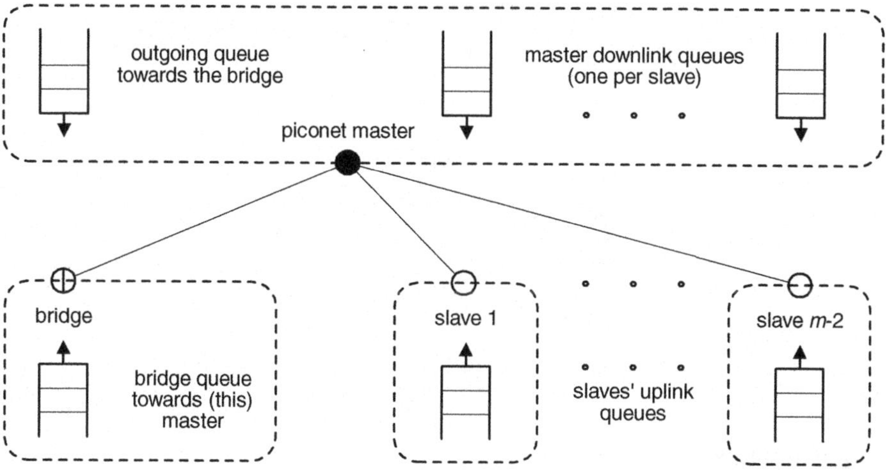

FIGURE 10.7
Portion of the queueing model of a scatternet – one piconet linked to one bridge.

10.4 The queueing model and traffic assumptions

Scatternet performance of both scatternet topologies may be analyzed using a queueing model, similar to the one that has been previously used to model the operation of a single piconet. In this model, each slave maintains an uplink queue, whereas the piconet master maintains one downlink queue per each slave in its piconet. At the same time, the master that communicates with bridge devices must maintain a separate output queue for each bridge. These queues hold packets with destinations in the other piconets until they are sent through the appropriate bridge. The portion of the scatternet model containing one piconet and a bridge is schematically shown in Fig. 10.7.

The bridge device maintains similar queues, one for each piconet in which it acts as a slave. Fig. 10.8 shows the queues in a bridge that connects three piconets.

For simplicity, we will initially consider simple scatternets with two piconets and a single bridge that links them, and evaluate their performance under all the three bridge scheduling approaches. These analyses for rendezvous-based, adaptive, and walk-in scheduling are presented in Chapters 11, 12, and 13, respectively. More complex scatternets will be analyzed in Chapters 13 and 14, respectively.

Our traffic model for the scatternet is similar to the one described in Section 2.4, and subsequently used to analyze the performance of simple piconets. We will assume that only ordinary slaves generate and receive any traffic, while the piconet masters and the bridge just route packets between their corresponding source and destination devices. This assumption is similar to the one we have previously used

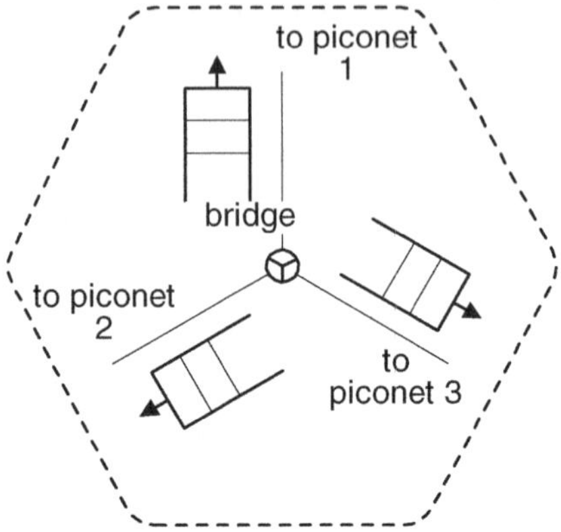

FIGURE 10.8
Portion of the queueing model of a scatternet – a bridge connecting three piconets.

in our analysis of piconet performance; it simplifies the analysis without restricting its generality. In fact, the case where piconet master(s) and/or bridge generate or consume packets may easily be accommodated by changing the corresponding packet arrival rates.

Slaves generate packets in bursts or batches, which correspond to application packets that are segmented into a number of Bluetooth baseband packets; more detailed discussion on segmentation and reassembly can be found in Section 2.3.

Packet burst arrivals follow a Poisson distribution with the arrival rate λ_{u1} for all slaves in piconet 1 and λ_{u2} for all slaves in piconet 2, while the length of the bursts is geometrically distributed with mean value of $\overline{B}$. (Modeling using single packets with Poisson arrivals has been shown to be inaccurate for real life data traffic [Paxson and Floyd, 1995].) The use of these distributions is actually a first approximation, since the exact characteristics of Bluetooth data traffic are unknown at this time. However, our analysis framework can easily accommodate any other distribution, provided its first and second moments are known.

Packets last one, three, or five time slots with probabilities p_1, p_3 and $p_5 = 1 - p_1 - p_3$, respectively [Bluetooth SIG, 2001*b*]. The corresponding probability generating function (PGF) is $G_p(z) = p_1 z + p_3 z^3 + p_5 z^5$, and the first and second moments of the packet length distribution are $\overline{L} = G'_p(1)$ and $\overline{L^2} = G''_p(1) + G'_p(1)$, respectively.

The length of the burst follows the probability distribution that may be described with a probability generating function (PGF) $G_b(z) = \sum_{k=0}^{\infty} b_k z^k$, where b_k is the probability that the burst will contain exactly k packets [Grimmett and Stirzaker, 1992]. We will also need the mean value of the burst length $\overline{B} = G'_b(1)$, while its second factorial moment is defined as $\overline{B^{(2)}} = E[B(B-1)] = G''_b(1)$. It is reasonable to assume that all slaves will use the same segmentation/reassembly mechanism, and consequently all packet bursts will have the same burst length and packet length distributions. The PGF for the duration of packet burst in time slots is then denoted as $R(z) = G_b(G_p(z))$ with the mean value $\overline{R} = \overline{B}\,\overline{L}$.

Traffic locality is the probability P_l that both the source and the destination of the packet burst are in the same piconet. (Of course, all packets within a single burst have the same destination.) The value of P_l is the same for all slaves. All packet-generating slaves within a piconet exhibit the same value for traffic locality P_l, i.e., the probability that both the source and destination of the burst will be in the same piconet. The probability that packet burst will be forwarded through the bridge is $1 - P_l$. For the packet bursts generated for the destinations which are in the same piconet, all the destinations are equally probable. Also, for the non-local traffic all destinations in the neighboring piconet are equally probable.

As data packets are queued at each intermediate node before being sent further on, the performance of the scatternet will be mainly determined by queueing delays in each of those queues. Three delay variables will be used as our main performance indicators. First is the access delay W_{ai} $(i = 1, 2)$, the time a data packet has to wait in the uplink queue of the source device before it is serviced. This parameter is important for both local and non-local traffic, as all packets, regardless of their destination, must wait in the corresponding uplink queue at the originating device. The other two are end-to-end delays from the moment the packet enters the uplink queue at the source, to the time it arrives at its destination device. We distinguish between end-to-end delay for local traffic, W_{ii} $(i = 1, 2)$, and the one for non-local traffic, W_{ij} $(i, j = 1, 2; i \neq j)$, as packets with non-local destinations have to pass through the bridge.

We finish our introductory presentation by noting that the Bluetooth specification does not define a protocol for forwarding or routing data from one piconet to another [Bluetooth SIG, 2003*b*]. In fact, because addressing schemes used in Bluetooth have only local significance, a bridging function cannot exist *at all* at the Bluetooth link layer. Instead, it has to be handled by the higher layers of the protocol stack, which is likely to incur additional cost in terms of performance. Still, the analysis of different bridge scheduling algorithms should provide important insights into their performance. It will give us the theoretical lower limits of bridge performance, thus providing a convenient benchmark against which the performance of actual networks may be assessed. A number of routing algorithms specifically targeting Bluetooth networks has been proposed [Albrecht, Frank, Martini, Schetelig, Vilavaara and Wenzel, 1999; Alzoubi et al., 2002; Bhagwat and Segall, 1999; Bose et al., 2001; Kapoor and Gerla, 2003; Lin et al., 2003; Prabhu and Chockalingam, 2002; Raman, Bhagwat and Seshan, 2001; Song, Li, Wang and Wang, 2003; Song et al., 2005; Stojmenovic, 2004*b*; Stojmenovic and Xu, 2001; Sun et al., 2002], but their detailed analysis is beyond the scope of this book.

11

Rendezvous-based bridge scheduling

We begin now our analysis of different bridge scheduling approaches and their impact on the performance of Bluetooth scatternets. In this chapter, we will analyze the performance of rendezvous-based bridge scheduling in a simple scatternet with two Bluetooth piconets linked through a bridge device. Both the Master/Slave (MS) and Slave/Slave (SS) topologies are discussed. In both cases, the piconet master polls its ordinary slaves according to the exhaustive and 1-limited service policy.

Our analysis is based on the theory of $M^{[x]}/G/1$ queues with vacations [Bertsekas and Gallager, 1991; Takagi, 1988; Takagi, 1991]. We first derive the probability distribution for bridge cycle time, under the assumption that this distribution is independent of the intra-piconet polling scheme. Next, we derive the probability distribution for piconet cycle time conditioned on the duration of bridge cycle time, under exhaustive and 1-limited intra-piconet scheduling policies. We investigate the impact of scatternet and traffic parameters on the mean values of access and end-to-end packet delays; in the latter case, we consider local (i.e., intra-piconet) and non-local (inter-piconet) traffic separately.

The chapter is structured as follows. Section 11.1 presents the queueing theoretic analysis of the scatternet with an MS bridge, followed by an analogous analysis of the scatternet with an SS bridge in Section 11.4. Analytical results are shown and summarized in Section 11.6.

11.1 MS bridge topology

Let m_1 and m_2 denote the numbers of members in piconets P_1 and P_2, respectively, and let the master of P_1 act also as the bridge toward piconet P_2. As before, we assume that only ordinary slaves generate and receive data packets; therefore, the number of packet-generating slaves will be $m_1 - 1$ and $m_1 - 2$ for piconets P_1 and P_2, respectively. The burst arrival rates for inter-piconet traffic will be $\lambda_{b12} = (m_1 - 1)\lambda_{u1}(1 - P_l)$ and $\lambda_{b21} = (m_2 - 2)\lambda_{u2}(1 - P_l)$, for traffic flows from P_1 to P_2 and vice versa, respectively.

According to Fig. 10.1(b), the bridge cycle time for the MS bridge under rendezvous bridge scheduling is $T_{cyc} = T_1 + T_r$. In this case, the bridge residence time in its own piconet, T_1, has a fixed value, while the bridge exchange time T_r is a random variable. The PGF for the bridge cycle time has the form

$$T_{cyc}(z) = z^{T_1} T_r(z) \tag{11.1}$$

where $T_r(z)$ is the PGF for the duration of the bridge exchange time.

We will first find the probability distributions for the bridge cycle time, followed by the piconet cycle time.

Bridge cycle time

We assume that the probability distribution of bridge cycle time is independent of intra-piconet scheduling scheme and therefore the same derivation holds for any intra-piconet scheduling policy. The duration of the exchange between the bridge and the master of P_2 can be calculated by summing up the durations of the transmission of the data packets that were queued when the exchange started. The packets that arrive during the current exchange will be queued until the start of the next exchange [Bluetooth SIG, 2001*b*]; therefore, this service discipline is gated policy [Takagi, 1991].

The probability of exactly l data packet bursts arriving to the outgoing queue of the bridge (i.e., in the direction from P_1 to P_2) during the bridge cycle of i slots, is

$$b_{l,12} = \sum_{i=0}^{\infty} P_{(T_{cyc}=i)} e^{-\lambda_{b12} i} \frac{(\lambda_{b12} i)^l}{l!} \tag{11.2}$$

where $P_{(T_{cyc}=i)}$ denotes the probability that the bridge cycle time lasts i slots. Then, the PGF for the number of packet burst arrivals to the bridge queue during the bridge cycle time is

$$\begin{aligned} B_{12}(y) &= \sum_{l=0}^{\infty} b_{l,12} y^l = \sum_{l=0}^{\infty} y^l \sum_{i=0}^{\infty} P_{(T_{cyc}=i)} e^{-\lambda_{b12} i} \frac{(\lambda_{b12} i)^l}{l!} \\ &= \sum_{i=0}^{\infty} P_{(T_{cyc}=i)} e^{-\lambda_{b12}(1-y)i} \\ &= T^*_{cyc}(\lambda_{b12} - y\lambda_{b12}) \end{aligned} \tag{11.3}$$

where $T^*_{cyc}(s) = \sum_{i=0}^{\infty} P_{(T_{cyc}=i)} e^{-si}$ denotes the moment generating function [Grimmett and Stirzaker, 1992] or discrete-time LST transform of the bridge cycle time. (In the discussions that follow, we will use the term LST transform.) The LST transform of the bridge cycle time can also be represented as

$$\begin{aligned} T^*_{cyc}(s) &= \sum_{i=0}^{\infty} P_{(T_{cyc}=i)} \sum_{n=0}^{\infty} \frac{(-si)^n}{n!} = \sum_{n=0}^{\infty} \frac{(-s)^n}{n!} \sum_{i=0}^{\infty} P_{(T_{cyc}=i)} i^n \\ &= \sum_{n=0}^{\infty} \frac{(-s)^n}{n!} T^{(n)}_{cyc} \end{aligned} \tag{11.4}$$

where $T_{cyc}^{(n)}$ denotes the n-th moment of the bridge cycle time:

$$T_{cyc}^{(n)} = (-1)^n \frac{d^n}{ds^n} T_{cyc}^*(s)|_{s=0}$$

Since the packet bursts are of random size described with the PGF $G_b(y)$, the PGF for the number of packet arrivals during the bridge cycle time is

$$A_{12}(y) = B_{12}(G_b(y)) = T_{cyc}^* \left(\lambda_{b12} - \lambda_{b12} G_b(y)\right) \tag{11.5}$$

Then, the probability that exactly k data packets will be sent from the bridge to the master of P_2 is

$$\begin{aligned} a_{k,12} &= \frac{1}{k!}\frac{d^k}{dy^k} A_{12}(0) \\ &= \sum_{l=0}^{\infty} \frac{1}{k!}\frac{d^k}{dy^k} (G_b(y))^l \Bigg|_{y=0} \sum_{i=0}^{\infty} P_{(T_{cyc}=i)} e^{-\lambda_{b12} i} \frac{(\lambda_{b12} i)^l}{l!} \\ &= \frac{1}{k!}\frac{d^k}{dy^k} \left(\sum_{i=0}^{\infty} P_{(T_{cyc}=i)} e^{-\lambda_{b12} i (1-G_b(y))} \right)\Bigg|_{y=0} \\ &= \frac{1}{k!}\frac{d^k}{dy^k} T_{cyc}^*(\lambda_{b12} - \lambda_{b12} G_b(y))\Bigg|_{y=0} \end{aligned} \tag{11.6}$$

We note that the value $a_{k,12}$ actually consists of the linear combination of the first k moments of the bridge cycle time distribution and the probability of no packet arrivals during the bridge cycle time, which is equal to $T_{cyc}^*(\lambda_{b12})$. By expanding (11.6), we obtain:

$$\begin{aligned} a_{k,12} = T_{cyc}^*(\lambda_{b12}) \sum_{n_1=1}^{k} \frac{d^{(n_1)}}{dy^{n_1}} G_b(0) \sum_{n_2=0}^{k-n_1} \frac{d^{(n_2)}}{dy^{n_2}} G_b(0) \cdots \sum_{n_k=0}^{k-n_1-n_2 \ldots -n_{k-1}} \frac{d^{(n_k)}}{dy^{n_k}} G_b(0) \\ \cdot \binom{k}{\max(n, n_1)} \lambda_{b12}^n (-1)^n T_{cyc}^{(n)} \end{aligned} \tag{11.7}$$

where $n = \sum_{i=1}^{k} \left\lceil \frac{n_i}{k} \right\rceil$, and we have assumed that $\frac{d^{(0)}}{dy^0} G_b(0) = 1$.

By the same token the probability of exchanging k data packets in the opposite direction (i.e., from the master of P_2 to the bridge) during one bridge cycle, is

$$a_{k,21} = \frac{1}{k!}\frac{d^k}{dy^k} T_{cyc}^* \left(\lambda_{b21} - \lambda_{b21} G_b(y)\right)|_{y=0} \tag{11.8}$$

The number of packets queued at the bridge may not be equal to the number of packets queued at the master of P_2; hence, empty (POLL or NULL) packets must be inserted to balance out the difference. This procedure will be referred to as *padding*. Also, the exchange ends with an empty frame, i.e., a POLL packet followed by a

NULL packet. Therefore, the PGF for the length of the exchange time between the bridge and the master of piconet P_2 can be written in the form

$$
\begin{aligned}
T_r(z) = & \sum_{k=0}^{\infty} a_{k,12} a_{k,21} G_p(z)^{2k} z^2 \\
& + \sum_{j=1}^{\infty} \sum_{k=0}^{\infty} (a_{k,12} a_{k+j,21} + a_{k+j,12} a_{k,21}) \cdot G_p(z)^{2k+j} z^{j+2}
\end{aligned}
\tag{11.9}
$$

The first term denotes the probability that both packet queues (the outgoing queue at the master and the corresponding outgoing queue at the bridge) have equal number of packets. The second term corresponds to the case when these queues have different lengths, in which case empty packets must be inserted in some frames in order to service all of the data packets. Finally, the term z^2 stands for the empty frame that terminates the exchange. By replacing z with e^{-s}, we are able to obtain the following equation related to the bridge cycle time

$$T^*_{cyc}(s) = e^{-T_1 s} T^*_r(s) \tag{11.10}$$

On the left hand side of (11.10), we have the LST transform of the bridge cycle time; on the right hand side, we have a function of its moments. Therefore, an approximate expression for $T^*_{cyc}(s)$ can be found by finding its first k moments by differentiating (11.10) with respect to s. For finding those k moments, together with $T^*_{cyc}(\lambda_{b12})$ and $T^*_{cyc}(\lambda_{b21})$, we need $k+2$ equations. Two of these can be obtained by substituting $s = \lambda_{b12}$ and $s = \lambda_{b21}$ in (11.10), while the remaining k equations are obtained from (11.10), through differentiation with respect to s, and truncation of moments of order higher than k. Once the moments are found, we can determine the coefficients $a_{k,12}$ and $a_{k,21}$, which means that PGFs $T_r(z)$ and $T_{cyc}(z)$ can be determined with the required accuracy.

Due to the presence of empty packets, the actual frame length will not be equal to twice the data packet length given by the PGF $G_p(z)^2$. The PGF for the frame length during the bridge exchange may be obtained from (11.9) as

$$
\begin{aligned}
F_b(z) = & \sum_{k=0}^{\infty} a_{k,12} a_{k,21} \left(G_p(z)^{2k} z^2 \right)^{1/(k+1)} \\
& + \sum_{j=1}^{\infty} \sum_{k=0}^{\infty} (a_{k,12} a_{k+j,21} + a_{k+j,12} a_{k,21}) \cdot \left(G_p(z)^{2k+j} z^{j+2} \right)^{1/(k+j+1)}
\end{aligned}
\tag{11.11}
$$

and the mean frame length for bridge-master exchange may be obtained as $\overline{F_b} = F_b'(1)$.

Piconet cycle time

In order to determine the probability distribution of the access delay, we need the mean duration of piconet service cycle, i.e., the time interval for a piconet master to service all of its slaves once. Due to the complexity of calculations, we will assume symmetric slaves with equal traffic loading. As noted above, one channel may be modeled as a pair of queues, for which the burst arrival rates will be $\lambda_{u1} = \lambda_{u2} = \lambda$ for the slave (uplink) queue, and $\lambda_{d1} = \lambda P_l + \lambda_{b21}/(m_1 - 1)$ for the corresponding downlink queue at the master. Due to the symmetry of the scatternet under rendezvous-based scheduling, we will consider one piconet, say, P_1, only.

In the local operation phase, a number of frames may be exchanged between the master and a single slave during a single visit to that slave. Those packets have been queued in the corresponding uplink and/or downlink queues during the previous piconet cycle. (As in the case of bridge exchanges, the number of packets in one direction may be different from the number of packets in the opposite one, and empty packets may have to be added to balance out the difference; furthermore, an empty packet will signalize that both queues are empty.) However, the previous piconet cycle may have been interrupted by the bridge exchange, and the number of queued packets depends on the probability distribution of the length of the bridge exchange, as well as on the probability distribution of the bridge cycle time.

To model the impact of bridge exchanges on the operation of the piconet, let us start by assuming that the bridge exchange has a fixed duration and derive the conditional probability distribution for the piconet cycle time. This result will be used, then, to derive the unconditional probability distribution of the piconet cycle time.

Let us consider the piconet cycle in P_1, in the presence of bridge exchanges which occur after every T_1 slots of local operation and last n slots (which, in turn, means that the bridge cycle time lasts for $n + T_1$ slots). Let $C_1^t|n$ denote the total number of slots in a piconet cycle, and let $C_1^t(z)|n$ denote its PGF. (At this moment, the notation of piconet cycle time is generic, i.e., it is not specifically tied to any intra-piconet scheduling policy.) Part of the piconet cycle will be used for the bridge exchange(s); therefore, let the "pure piconet cycle" $C_1^p|n$ denote the number of slots spent exclusively on communication with ordinary slaves between the beginning of the visit to the slave j (where $j = 1 \,..\, m_1 - 1$) and the end of the visit to the slave $(j + m_1 - 1) \bmod m_1$. We will denote its conditional PGF as $C_1^p(z)|n = \sum_{i=0}^{\infty} r_i^{(n)} z^i$ (where (n) should be considered as a superscript, rather than as power).

Also, let $S_1^p|n$ denote the 'pure' single channel service time, i.e., the time to empty both uplink and downlink queues for one particular slave channel without counting the slots used for the bridge exchange(s), conditioned upon the bridge exchange time being equal to n slots. The corresponding conditional PGF will be denoted as $S_1^p(z)|n$.

Let the pure piconet cycle last i slots with the probability $r_i^{(n)}$. Now consider a super-cycle that consists of $T_1 i$ time slots; during this super-cycle, all combinations of relative positions of beginnings of piconet cycle and local operation (i.e., master residence) times will be repeated k times, where $k = 1 \,..\, (T_1 i)/lcm(T_1, i)$ and $lcm(T_1, i)$ is the least common multiple for the current value of piconet cycle i and the master residence time in piconet P_1, T_1. Obviously, the super-cycle will contain $n_p = T_1$ piconet cycles (without interruptions) and $n_b = i$ master residence times.

Then, the piconet cycle can be interrupted either $\left\lceil \frac{i}{T_1} \right\rceil$ times or $\left\lfloor \frac{i}{T_1} \right\rfloor$ times (the two values will be the same if $\frac{i}{T_1}$ is an integer). The probabilities of these two events are

$$pp_{i,\lceil \frac{i}{T_1} \rceil} = \frac{n_b - n_p \left\lfloor \frac{n_b}{n_p} \right\rfloor}{n_p} = 1 - pp_{i,\lfloor \frac{i}{T_1} \rfloor} \tag{11.12}$$

The PGF for the number of bridge exchanges within one piconet cycle may, then, be expressed as

$$I(z) = \sum_{i=0}^{\infty} r_i^{(n)} \left(pp_{i,\left\lfloor \frac{i}{T_1} \right\rfloor} z^{\left\lfloor \frac{i}{T_1} \right\rfloor} + pp_{i,\left\lceil \frac{i}{T_1} \right\rceil} z^{\left\lceil \frac{i}{T_1} \right\rceil} \right) \tag{11.13}$$

while the PGF for the number of slots inserted in the piconet cycle due to the bridge exchange has the form $I(z^n)$. Then, the PGF for the total length of the piconet cycle in the presence of bridge exchanges that occur at every T_1 slots of "pure" piconet operation and last for n slots, is

$$C_1^t(z)|n = \sum_{i=0}^{\infty} r_i^{(n)} \left(pp_{i,\left\lfloor \frac{i}{T_1} \right\rfloor} z^{i+n\left\lfloor \frac{i}{T_1} \right\rfloor} + pp_{i,\left\lceil \frac{i}{T_1} \right\rceil} z^{i+n\left\lceil \frac{i}{T_1} \right\rceil} \right) \tag{11.14}$$

Exhaustive polling

In order to determine $C_1^p(z)|n$, we have to find the pure channel service time. The probability that k data packets are sent from the slave to the master in piconet P_1 without taking into account the packets from the bridge exchange, may be obtained as

$$a_{k,u1}|n = \frac{1}{k!} \frac{d^k}{dy^k} C_1^{t*} \left(\lambda_{u1} - \lambda_{u1} G_b(y)\right)\big|_{y=0} |n \tag{11.15}$$

where $C_1^{t*}(\lambda_{u1} - \lambda_{u1} G_b(y))|n$ denotes the PGF for the number of data packet arrivals in the slave uplink queue during the total piconet cycle time. By the same token, the equivalent conditional probability that k data packets are sent from the master to the slave in piconet P_1 is

$$a_{k,d1}|n = \frac{1}{k!} \frac{d^k}{dy^k} C_1^{t*} \left(\lambda_{d1} - \lambda_{d1} G_b(y)\right)\big|_{y=0} |n \tag{11.16}$$

By combining these elements together, the conditional PGF for the pure channel service time becomes

$$\begin{aligned} S_1^p(z)|n &= \sum_{k=0}^{\infty} (a_{k,u1}|n)(a_{k,d1}|n) G_p(z)^{2k} z^2 \\ &+ \sum_{j=1}^{\infty} \sum_{k=0}^{\infty} \left((a_{k,u1}|n)(a_{k+j,d1}|n) + (a_{k+j,u1}|n)(a_{k,d1}|n)\right) G_p(z)^{2k+j} z^{(j+2)} \end{aligned} \tag{11.17}$$

Finally, the conditional PGF for the pure piconet cycle time becomes

$$C_1^p(z)|n = (S_1^p(z)|n)^{m_1-1} \tag{11.18}$$

In order to solve the last equation we have to truncate the $C_1^p(z)|n$ to i_{max} members $r_i^{(n)}$. This is an approximate solution – but it should be sufficient for large i_{max}. Then, we have to set i_{max} equations in the form $r_i^{(n)} = \frac{1}{i!}\frac{d^i}{dz^i}C_1^p(0)|n$. In each equation (say, the i-th), we will have the term $r_i^{(n)}$ on the left side and the function of all $r_i^{(n)}$ mass probabilities on the right side. The resulting system of equations can be solved using numerical solvers (e.g., Waterloo Maple).

When the PGF for the pure piconet cycle time is known, the PGF for the total piconet cycle time can be found from (11.14). Then, the unconditional PGF for the total piconet cycle time can be obtained by using the mass probabilities for the bridge exchange distribution from the PGF $T_r(z) = \sum_{n=0}^{\infty} p_n z^n$, which is actually derived in the form given by (11.9). This unconditional PGF of the total piconet cycle time can be approximated with

$$C_1^t(z) = \sum_{n=0}^{n_{max}} p_n C_1^t(z)|n \tag{11.19}$$

where n_{max} is some sufficiently large value that depends on the value of T_1 and the intensity of inter-piconet traffic.

The PGF for the frame length for the regular slave-master exchange will be

$$F_s(z) = \sum_{n=0}^{n_{max}} p_n F_s(z)|n \tag{11.20}$$

which differs from $G_p(z)^2$, due to the presence of empty slots to balance the unmatched data packets. The values p_n are mass probabilities that bridge exchange lasts n time slots. The conditional PGF for the frame time is:

$$\begin{aligned} F_s(z)|n = &\sum_{k=0}^{k_{max}} (a_{k,u1}|n)(a_{k,d1}|n)(G_p(z)^{2k}z^2)^{1/(k+1)} \\ &+ \sum_{j=1}^{j_{max}} \sum_{k=0}^{k_{max}} \left((a_{k,u1}|n)(a_{k+j,d1}|n) + (a_{k+n,u1}|n)(a_{k,d1}|n)\right) \\ &\cdot (G_p(z)^{2k+j}z^{j+2})^{1/(k+j+1)} \end{aligned} \tag{11.21}$$

where k_{max} and j_{max} are determined according to the accuracy required. The mean frame length will be $\overline{F_s} = F_s'(1)$.

From the viewpoint of a particular slave, the master is not available—i.e., takes a vacation—during the time it services other slaves. The vacation lasts until the next visit to the slave; if the slave queue is empty at that time, the master will immediately start a new vacation. The vacation time is, then, the time while the master is busy servicing other slave queues. The duration of the vacation period V_1 may be described with the following PGF

$$V_1(z) = \sum_{n=0}^{n_{max}} p_n (S_1^t(z)|n)^{m_1-2} \tag{11.22}$$

and its first and second moments are $\overline{V_1} = V_1'(1)$ and $\overline{V_1^2} = V_1''(1) + V_1'(1)$, respectively.

1-limited polling

In order to determine $C_1^p(z)|n$, we have to find the pure channel service time. The probability that zero data packets (i.e., a single NULL packet) are sent from the slave to the master, without taking into account the packets from the bridge exchange, may be obtained as

$$a_{0,u1}|n = 1 - \lambda_{u1}\overline{B}\,\overline{C_1^t|n} \tag{11.23}$$

The probability that one data packet is sent from the slave to the master in a polling cycle is then:

$$a_{1,u1}|n = \lambda_{u1}\overline{B}\,\overline{C_1^t|n} \tag{11.24}$$

By the same token, the equivalent conditional probabilities that one POLL packet or one data packet is sent from the master to the slave are respectively equal to:

$$\begin{aligned} a_{0,d1}|n &= 1 - \lambda_{d1}\overline{B}\,\overline{C_1^t|n} \\ a_{1,d1}|n &= \lambda_{d1}\overline{B}\,\overline{C_1^t|n} \end{aligned} \tag{11.25}$$

By combining these elements together, the conditional PGF for the pure channel service time becomes

$$\begin{aligned} S_1^p(z)|n = {} & (a_{0,u1}|n)(a_{0,d1}|n)z^2 \\ & + \left((a_{1,u1}|n)(a_{0,d1}|n) + (a_{0,u1}|n)(a_{1,d1}|n)\right) G_p(z)z \\ & + (a_{1,u1}|n)(a_{1,d1}|n)G_p(z)^2 \end{aligned} \tag{11.26}$$

Finally, the conditional PGF for the pure piconet cycle time becomes

$$C_1^p(z)|n = (S_1^p(z)|n)^{m_1-1} \tag{11.27}$$

and we can solve it in the similar way to the one indicated for the exhaustive service.

When the PGF for the pure piconet cycle time is known, the PGF for the total piconet cycle time can be found from (11.14). Then, the unconditional PGF for the total piconet cycle time can be obtained by using the mass probabilities for the bridge exchange distribution from the PGF $T_r(z) = \sum_{n=0}^{\infty} p_n z^n$, which is actually derived in the form given by (11.9). This unconditional PGF of the total piconet cycle time can be approximated with

$$C_1^t(z) = \sum_{n=0}^{n_{max}} p_n C_1^t(z)|n \tag{11.28}$$

where n_{max} is some sufficiently large value that depends on the value of T_1 and the intensity of inter-piconet traffic.

The PGF for the frame length for the regular slave-master exchange in the case of 1-limited service is the same as the channel time:

$$F_s(z) = S_1^p(z) \tag{11.29}$$

The duration of the vacation period V_1 experienced by the slave has the same PGF of the same form as (11.22).

11.2 Packet delays: the MS bridge case

We are now ready to calculate access delay and end-to-end delay for local and non-local traffic for two piconets interconnected by the MS bridge. Delay components for both kinds of delay are shown in Figure 11.1.

Exhaustive polling

By substituting e^{-s} in place of z in the previously derived PGFs, we obtain the LSTs for individual components of the access delay. The LST for the access delay at the slave in a Bluetooth piconet is obtained from the analysis of exhaustive FCFS system with vacations and batch arrivals [Takagi, 1991, Chapter 2, Equation 3.20], and it has the form

$$W_{a1}^*(s) = \frac{1 - V_1^*(s)}{s\overline{V_1}} \cdot \frac{1 - G_b(F_s^*(s))}{\overline{B}\,(1 - F_s^*(s))} \cdot \frac{s(1 - \lambda_{u1}\,\overline{B}\,\overline{F_s})}{s - \lambda_{u1} + \lambda_{u1} G_b(F_s^*(s))} \tag{11.30}$$

The mean access delay is obtained as

$$\begin{aligned} \overline{W_{a1}} &= -(W_{a1}^*)'(0) \\ &= \frac{\lambda_{u1}\overline{B}\,\overline{F_s^2}}{2(1 - \lambda_{u1}\,\overline{B}\,\overline{F_s})} + \frac{\overline{B^{(2)}}\,\overline{F_s}}{2\overline{B}(1 - \lambda_{u1}\overline{B}\,\overline{F_s})} + \frac{\overline{V_1^2}}{2\overline{V_1}} \end{aligned} \tag{11.31}$$

where $\overline{F_s^2} = F_s'(1) + F''s(1)$ denotes the second moment of the frame length distribution during ordinary master-slave exchange.

Note that the burstiness of the traffic is essentially preserved under exhaustive service scheduling, because the entire burst from the slave is transferred without

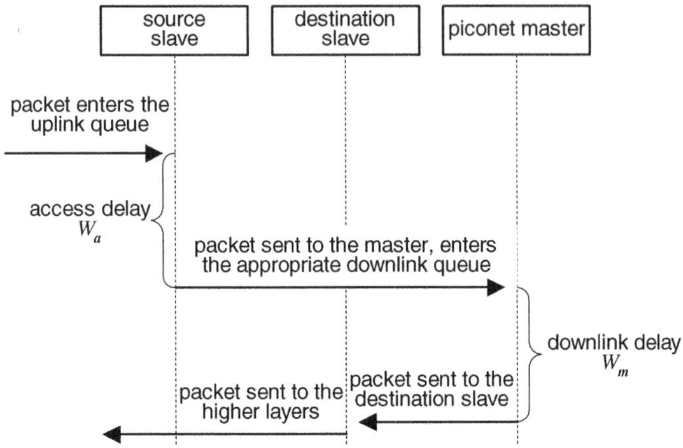

(a) Local traffic.

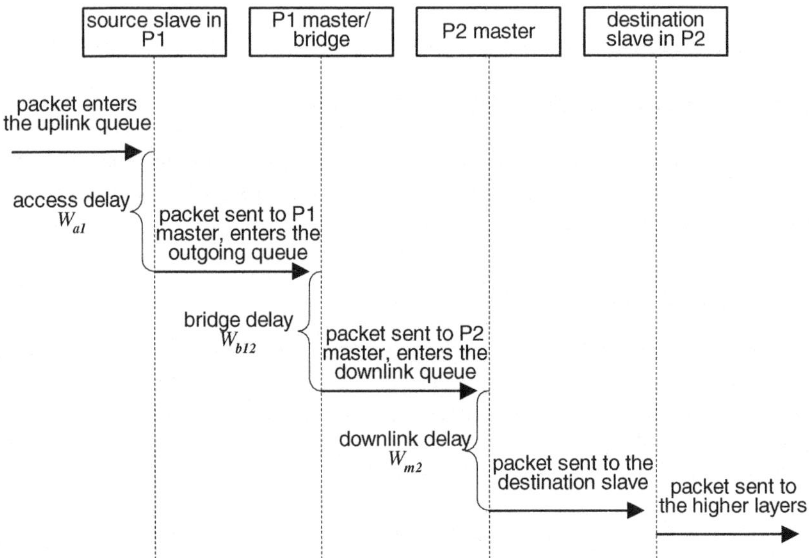

(b) Non-local traffic.

FIGURE 11.1
Delay components in the scatternet with an MS bridge.

interruption to the corresponding downlink queue. Occasionally, a burst might be interrupted due to the bridge exchange, but even in this case the transfer could be resumed after the exchange, and such burst will actually reappear in the corresponding downlink queue. Therefore, the queueing delay at the master is equal to the access delay at the slave, $W^*_{m1}(s) = W^*_{a1}(s)$, which may be obtained from (11.30).

The calculation of the end-to-end delay is slightly more involved, as two distinct cases may be observed. If both the source and destination nodes of a packet are in the same piconet, the total delay is the sum of two components, the access delay at the slave and the queueing delay at the master. The LST of this delay is $W_{11}^*(s) = W_{a1}^*(s)W_{m1}^*(s)$, and the mean value of the end-to-end delay for local traffic is equal to

$$\overline{W_{11}} = \overline{W_{a1}} + \overline{W_{m1}} \tag{11.32}$$

An analogous expression holds for piconet P_2.

Packets may also go from the source in one piconet to the destination in the other one. Such packets will have to pass through the bridge, which incurs an additional queueing delay. Since the packets in the bridge are served exhaustively, the corresponding LST for the bridge queueing delay (for packets going from P_1 to P_2) under rendezvous-based bridge scheduling is:

$$W_{b12}^*(s) = \frac{1 - T_1^*(s)}{s\overline{T_1}} \cdot \frac{s(1 - \lambda_{b12}\overline{B}\,\overline{F_b})}{s - \lambda_{b12} + \lambda_{b12}G_b(F_b^*(s))} \cdot \frac{1 - G_b(F_b^*(s))}{\overline{B}(1 - F_b^*(s))} \tag{11.33}$$

where $T_1^*(s) = \overline{T_1}/s$ is the LST of the vacation time taken after serving the bridge's and master's queue toward the bridge.

The mean bridge queueing delay for packets going from P_1 to P_2 under rendezvous-based scheduling is

$$\overline{W_{b12}} = \frac{\lambda_{b12}\overline{B}\,\overline{F_b^2}}{2(1 - \lambda_{b12}\overline{B}\,\overline{F_b})} + \frac{\overline{B^{(2)}}\,\overline{F_b}}{2\overline{B}(1 - \lambda_{b12}\overline{B}\,\overline{F_b})} + \frac{\overline{T_1^2}}{2\overline{T_1}} \tag{11.34}$$

where $\overline{F_b^2} = G''_{fb}(1) + G'_{fb}(1)$ denotes the second moment of frame length distribution during bridge master exchange. Also, since T_1 is constant, the third component of the previous expression is only $T_1/2$.

By the same token, packets going in the opposite direction (i.e., from P_2 to P_1) will experience the queueing delay similar to (11.33).

Overall, the mean queueing end-to-end delay time will be

$$\begin{aligned} \overline{W_{12}} &= \overline{W_{a1}} + \overline{W_{b12}} + \overline{W_{m2}} \\ \overline{W_{21}} &= \overline{W_{a2}} + \overline{W_{b21}} + \overline{W_{m1}} \end{aligned} \tag{11.35}$$

for packets going from P_1 to P_2 and from P_2 to P_1, respectively.

1-limited polling

When the PGFs for the piconet cycle time, frame time, and vacation time are known, we may calculate the LST for distribution of the waiting time (access delay) at the slave queue [Takagi, 1991]:

$$W_{a1}^*(s) = \frac{1 - V_1^*(s)}{s\overline{V_1}} \cdot \frac{s(1 - \lambda_{u1}\overline{B}\,\overline{C_1^t})}{s - \lambda_{u1} + \lambda_{u1}G_b(C_1^{t*}(s))} \cdot \frac{1 - G_b(C_1^{t*}(s))}{\overline{B}\,(1 - C_1^{t*}(s))} \tag{11.36}$$

where $V_1^*(s)$ and $C_1^{t*}(s)$ denote the LST of the vacation time and cycle time probability distributions, respectively. The average access delay at the slave is then calculated as $\overline{W_{a1}} = -(W_{a1}^*)'(0)$, and its value is:

$$\overline{W_{a1}} = \frac{\lambda_{u1}\overline{B}\,\overline{(C_1^t)^2}}{2(1-\lambda_{u1}\,\overline{B}\,\overline{C_1^t})} + \frac{\overline{B^{(2)}}\,\overline{C_1^t}}{2\overline{B}(1-\lambda_{u1}\overline{B}\,\overline{C_1^t})} + \frac{\overline{V_1^2}}{2\overline{V_1}} \tag{11.37}$$

where $\overline{B^{(2)}} = E[B(B-1)] = G_b''(1)$ denotes the second factorial moment of the packet burst length distribution.

Analogous expressions may be obtained for piconet P_2, except that the number of packet-generating slaves should be set to $m_2 - 2$.

It may be interesting to analyze the behavior of the mean access delay when the aggregate packet arrival rate is kept constant, $\lambda_{u1}\overline{B} = \text{const}$, while the mean burst size $\overline{B}$ is variable. First, the mean value of exchange time, $\overline{T_r}$, is proportional to $\lambda_{u1}\overline{B}$ rather than on either of these separately. Therefore, it will be constant when the aggregate packet arrival rate is constant, even though the mean burst size changes. Second, the mean piconet cycle time $\overline{C_1^t}$ is also dependent on the product of λ_{u1} and $\overline{B}$, and therefore constant as well. Consequently, the mean vacation time is also constant.

Having found this, let us take a closer look at the three terms in (11.37). Since we have assumed a geometric distribution of packet burst length, the second factorial moment is $\overline{B^{(2)}} = G_b''(1) = 2\overline{B}(\overline{B}-1)$, therefore $\overline{B^{(2)}}/\overline{B} = 2(\overline{B}-1)$.

With that in mind, $C_1''^t(z)$ may be expressed as a linear combination of products of components which contain zeroth, first and second derivatives of PGFs for channel time and bridge exchange time. When $z = 1$, all components will be linear functions of packet arrival rate, e.g., $k_1 + k_2\lambda_{u1}\overline{B}$ (where k_1 and k_2 are constants), except for the second derivative $T_r''(1)$, which contains the second derivative $G_b''(1)$ and has the form $k_3 + k_4\lambda_{u1}\overline{B}^2$. Since $\lambda_{u1}\overline{B} = \text{const}$, this means that $T_r''(1)$ is a linear function of $\overline{B}$. Overall, $\overline{C_1^{2t}}$ will be a linear function of $\overline{B}$.

By the same token, the second moment of vacation time $\overline{V_1^2}$ can be shown to have the form $K_5 + K_6\lambda_{u1}\overline{B}^2$, and is therefore a linear function of $\overline{B}$.

In summary, the numerators of all three fractions from (11.37) are linear functions of $\overline{B}$, and the denominators of all three depend on the packet arrival rate $\lambda_{u1}\overline{B}$ which is constant. Therefore, we may conclude that the mean access delay time $\overline{W_{a1}}$ is a linear function of mean burst size $\overline{B}$ – or, in plain words, the burstier the traffic, the longer the access delay.

The calculation of end-to-end delay for 1-limited intra-piconet scheduling is similar to the case of exhaustive scheduling given in 11.2 but is slightly more complicated due to the change of the burst size caused by the scheduling policy. The burstiness of the traffic in the downlink (master) queue will differ from that of the traffic in the slave (uplink) queue because the bursts from different source slaves with the same destination will become interleaved in the same downlink queue, which will in turn lead to an equivalent decrease in burst length. Exact analysis of this phenomenon is fairly involved, and we will only present an approximate model for the decrease of the burst length $\overline{B} = 1/p_B$, where p_B is the parameter of geometric distribution [Grimmett and Stirzaker, 1992]. Since two slaves in a piconet, say, P_1, have the same local destination with the probability $P_l^2/(m_1 - 2)^2$, the probability that the two bursts will not overlap in time (and, consequently, that they will not be interlaced in the same downlink queue) will be $1 - P_l^2/(m_1 - 2)^2$. Therefore, given the parameter p_B of the geometric distribution of burst length at the source slave, the equivalent parameter p_{Bm} of the packet burst at the master (downlink) queue will be

$$p_{Bm} = p_B \left(1 - \frac{P_l^2}{(m_1 - 2)^2}\right) + 1 \cdot \frac{P_l^2}{(m_1 - 2)^2} \tag{11.38}$$

The new equivalent mean burst length will be $\overline{B_m} = 1/p_{Bm}$. In order to maintain the same server utilization under decreased burst length, the burst arrival rate has to be scaled so that $\lambda_{d1m}\overline{B_m} = \lambda_{d1}\overline{B}$.

The mean queueing delays in master downlink queues in piconets P_1 and P_2, respectively, are equal to

$$\begin{aligned} \overline{W_{m1}} &= \frac{\lambda_{d1m}\overline{B_m}\ \overline{X_{c1}^2}}{2(1 - \lambda_{d1m}\overline{B_m}\ \overline{C_1^t})} + \frac{\overline{B_m^{(2)}}\ \overline{C_1^t}}{2\overline{B_m}(1 - \lambda_{d1m}\overline{B_m}\ \overline{C_1^t})} + \frac{\overline{V_1^2}}{2\overline{V_1}} \\ \overline{W_{m2}} &= \frac{\lambda_{d2m}\overline{B_m}\ \overline{(C_2^t)^2}}{2(1 - \lambda_{d2m}\overline{B_m}\ \overline{C_2^t})} + \frac{\overline{B_m^{(2)}}\ \overline{C_2^t}}{2\overline{B_m}(1 - \lambda_{d2m}\overline{B_m}\ \overline{C_2^t})} + \frac{\overline{V_2^2}}{2\overline{V_2}} \end{aligned} \tag{11.39}$$

Mean values of the end-to-end delay for local traffic, bridge delay, and end-to-end delay for non-local traffic are the same as calculated in Section 11.2.

Stability considerations

Stability of scatternet operation means that all the relevant queues are stable, i.e., that the mean rate of servicing the packets in each queue is larger than the mean packet arrival rate. This applies to uplink queues at each slave, downlink queues at the master, and the bridge and master outgoing queues alike.

Stability conditions for the slave uplink queues can be derived as follows. Let us assume that the number of serviced packets per slave in one piconet cycle is limited to M, in which case the mean number of packet arrivals in the uplink queue during the same interval must be less than M. If we consider uplink slave's queue and downlink master's queue in parallel and let $M \to \infty$, we get:

$$(\lambda_{u1} + \lambda_{d1})\overline{BL}(m_1 - 2)\sum_{n=0}^{\infty} p_n \left(1 + \frac{n}{n + T_1}\right) < 1 \tag{11.40}$$

This stability criterion is tighter than the one given by the equations for the access delay,

$$\begin{aligned} \lambda_{u1}\ \overline{B}\ \overline{F_s} &< 1 \\ \lambda_{d1}\ \overline{B}\ \overline{F_s} &< 1 \end{aligned} \tag{11.41}$$

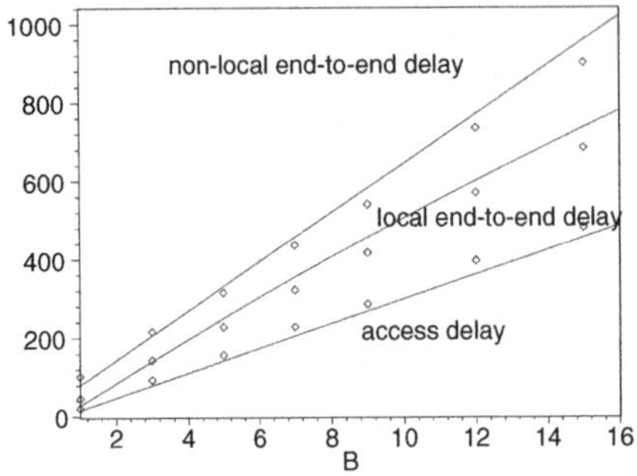

(a) Limited service scheduling.

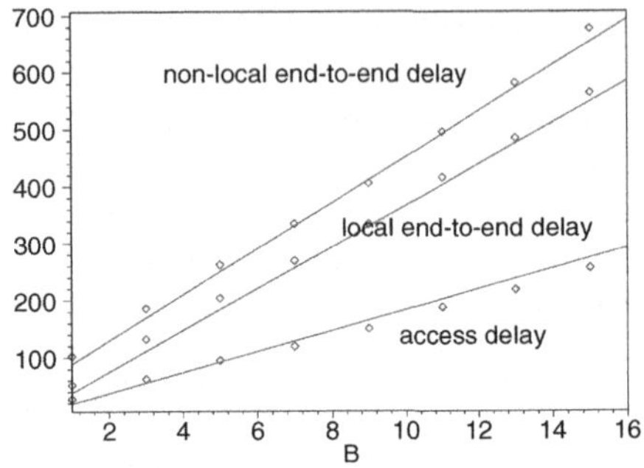

(b) Exhaustive service scheduling.

FIGURE 11.2
MS bridge: delays as functions of mean packet burst length under constant aggregate packet arrival rate. (From J. Mišić and V. B. Mišić, "Bridges of Bluetooth county: topologies, scheduling, and performance," *IEEE J. Select. Areas Commun. – Wireless Series* **21**(2):240–258, ©2004 IEEE. Reprinted with permission.)

which does not take into account the increase of the vacation time caused by the increase in packet burst arrival rates.

Under rendezvous-based scheduling, the stability condition for the bridge is

$$\lambda_{b12}\,\overline{B}\,\overline{F_b} < 1 \tag{11.42}$$

A similar set of conditions should hold in P_2. Note that $\overline{F_s}$ and $\overline{F_b}$ depend on the symmetry of the uplink and downlink packet burst arrival rates; their maximum value is $2\overline{L}$. The last stability condition can be rewritten as $(m_1-1)\lambda_{u1}(1-P_l)\overline{B}\,\overline{F_b} < 1$. Therefore, under low intensity of inter-piconet traffic the stability of slave queues and master downlink queues will be more critical than the stability of bridge queue and master queue toward the bridge. Under high intensity of inter-piconet traffic, the opposite holds.

11.3 Performance of the MS bridge

In order to assess the impact of various scatternet and traffic parameters on performance indicators such as access delay and end-to-end-delay, we have plotted analytical solutions for access and end-to-end delays. For all measurements, both piconets were assumed to have six packet-generating slaves ($m_1 = 7, m_2 = 8$), mean packet length was $\overline{L} = 3$ with $p_1 = p_3 = p_5 = 1/3$, and traffic locality was $P_l = 0.9$.

Analytical results were subsequently verified with a Bluetooth piconet simulator, built using the object-oriented Petri Net-based simulation engine Artifex by RSoft-Design, Inc. [RSoft Design, Inc., 2003], running on a Linux platform. The simulator

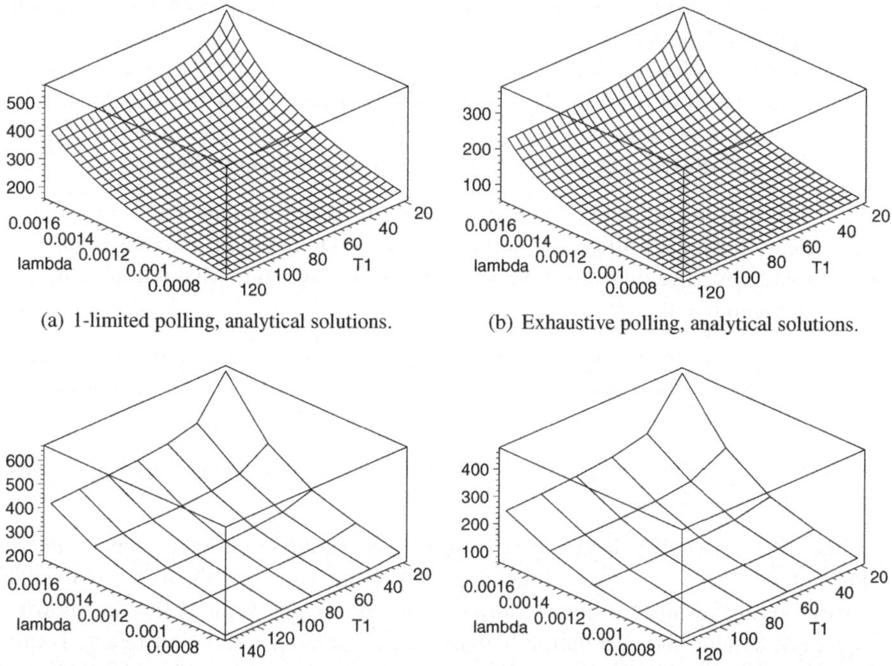

(a) 1-limited polling, analytical solutions.

(b) Exhaustive polling, analytical solutions.

(c) 1-limited polling, simulation results.

(d) Exhaustive polling, simulation results.

FIGURE 11.3
MS bridge: mean access delay as a function of burst arrival rate and time between bridge exchanges T_1. (From J. Mišić and V. B. Mišić, "Bridges of Bluetooth county: topologies, scheduling, and performance," *IEEE J. Select. Areas Commun. – Wireless Series* **21**(2):240–258, ©2004 IEEE. Reprinted with permission.)

operates at a MAC level, and it contains separate classes for slave devices with bursty packet generator, piconet masters (one of which contains the bridge logic), and a top-level scatternet class that integrates all other classes and provides measurement capabilities. In order to eliminate possible transient effects at system start-up, all measurements were taken after an initial warm-up delay needed to bring the system to a steady state. Since the simulation results correspond quite well to those obtained analytically, they will be presented and discussed together.

The impact of traffic burstiness is depicted by the diagrams in Fig. 11.2, where different delay variables are plotted as functions of the mean length of the burst, while keeping the uplink packet arrival rate per all slaves constant and equal to $\lambda\overline{B} = 0.015$. Continuous lines denote analytical solutions, while the diamonds show results obtained by simulation. The following observations may be made:

- As predicted, an increase in mean burst length corresponds to a nearly linear increase in all delay variables. In other words, the delays are almost linear functions of traffic burstiness, and smaller delays could be achieved by keeping the burst length as small as possible, which may be accomplished through appropriate segmentation mechanisms. Packet segmentation is another topic currently missing from the Bluetooth specification [Bluetooth SIG, 2001*b*] and, as such, it presents an interesting research topic; some preliminary results are reported in [Kalia et al., 1999; Kalia, Bansal and Shorey, 2000].

- The end-to-end delay for non-local traffic is higher than the corresponding delay for local traffic, as could be expected – non-local traffic has to pass one extra hop. The difference, however, does increase somewhat with the increase in mean burst length.
- Delays can reach rather high values: for values of mean burst length of 15 and above, end-to-end delays can reach values in the range of thousands of Bluetooth time slot intervals T (remember that $T = 0.625\mu s$).
- All three delay variables are smaller by about 30% when exhaustive service scheduling is used.

Fig. 11.3 shows the dependency of mean access delay on burst arrival rate and time interval between bridge exchanges, for mean burst length of $\overline{B} = 10$. The corresponding results for mean end-to-end delay for non-local traffic are shown in Fig. 11.4. Again, the correlation between analytical and simulation results is quite good, both in shape and in absolute values. Also, there appears to be a broad but not very pronounced minimum in the end-to-end delay for non-local traffic. Finally, it should be noted that mean end-to-end delay for non-local (i.e., inter-piconet) traffic can reach values in the range of seconds; similar results have been obtained for file transmissions in the single piconet case [Capone et al., 2001]. Although such long delays might seem unacceptable for individual packets, they may still be quite manageable in the context of file transmissions.

Mean non-local end-to-end delays are shown from a slightly different perspective in Fig. 11.5: i.e., as a function of traffic locality P_l and time between bridge exchanges T_1. Both analytical solutions (upper row) and simulation results (bottom row) show that the dependency is similar in shape to the previous one. This similarity should come as no surprise, as the decrease of the probability of local destinations actually leads to an increase of the number of packets that have to be routed through the bridge. This increase in turn leads to longer bridge exchanges and less time for servicing local slaves, hence the longer delays. Again, the exhaustive polling performs better than its 1-limited counterpart.

The relative advantage of exhaustive over 1-limited polling may be illustrated through the ratios of end-to-end delays for local and non-local traffic, as shown in Fig. 11.6. (Ratios of access delays has been omitted for brevity, and because they are less interesting anyway.) As can be seen, exhaustive service has about 15 to 20% advantage for a wide range of values of independent parameters. Limited service offers comparable or smaller delays only at very small values of T_1 and high values of λ – which are not very likely in practice. Non-local end-to-end delay is always better under exhaustive polling.

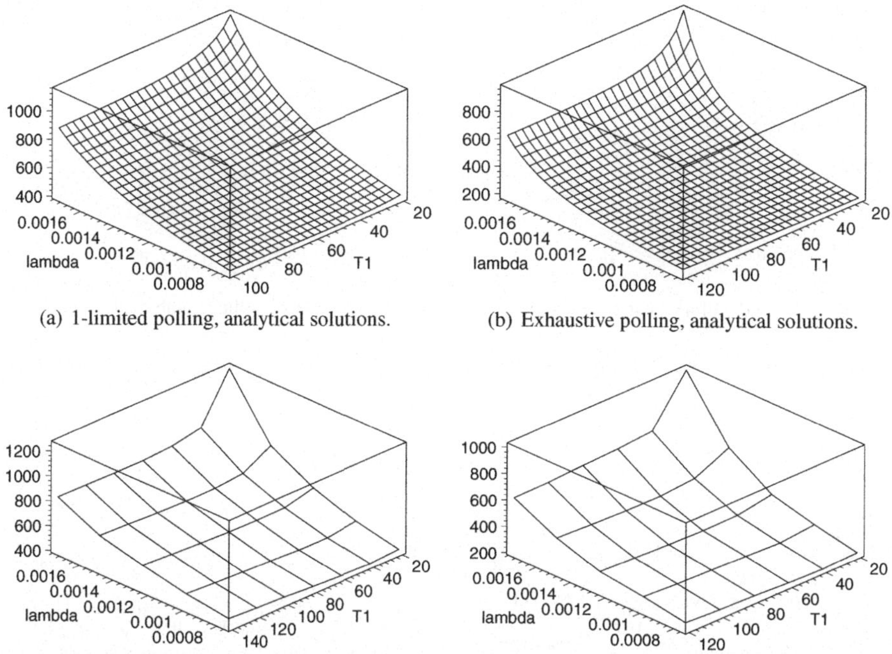

(a) 1-limited polling, analytical solutions.

(b) Exhaustive polling, analytical solutions.

(c) 1-limited polling, simulation results.

(d) Exhaustive polling, simulation results.

FIGURE 11.4
MS bridge: mean end-to-end delay for non-local traffic as a function of burst arrival rate and time between bridge exchanges T_1. (From J. Mišić and V. B. Mišić, "Bridges of Bluetooth county: topologies, scheduling, and performance," *IEEE J. Select. Areas Commun. – Wireless Series* **21**(2):240–258, ©2004 IEEE. Reprinted with permission.)

11.4 SS bridge topology

As this topology is symmetrical with respect to the piconets it contains, it suffices to consider the packet exchange between the bridge and one of piconet masters only. Under rendezvous-based scheduling, the bridge residence time in each piconet is fixed, and there is no guarantee that both the bridge queue and the outgoing queue of the corresponding piconet master will be completely empty at the end of the exchange. The bridge cycle time is equal to $T_1 + T_2$, and the PGF for the bridge cycle time in this case is

$$T_{cyc}(z) = z^{T_1+T_2} \tag{11.43}$$

We start by finding the probability distributions for the bridge exchange time and piconet cycle time. To facilitate the derivation, we will first assume that the bridge

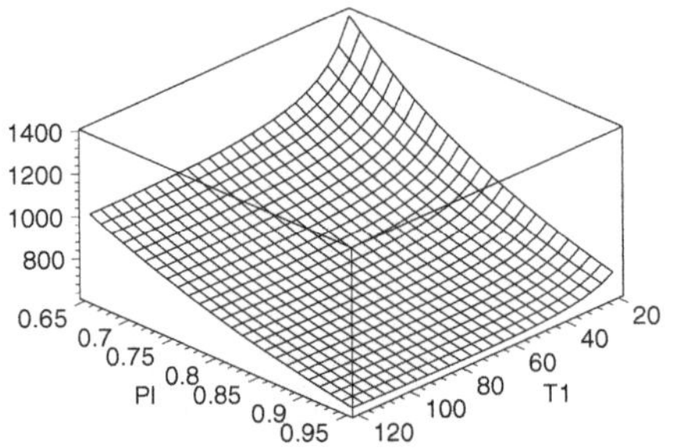

(a) 1-limited polling, analytical solutions.

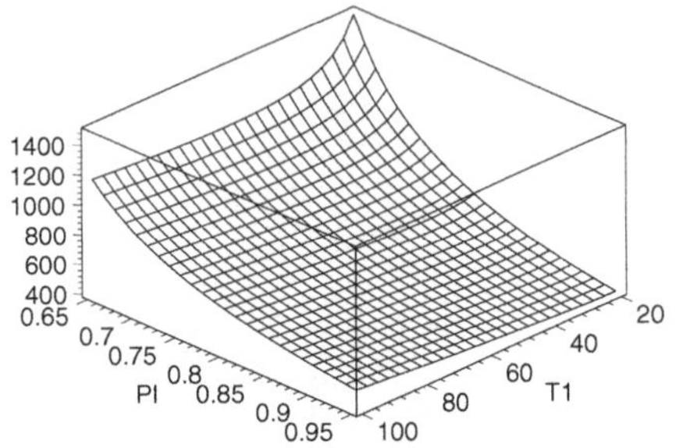

(b) Exhaustive polling, analytical solutions.

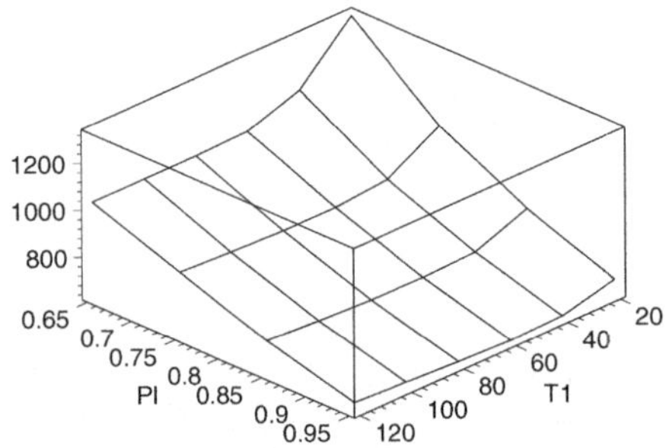

(c) 1-limited polling, simulation results.

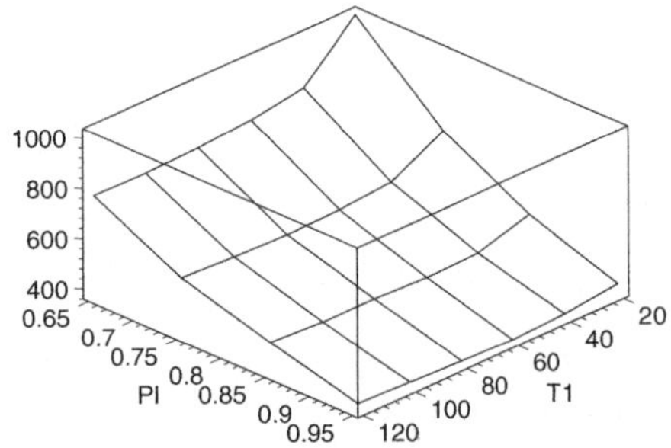

(d) Exhaustive polling, simulation results.

FIGURE 11.5
MS bridge: mean end-to-end delay for non-local traffic as a function of traffic locality, P_l, and time between bridge exchanges, T_1. (From J. Mišić and V. B. Mišić, "Bridges of Bluetooth county: topologies, scheduling, and performance," *IEEE J. Select. Areas Commun. – Wireless Series* **21**(2):240–258, ©2004 IEEE. Reprinted with permission.)

exchange is limited to T_1 (T_2) time slots, then evaluate its impact on the piconet cycle time, and finally derive the probability distribution for the piconet cycle time.

Bridge exchange time

We assume that the PDF of the bridge exchange time is independent of the intra-piconet polling scheme. This problem is similar to the problem of M/G/1 queue with vacations and gated time-limited service, which has been considered in [Leung and Eisenberg, 1989]. However, the results presented there cannot be directly used because of the coupling of bridge and master queues, and the existence of POLL/NULL packets. Instead, we have to proceed step by step, from packet bursts through packets to slots.

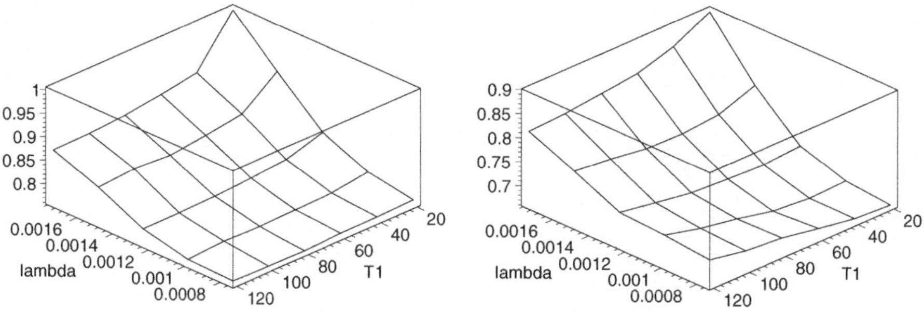

(a) Ratio of local end-to-end delays.

(b) Ratio of non-local end-to-end delays.

FIGURE 11.6
MS bridge: ratios of mean delays for exhaustive vs. 1-limited polling as functions of burst arrival rate and time between bridge exchanges, T_1 – simulation results only. (From J. Mišić and V. B. Mišić, "Bridges of Bluetooth county: topologies, scheduling, and performance," *IEEE J. Select. Areas Commun. – Wireless Series* **21**(2):240–258, ©2004 IEEE. Reprinted with permission.)

The probability that exactly l data packet bursts will be sent from P_1 to P_2 is

$$b_{l,12} = \sum_{i=0}^{\infty} P_{(T_{cyc}=i)} e^{-\lambda_{b12} i} \frac{(\lambda_{b12} i)^l}{l!} \tag{11.44}$$

where $P_{(T_{cyc}=i)}$ denotes the probability that the bridge cycle time is equal to k slots. The PGF for the number of packet burst arrivals to the bridge queue during the bridge cycle time is

$$\begin{aligned} B_{12}(y) &= \sum_{l=0}^{\infty} b_{l,12} y^l = \sum_{l=0}^{\infty} y^l \sum_{i=0}^{\infty} P_{(T_{cyc}=i)} e^{-\lambda_{b12} i} \frac{(\lambda_{b12} i)^l}{l!} \\ &= \sum_{i=0}^{\infty} P_{(T_{cyc}=i)} e^{-\lambda_{b12}(1-y)i} \\ &= T^*_{cyc}(\lambda_{b12} - y\lambda_{b12}) \end{aligned} \tag{11.45}$$

where $T^*_{cyc}(s) = \sum_{i=0}^{\infty} P_{(T_{cyc}=i)} e^{-si}$ denotes the moment generating function [Grimmett and Stirzaker, 1992] or discrete-time LST transform of the bridge cycle time.

Since the packet bursts are of random size described with the PGF $G_b(y)$, the PGF for the number of packet arrivals during the bridge cycle time is

$$A_{12}(y) = B_{12}(G_b(y)) = T^*_{cyc}\left(\lambda_{b12} - \lambda_{b12} G_b(y)\right) \tag{11.46}$$

The probability that exactly k data packets will be sent from the bridge to the master of P_2 is

$$
\begin{aligned}
a_{k,12} &= \frac{1}{k!}\frac{d^k}{dy^k}A_{12}(0) \\
&= \sum_{l=0}^{\infty}\frac{1}{k!}\frac{d^k}{dy^k}(G_b(y))^l_{|y=0}\sum_{i=0}^{\infty}P_{(T_{cyc}=i)}e^{-\lambda_{b12}i}\frac{(\lambda_{b12}i)^l}{l!} \\
&= \frac{1}{k!}\frac{d^k}{dy^k}\left(\sum_{i=0}^{\infty}P_{(T_{cyc}=i)}e^{-\lambda_{b12}i(1-G_b(y))}\right)_{|y=0} \\
&= \frac{1}{k!}\frac{d^k}{dy^k}T^*_{cyc}(\lambda_{b12}-\lambda_{b12}G_b(y))_{|y=0}
\end{aligned}
\tag{11.47}
$$

We note that the value $a_{k,12}$ is a linear combination of the first k moments of the bridge cycle time distribution and the probability of no packet arrivals during the bridge cycle, which is equal to $T^*_{cyc}(\lambda_{b12})$. Expanding the expression (11.47), we obtain

$$
\begin{aligned}
a_{k,12} = T^*_{cyc}(\lambda_{b12})&\binom{k}{\max(n,n_1)}\lambda^n_{b12}(-1)^n T^{(n)}_{cyc} \\
&\cdot\sum_{n_1=1}^{k}\frac{d^{(n_1)}}{dy^{n_1}}G_b(0)\sum_{n_2=0}^{k-n_1}\frac{d^{(n_2)}}{dy^{n_2}}G_b(0)\cdots\sum_{n_k=0}^{k-n_1-n_2\ldots-n_{k-1}}\frac{d^{(n_k)}}{dy^{n_k}}G_b(0)
\end{aligned}
\tag{11.48}
$$

where $n = \sum_{i=1}^{k}\left\lceil\frac{n_i}{k}\right\rceil$, and the k-th moment of the bridge cycle time distribution is defined as

$$
T^{(k)}_{cyc} = \sum_{n=0}^{\infty}P_{(T_{cyc}=n)}n^k = (-1)^k\frac{d^k}{ds^k}T^*_{cyc}(s)\bigg|_{s=0} \tag{11.49}
$$

For completeness, we also assume that $\frac{d^{(0)}}{dy^0}G_b(0) = 1$ in (11.48).

By the same token, the probability of sending k data packets from the master of P_2 to the bridge in a single bridge cycle is

$$
a_{k,21} = \frac{1}{k!}\frac{d^k}{dy^k}T^*_{cyc}\left(\lambda_{b21}-\lambda_{b21}G_b(y)\right)\bigg|_{y=0} \tag{11.50}
$$

As before, whenever the numbers of exchanged data packets in two directions differ, POLL or NULL packets must be inserted, and the bridge exchange terminates with an empty frame. The PGF for the total number of slots that have been generated during the bridge cycle is

$$
\begin{aligned}
T_r(z) &= \sum_{k=0}^{\infty}a_{k,12}a_{k,21}G_p(z)^{2k}z^2 \\
&+\sum_{j=1}^{\infty}\sum_{k=0}^{\infty}(a_{k,12}a_{k+j,21}+a_{k+j,12}a_{k,21})\cdot G_p(z)^{2k+j}z^{j+2}
\end{aligned}
\tag{11.51}
$$

The first term in (11.51) denotes the probability that both packet queues (the outgoing queue at the master and the corresponding outgoing queue at the bridge) have the same number of packets. The second term corresponds to the case where these queues have different lengths, in which case empty packets must be inserted in some frames in order to service all of the data packets. The term z^2 in (11.51) stands for the POLL/NULL frame that terminates the exchange. Therefore, the probability that the exchange takes exactly i slots is $c_i = \frac{1}{i!}\frac{d^i}{dz^i}T_r(0)$.

Now, the exchange may be terminated because of the pending rendezvous before all the queued packets are actually exchanged. The 'leftover' packets will have to be serviced in the next exchange, together with the 'fresh' ones that have arrived during the bridge residence in the other piconet. The probability q_i of having a total of i slots worth of queued packets at the beginning of an exchange (the 'unfinished exchange work') is

$$q_i = \sum_{j=0}^{T_1-1} q_j c_i + \sum_{j=T_1}^{T_1+i} q_j c_{i-j+T_1} \tag{11.52}$$

By multiplying both sides with z^i and summing over i, we obtain

$$Q(z) = \sum_{i=0}^{\infty} q_i z^i = \sum_{i=0}^{\infty} z^i \sum_{j=0}^{T_1-1} q_j c_i + \sum_{i=0}^{\infty} z^i \sum_{j=T_1}^{T_1+i} q_j c_{i-j+T_1} \tag{11.53}$$

By manipulating the order of summation, we obtain

$$Q(z) = \frac{z^{T_1} T_r(z) \sum_{j=0}^{T_1-1} q_j (1 - z^j)}{z^{T_1} - T_r(z)} \tag{11.54}$$

In order to find q_j, $j = 0 \mathinner{..} T_1 - 1$, we follow the approach outlined by Takagi [1991] for solving G-limited systems. Namely, since $Q(z)$ has to be analytic function of $|z| \in (0, 1)$, the denominator and numerator must have identical roots. By Rouché's theorem [Bak and Newman, 1982], the denominator of $Q(z)$ has exactly T_1 zeros. One of the zeros is $z_0 = 1$, and the remaining $T_1 - 1$ can be determined by applying Lagrange's theorem [Whittaker and Watson, 1952]:

$$z_m = \sum_{n=1}^{\infty} \frac{e^{2\pi mn\sqrt{-1}/T_1}}{n!} \frac{d^{n-1}}{dz^{n-1}} \left(T_r(z)^{n/T_1} \right) |_{z=0} \tag{11.55}$$

for $m = 1, 2, \mathinner{..} T_1 - 1$. Having found these zeros, we can set a total of $T_1 - 1$ equations by substituting the values for z_m in the numerator of $Q(z)$:

$$z_m^{T_1} \sum_{j=0}^{T_1-1} q_j (1 - z_m^j) T_r(z_m) = 0 \tag{11.56}$$

The last equation is obtained from the condition $Q(1) = 1$ which follows from applying l'Hôpital's rule to $Q(z)$. In practice, the summation in (11.55) can go only up to some finite value of n, and the small imaginary parts can be neglected in the solutions.

Once we know the probability distribution of unfinished bridge exchange work at the beginning of the bridge exchange, the PGF for the duration of the bridge exchange becomes

$$T_{rl}(z) = \sum_{n=0}^{T_1} p_n z^n = \sum_{i=0}^{T_1-1} q_i z^i + \sum_{i=T_1}^{\infty} q_i z^{T_1} \tag{11.57}$$

The PGF for length of the frame is

$$\begin{aligned} F_b(z) &= \sum_{k=0}^{\infty} a_{k,12} a_{k,21} \left(G_p(z)^{2k} z^2\right)^{1/(k+1)} \\ &+ \sum_{j=1}^{\infty} \sum_{k=0}^{\infty} (a_{k,12} a_{k+j,21} + a_{k+j,12} a_{k,21}) \cdot \left(G_p(z)^{2k+j} z^{j+2}\right)^{1/(k+j+1)} \end{aligned} \tag{11.58}$$

As before, the frame length is not equal to twice the data packet size because of the need to use empty packets to balance the difference in the numbers of data packets, as well as because of the terminating empty frame.

In order to determine the probability distribution for the delay from the moment the packet arrives to master queue toward the bridge, to the moment when it leaves the queue, we will use the principle of stochastic decomposition property. This principle states (in the case of work conserving disciplines) that the amount of work at an arbitrary time epoch in the M/G/1 vacation model, is distributed as the sum of the amount of work at an arbitrary epoch in an M/G/1 queue without vacations, and the amount of work at an arbitrary epoch during a vacation. Therefore, the LST for the packet delay at the bridge queue may be expressed as

$$\begin{aligned} W^*_{m1b}(s) &= \frac{s(1-\lambda_{b12}\overline{F_b B})}{s-\lambda_{b12}+\lambda_{b12} G_b(F^*_b(s)} \cdot \frac{1-G_b(F^*_b(s))}{\overline{B}(1-F^*_b(s))} \\ &\cdot \frac{1-V^*_T(s)}{s\overline{V_T}} \cdot \frac{Q\left(1-\dfrac{s}{\lambda_{b12}\overline{B}-s\overline{B}+s}\right)}{V^*_T(s)} \end{aligned} \tag{11.59}$$

where $V^*_T(s) = \dfrac{T^*_{cyc}(s)}{T^*_{rl}(s)}$. The first term in this expression corresponds to the time needed to serve the first packet in the burst in the $M^{[x]}/G/1$ system. The second term corresponds to the time needed to serve the given target packet in the burst. The third term corresponds to the time needed to serve packets which arrive during the vacation, but before the target burst. Finally, the last term corresponds to time needed to serve packets which were already in the uplink queue when the vacation has started.

Mean delay in the master queue toward the bridge is $\overline{W_{m1b}} = -W^{*'}_{m1b}(0)$, which amounts to

$$\overline{W_{mb1}} = \frac{\lambda_{b12}\overline{B}(\overline{F_b^2})}{(1-\lambda_{b12}\overline{F_b}\overline{B})} + \frac{\overline{B^{(2)}}\,\overline{F_b}}{2\overline{B}(1-\lambda_{b12}\overline{F_b}\overline{B})} + \frac{\overline{V_T^2}}{2\overline{V_T}} - \overline{V_T} + \frac{Q'(1)}{\lambda_{b12}\overline{B}Q(1)} \tag{11.60}$$

where $\overline{V_T^2} = V_T^{*''}(0)$.

The delay in the bridge queue toward P_2 is equal to the delay in the master queue toward the bridge – but in piconet P_1: $W^*_{m1b}(s) = W^*_{a1b}(s)$.

Piconet cycle time

In order to determine the mean access delay, we need the mean duration of piconet service cycle, the time interval for the piconet master to service all of its slaves once. Again, due to the symmetry of the topology, it suffices to consider just one master-slave channel in one piconet, modeled as the pair of queues with burst arrival rates of $\lambda_{u1} = \lambda$ and $\lambda_{d1} = \lambda P_l + \lambda_{b21}/(m_1 - 1)$ for the uplink and downlink queue, respectively.

In this section we will use the approach similar to the one used for MS bridge in Section 11.1. Let the PGF of the pure number of slots in the piconet cycle in P_1, conditioned on the bridge exchanges of length n, be $C_1^p(z)|n = \sum_{i=0}^{\infty} r_i^{(n)} z^i$. The PGF of the total piconet time (which includes bridge exchanges) will be denoted as $C_1^t(z)$ and derived later. Also, let $S_1^p|n$ denote the single channel service time (i.e., the time to empty both uplink and downlink queues for one particular slave channel without counting the exchange slots with the bridge), conditioned with the bridge exchange time equal to n slots; the corresponding conditional PGF will be denoted as $S_1^p(z)|n$.

We should note that, within one bridge cycle of length T_1+T_2, the local time which the master devotes to its ordinary slaves is $T_l = T_1 + T_2 - n$. Let us assume that a pure piconet cycle lasts for i slots with probability $r_i^{(n)}$. Now, consider a super-cycle that consists of $T_l i$ time slots. During this super-cycle, all combinations of relative positions of beginnings of the piconet cycle and the beginnings of the master's local operation will be repeated k times, where $k = 1..(T_l i)/lcm(T_l, i)$ and $lcm(T_l, i)$ is the least common multiple for the current value of piconet cycle i and the duration of master's local operation time T_l. Obviously, the super-cycle will contain $n_p = T_l$ piconet cycles without bridge interruptions, as well as $n_b = i$ intervals of master's local operation.

The piconet cycle can be interrupted either $\left\lceil \frac{i}{T_l} \right\rceil$ times or $\left\lfloor \frac{i}{T_l} \right\rfloor$ times; those two values will be the same if $\frac{i}{T_l}$ is an integer. The probability that the piconet cycle will be interrupted $\left\lceil \frac{i}{T_l} \right\rceil$ times is $pp_{i,\left\lceil \frac{i}{T_l} \right\rceil} = \dfrac{n_b - n_p \left\lfloor \frac{n_b}{n_p} \right\rfloor}{n_p}$. By the same token, the probability that the piconet cycle will be interrupted $\left\lfloor \frac{i}{T_l} \right\rfloor$ times is $pp_{i,\left\lfloor \frac{i}{T_l} \right\rfloor} = 1 - \frac{n_b - n_p \left\lfloor \frac{n_b}{n_p} \right\rfloor}{n_p}$.

Then, the PGF for the total length of the piconet cycle in the presence of bridge exchanges, which occur after every T_l slots of local operation and last for n slots, is

$$C_1^t(z)|n = \sum_{i=0}^{\infty} r_i^{(n)} \left(pp_{i,\left\lfloor \frac{i}{T_l} \right\rfloor} z^{i+n\left\lfloor \frac{i}{T_l} \right\rfloor} + pp_{i,\left\lceil \frac{i}{T_l} \right\rceil} z^{i+n\left\lceil \frac{i}{T_l} \right\rceil} \right) \tag{11.61}$$

Exhaustive polling

Let us now determine $C_1^p(z)|n$. The probability that exactly k packets are exchanged between the slave and the master (without taking into account the packets from the bridge exchange), may be obtained as

$$a_{k,u1}|n = \frac{1}{k!} \frac{d^k}{dy^k} C_1^{t*}(\lambda_{u1} - \lambda_{u1} G_b(y))|_{y=0}|n \tag{11.62}$$

where $C_1^{t*}(\lambda_{u1} - \lambda_{u1} G_b(y))|n$ denotes the PGF for the number of packet arrivals in the slave uplink queue during the total piconet cycle time. By the same token, the conditional probability that k packets are exchanged between the master and the slave are

$$a_{k,d1}|n = \frac{1}{k!} \frac{d^k}{dy^k} C_1^{t*}(\lambda_{d1} - \lambda_{d1} G_b(y))|_{y=0}|n \tag{11.63}$$

By combining all these elements together, the conditional PGF for the channel service time becomes

$$\begin{aligned} S_1^p(z)|n &= \sum_{k=0}^{\infty} (a_{k,u1}|n)(a_{k,d1}|n) G_p(z)^{2k} z^2 \\ &+ \sum_{j=1}^{\infty} \sum_{k=0}^{\infty} \left((a_{k,u1}|n)(a_{k+j,d1}|n) + (a_{k+j,u1}|n)(a_{k,d1}|n) \right) G_p(z)^{2k+j} z^{(j+2)} \end{aligned} \tag{11.64}$$

Finally, the conditional PGF of the piconet cycle time becomes

$$C_1^p(z)|n = (S_1^p(z)|n)^{m_1-2} \tag{11.65}$$

In order to solve it, we will have to truncate the $C_1^p(z)|n$ to i_{max} members $r_i^{(n)}$; this is an approximate solution, but it should provide satisfactory results for large i_{max}. Then, we have to set i_{max} equations in the form $r_i^n = \frac{1}{i!} \frac{d^i}{dz^i} C_1^p(z)|n|z = 0$. In each equation i, on the left side we will have $r_i^{(n)}$ and on the right side we will have function of all $r_i^{(n)}$ mass probabilities.

When the PGF for the pure piconet cycle is known, the PGF for the total piconet cycle can be found using (11.61). The unconditional PGF for the total piconet cycle time can be obtained by using the mass probabilities for the bridge exchange distribution from the PGF $T_r(z) = \sum_{n=0}^{\infty} p_n z^n$, which is actually derived in the form given in (11.51). This unconditional PGF of the total piconet cycle time can then be approximated with

$$C_1^t(z) = \sum_{n=0}^{n_{max}} p_n C_1^t(z)|n \tag{11.66}$$

where n_{max} is a sufficiently large value which depends on the intensity of inter-piconet traffic.

The PGF for the frame length in the regular master-slave exchange will be

$$F_s(z) = \sum_{n=0}^{T_1} p_n F_s(z)|n \tag{11.67}$$

where p_n are mass probabilities that the bridge exchange lasts n time slots, and

$$\begin{aligned} F_s(z)|n = & \sum_{k=0}^{k_{max}} (a_{k,u1}|n)(a_{k,d1}|n)(G_p(z)^{2k} z^2)^{1/(k+1)} \\ & + \sum_{j=1}^{j_{max}} \sum_{k=0}^{k_{max}} \left((a_{k,u1}|n)(a_{k+j,d1}|n) + (a_{k+n,u1}|n)(a_{k,d1}|n)\right) \\ & \cdot (G_p(z)^{2k+j} z^{j+2})^{1/(k+j+1)} \end{aligned} \tag{11.68}$$

where k_{max} and j_{max} are determined according to the accuracy required. The mean frame length will be $\overline{F_s} = F_s'(1)$.

As in the case of the MS bridge, the vacation time is the time while the master is busy servicing other slave queues. The duration of the vacation period V_1 may be described with the following PGF

$$V_1(z) = \sum_{n}^{T_1} p_n (S_1^t(z)|n)^{m_1 - 2} \tag{11.69}$$

and its first and second moments are $\overline{V_1} = V_1'(1)$ and $\overline{V_1^2} = V_1''(1) + V_1'(1)$, respectively.

1-limited polling

For the case of 1-limited intra-piconet polling we also have to determine $C_1^p(z)|n$. The probabilities that only NULL packet, or only one data packet is exchanged between the slave and the master (without taking into account the packets from the bridge exchange), may be obtained as

$$\begin{aligned} a_{0,u1}|n &= 1 - \lambda_{u1}\overline{B}\,\overline{C_1^t|n} \\ a_{1,u1}|n &= \lambda_{u1}\overline{B}\,\overline{C_1^t|n} \end{aligned} \tag{11.70}$$

By the same token, the conditional probabilities that only POLL packet or data packet is exchanged between the master and the slave respectively are:

$$\begin{aligned} a_{0,d1}|n &= 1 - \lambda_{d1}\overline{B}\,\overline{C_1^t|n} \\ a_{1,d1}|n &= \lambda_{d1}\overline{B}\,\overline{C_1^t|n} \end{aligned} \tag{11.71}$$

By combining these elements together, the conditional PGF for the pure channel service time becomes

$$\begin{aligned} S_1^p(z)|n = {} & (a_{0,u1}|n)(a_{0,d1}|n)z^2 \\ & +((a_{1,u1}|n)(a_{0,d1}|n) + (a_{0,u1}|n)(a_{1,d1}|n))G_p(z)z \\ & +(a_{1,u1}|n)(a_{1,d1}|n)G_p(z)^2 \end{aligned} \tag{11.72}$$

Finally, the conditional PGF for the pure piconet cycle time becomes

$$C_1^p(z)|n = (S_1^p(z)|n)^{m_1-1} \tag{11.73}$$

and we can solve it in the similar way to the one indicated for the exhaustive service.

When the PGF for the pure piconet cycle time is known, the PGF for the total piconet cycle time can be found from (11.14). Then, the unconditional PGF for the total piconet cycle time can be obtained by using the mass probabilities for the bridge exchange distribution from the PGF $T_r(z) = \sum_{n=0}^{\infty} p_n z^n$, which is actually derived in the form given by (11.9). This unconditional PGF of the total piconet cycle time can be approximated with

$$C_1^t(z) = \sum_{n=0}^{n_{max}} p_n C_1^t(z)|n \tag{11.74}$$

where n_{max} is some sufficiently large value that depends on the value of T_1 and the intensity of inter-piconet traffic.

The PGF for the frame length for the regular slave-master exchange in the case of 1-limited service is the same as the channel time i.e.:

$$F_s(z) = S_1^p(z) \tag{11.75}$$

The duration of the vacation period V_1 experienced by the slave has the same PGF as given in expr. (11.69).

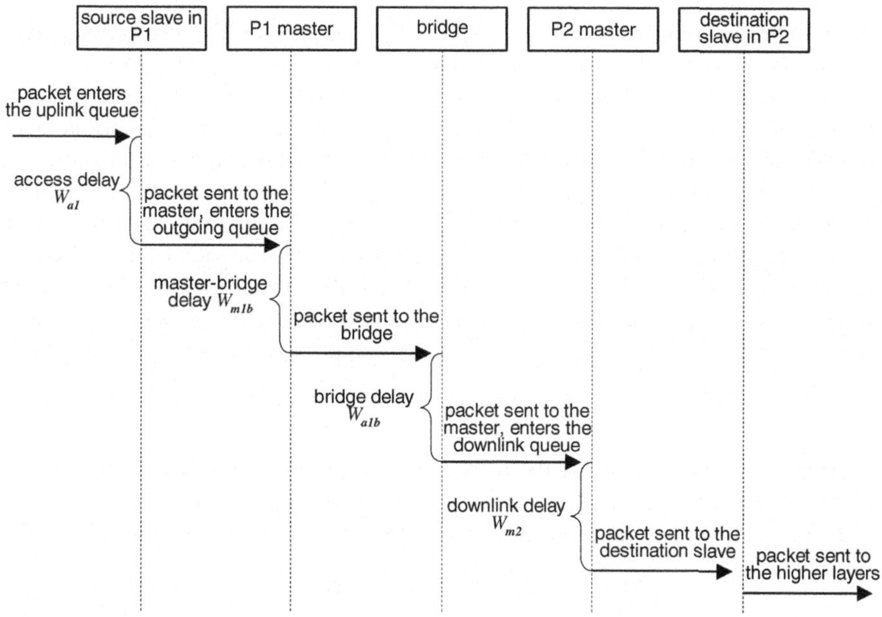

FIGURE 11.7
Components of non-local end-to-end delays in the scatternet with an SS bridge.

11.5 Packet delays: the SS bridge case

In this section we will calculate access delay and end-to-end delay for local and non-local traffic for two piconets interconnected by the SS bridge.

In case of exhaustive intra-piconet polling, we can obtain the corresponding LST for the access delay in piconet P_1 as

$$W^*_{a1}(s) = \frac{1 - V_1^*(s)}{s\overline{V_1}} \cdot \frac{1 - G_b(F_s^*(s))}{\overline{B}\,(1 - F_s^*(s))} \cdot \frac{s(1 - \lambda_{u1}\overline{B}\,\overline{F_s})}{s - \lambda_{u1} + \lambda_{u1} G_b(F_s^*(s))} \tag{11.76}$$

Mean access delay can now be obtained as $\overline{W_{a1}} = -(W^*_{a1})'(0)$, which evaluates to

$$\overline{W_{a1}} = \frac{\lambda_{u1}\overline{B}\,\overline{F_s^2}}{2(1 - \lambda_{u1}\,\overline{B}\,\overline{F_s})} + \frac{\overline{B^{(2)}}\,\overline{F_s}}{2\overline{B}(1 - \lambda_{u1}\overline{B}\,\overline{F_s})} + \frac{\overline{V_1^2}}{2\overline{V_1}} \tag{11.77}$$

The LST for the delay at the master has the same form as for the access delay, $W^*_{a1}(s) = W^*_{m1}(s)$. The LSTs for both access and master delay in the piconet P_2 can be calculated from the expressions similar to (11.76).

As before, local and non-local end-to-end delays have to be considered separately. The local traffic will experience two delays only, the access delay and the queueing delay in the downlink queue: $\overline{W_{11}} = \overline{W_{a1}} + \overline{W_{m1}}$ and $\overline{W_{22}} = \overline{W_{a2}} + \overline{W_{m2}}$, for piconets P_1 and P_2, respectively. These components were shown in Fig. 11.1(a).

Non-local traffic will have as much as four hops to make and four delays to experience: access delay, delay in the master bridging queue, delay in the bridge queue toward the master of the other piconet, and finally the delay in that master's downlink queue; these components are shown in Fig. 11.7. The first and last component are the same as in the case of local traffic. The LST of the distribution of the waiting time at the bridge queue of P_1 master is given by (11.59), and the LST for the delay in the bridge queue $W^*_{a1b}(s) = W^*_{m1b}(s)$. Then, the overall LST of the probability distribution of end-to-end queueing delay for non-local traffic going from P_1 to P_2 may be written as $W^*_{12}(s) = W^*_{a1}(s)W^*_{m1b}(s)W^*_{a1b}(s)W^*_{m2}(s)$, and its mean value is

$$\overline{W_{12}} = \overline{W_{a1}} + \overline{W_{m1b}} + \overline{W_{a1b}} + \overline{W_{m2}} \tag{11.78}$$

Analogous expressions for the packets flowing in the opposite direction may be obtained with ease, due to the inherent symmetry of the scatternet.

In case of 1-limited intra-piconet polling, the LST for the access delay at the slave queue is

$$W^*_{a1}(s) = \frac{1 - V^*_1(s)}{s\overline{V_1}} \cdot \frac{s(1 - \lambda_{u1}\overline{B}\,\overline{C^t_1})}{s - \lambda_{u1} + \lambda_{u1}G_b(C^{t*}_1(s))} \cdot \frac{1 - G_b(C^{t*}_1(s))}{\overline{B}\,(1 - C^{t*}_1(s))} \tag{11.79}$$

where $V^*_1(s)$ and $C^{t*}_1(s)$ denote the LST of the vacation time and total cycle time probability distributions, respectively. The average access delay at the slave is then calculated as $\overline{W_{a1}} = -W^{*'}_{a1}(0)$, and its value is:

$$\overline{W_{a1}} = \frac{\lambda_{u1}\overline{B}\,\overline{(C^t_1)^2}}{2(1 - \lambda_{u1}\overline{B}\,\overline{C^t_1})} + \frac{\overline{B^{(2)}}\,\overline{C^t_1}}{2\overline{B}(1 - \lambda_{u1}\overline{B}\,\overline{C^t_1})} + \frac{\overline{V^2_1}}{2\overline{V_1}} \tag{11.80}$$

The delay in the downlink queue at the master can be calculated by taking into account the change in burstiness because of burst interleaving in downlink queues, as explained in Sec. 11.2 and its LST is:

$$W^*_{m1}(s) = \frac{1 - V^*_1(s)}{s\overline{V_1}} \cdot \frac{s(1 - \lambda_{d1}\overline{B_m}\,\overline{C^t_1})}{s - \lambda_{d1} + \lambda_{d1}G_{bm}(C^{t*}_1(s))} \cdot \frac{1 - G_{bm}(C^{t*}_1(s))}{\overline{B_m}\,(1 - C^{t*}_1(s))} \tag{11.81}$$

The mean end-to-end delay for local traffic will be the sum of access and master delay.

As mentioned above, non-local traffic will have as much as four hops to make and four delays to experience, as shown in Fig. 11.7. The first and the last component have already been calculated. The LST for the waiting time at the bridge queue of P_1 master is given by (11.59), and the LST for the delay in the bridge queue $W^*_{a1b}(s) = W^*_{m1b}(s)$.

The overall LST for the distribution of end-to-end delay of non-local traffic going from P_1 to P_2 has the form given in (11.78). Analogous expressions for the packets flowing in the opposite direction may be obtained with ease, due to the inherent symmetry of the scatternet. Both mean access delay and mean end-to-end delay may be shown to be linear functions of the mean burst length when the packet arrival rate (i.e., burst arrival rate times mean burst length) is kept constant.

Stability considerations

The stability conditions for the slave uplink queues can be derived as follows. If the number of serviced packets per slave in one piconet cycle is limited to M, the average number of packet arrivals in the slave's queue during the piconet cycle time must be less than M. Considering the uplink and downlink queue in parallel and letting $M \to \infty$, we get

$$(\lambda_{u1} + \lambda_{d1})\overline{B}(m_1 - 2)\left(1 + \frac{\overline{T_{rl}}}{T_{cyc}}\right)\overline{L} < 1 \tag{11.82}$$

Bridge stability is determined from the condition that the mean duration of the bridge exchange must be shorter than the bridge residence time in the piconet, $\overline{T_r} < T_1$.

Note that the different bridge cycle times will lead to different values of $\overline{F_s}$ and $\overline{F_b}$, and the actual stability limits for the SS bridge topology will differ from those for the MS bridge topology. However, the upper bound for both $\overline{F_s}$ and $\overline{F_b}$ is still $2\overline{L}$, which is sufficient to derive the limits on the total burst packet arrival rate at the bridge.

11.6 Performance of the SS bridge

As before, we have plotted the analytical solutions for different delay variables, and verified those solutions through simulation. The following parameter values were used, unless otherwise specified: each piconet had six active slaves, i.e., $m_1 = m_2 = 8$; burst arrival rate per slave, if fixed, was $\lambda_{u1} = \lambda = 0.0015$; mean packet length was $\overline{L} = 3$, with $p_1 = p_2 = p_3 = 1/3$; traffic locality was $P_l = 0.9$; and times between bridge exchanges, if fixed, were $T_1 = T_2 = 100T$, where $T = 0.625\mu s$ is the time slot of the Bluetooth clock.

In order to assess the impact of traffic burstiness, we have calculated and plotted the delay times as functions of mean burst size $\overline{B}$, as shown in Fig. 11.8. In both diagrams, analytical solutions are shown as continuous lines, while diamonds stand for results obtained through simulation. As predicted, the delays are nearly linear functions of mean burst size for both polling schemes, but delays are significantly lower when exhaustive polling is used.

The dependency of mean access delay on burst arrival rate λ_{u1} and time between bridge exchanges T_1 (we assume that $T_2 = T_1$) is shown in Fig. 11.9. Exhaustive

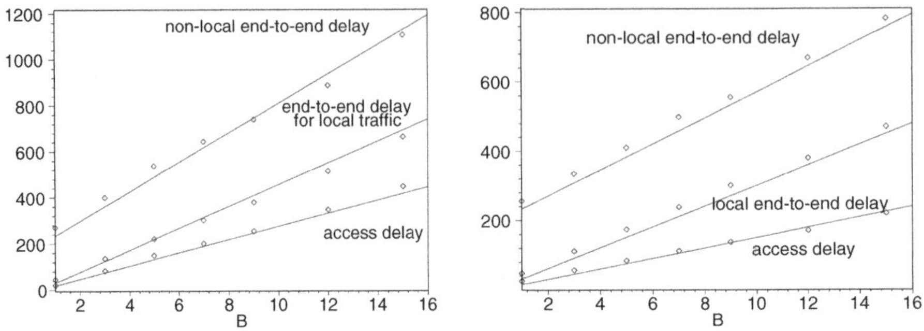

(a) 1-limited polling. (b) Exhaustive polling.

FIGURE 11.8
SS bridge: delays as functions of mean packet burst length, under constant aggregate packet arrival rate. (From J. Mišić and V. B. Mišić, "Bridges of Bluetooth county: topologies, scheduling, and performance," *IEEE J. Select. Areas Commun. – Wireless Series* **21**(2):240–258, ©2004 IEEE. Reprinted with permission.)

polling can be seen to offer a distinct advantage over the 1-limited polling, even though the shapes of the delay dependency as burst arrival rate and time T_1 change are quite similar. Note that very small values of T_1 lead to an increase in delay times. Again, the agreement between analytical solutions and simulation results is very good.

The next set of diagrams (Fig. 11.10) show the dependency of mean non-local end-to-end delay on burst arrival rate λ_{u1} and time between bridge exchanges T_1 (we assume that $T_2 = T_1$). We do not show the local end-to-end delay, since it would provide little extra information except for the fact that it is lower than the its non-local counterpart by about $T_1 + T_2$. As before, both polling schemes exhibit similar results; as before, exhaustive polling offers a distinct advantage over 1-limited polling; and as before, the agreement between analytical solutions and simulation results is very good.

Note that the end-to-end delay for non-local traffic increases with T_1, the time between bridge exchanges, especially at lower burst arrival rates, unlike its MS bridge counterpart where the rate of increase is barely noticeable. This is due to the fact that bridge exchanges (or, rather, the time instants in which they start) are spaced exactly T_1 apart, hence the non-local end-to-end delay virtually contains $2T_1$ as an additive component.

Fig. 11.11 shows the dependency of mean non-local end-to-end delay on traffic locality P_l and time between bridge exchanges T_1. As in the case of the scatternet with a MS bridge, this dependency is similar in shape to the previous one (end-to-end delay vs. burst arrival rate and time between bridge exchanges). Again, exhaustive polling performs noticeably better than 1-limited polling.

Finally, Fig. 11.12 shows the ratios of end-to-end delay times under exhaustive polling to those obtained under 1-limited polling, using simulation results. As can

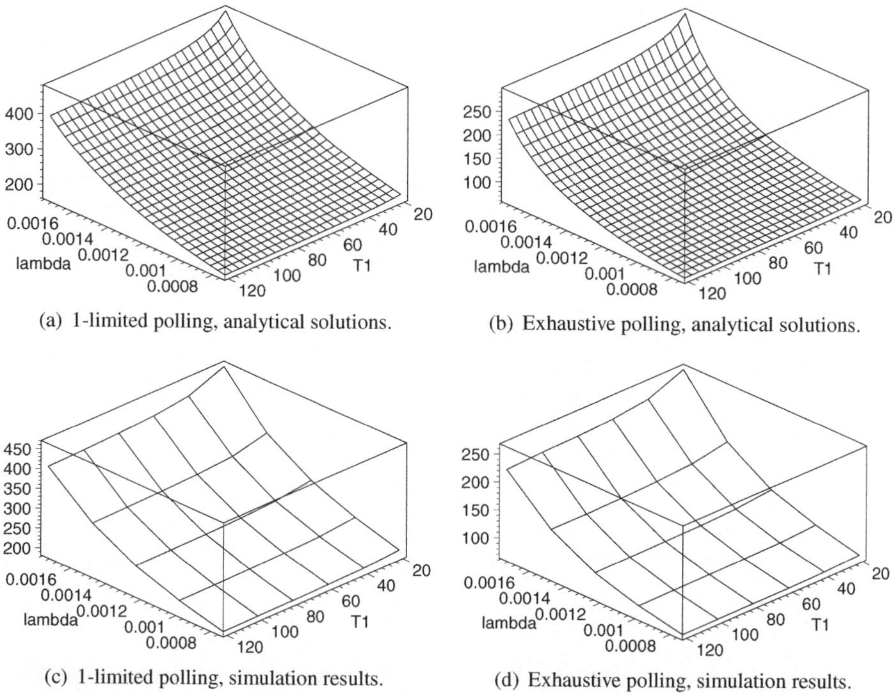

(a) 1-limited polling, analytical solutions.
(b) Exhaustive polling, analytical solutions.
(c) 1-limited polling, simulation results.
(d) Exhaustive polling, simulation results.

FIGURE 11.9

SS bridge: mean access delay as a function of burst arrival rate, $\lambda_{u1} = \lambda$, and time between bridge exchanges, T_1. (From J. Mišić and V. B. Mišić, "Bridges of Bluetooth county: topologies, scheduling, and performance," *IEEE J. Select. Areas Commun. – Wireless Series* **21**(2):240–258, ©2004 IEEE. Reprinted with permission.)

be seen, the exhaustive polling scheme outperforms 1-limited polling one by about 10 to 20% in terms of access delay, and about 13 to 17% in terms of end-to-end delay for non-local traffic, in the entire range of values of $\lambda_{u1} = \lambda$ and T_1 displayed on the diagram.

The results of our analysis may be summarized as follows.

First, performance of Bluetooth networks, expressed in terms of mean access delay and mean end-to-end delay, shows monotonic behavior with respect to different traffic parameters such as packet or burst arrival rate, traffic locality, and mean burst length. The dependence is by no means linear, though, and sharp increases of delay times may be experienced when one or more of the following conditions happen: high burst arrival rates, low probability of local traffic, and short intervals between bridge exchanges. Such behaviour could have been expected because the delays in the network are actually queueing delays.

Regardless of the topology and/or scheduling policy, delays are not too sensitive on the time interval between bridge exchanges. There is a caveat, though: this time

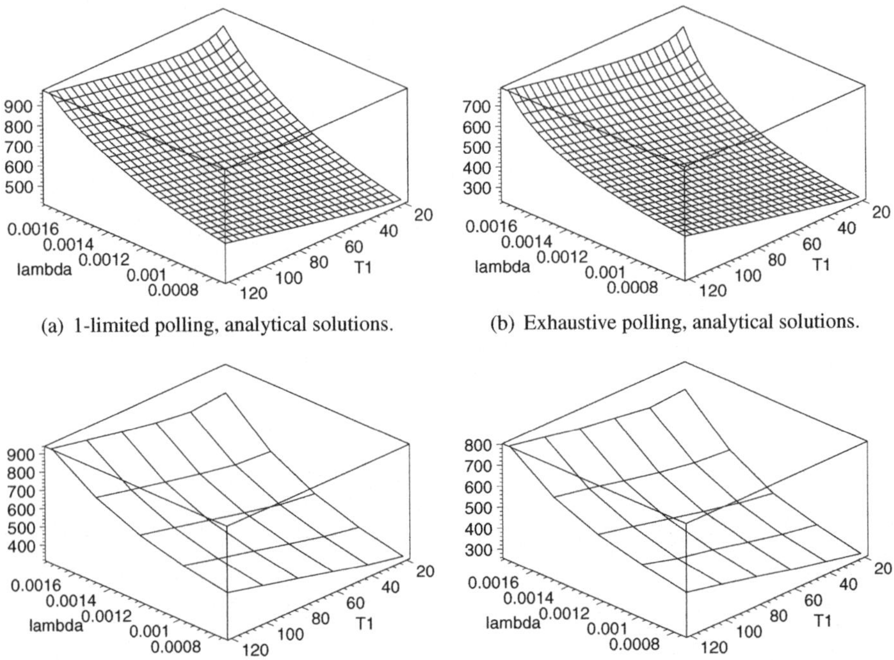

(a) 1-limited polling, analytical solutions.

(b) Exhaustive polling, analytical solutions.

(c) 1-limited polling, simulation results.

(d) Exhaustive polling, simulation results.

FIGURE 11.10

SS bridge: mean end-to-end delay for non-local traffic as a function of burst arrival rate, $\lambda_{u1} = \lambda$, and time between bridge exchanges, T_1. (From J. Mišić and V. B. Mišić, "Bridges of Bluetooth county: topologies, scheduling, and performance," *IEEE J. Select. Areas Commun. – Wireless Series* **21**(2):240–258, ©2004 IEEE. Reprinted with permission.)

interval should not be too small, otherwise there is not enough time to exchange any meaningful number of packets. Satisfactory results should be obtained with values in the range over 40 to $60T$. In general, the access delay and local end-to-end delay slowly decrease when T_1 increases, while the non-local end-to-end delay increases at a slightly higher rate than the other two. (This rate is higher in the topology with an SS bridge, for reasons explained in Sec. 11.6, but it tends to flatten out at higher burst arrival rates.) Still, the gradients are quite low and the exact value of T_1 does not seem to be critical for scatternet performance, although this issue probably deserves more attention in future research.

Second, with regard to scatternet topology, we may conclude that:

- Mean access delay is lower in the scatternet topology with an SS bridge, due to the fact that both piconet masters can spend more time servicing the slaves in their respective piconets. The same holds for mean end-to-end delay for local (i.e., intra-piconet) traffic.

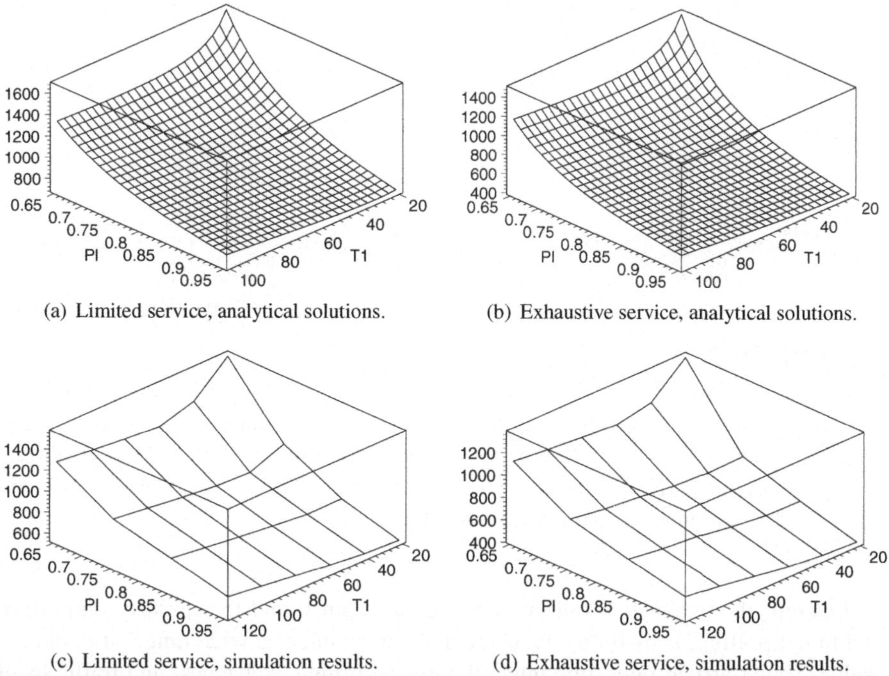

FIGURE 11.11

SS bridge: mean end-to-end delay for non-local traffic as a function of traffic locality, P_l, and time between bridge exchanges, T_1. (From J. Mišić and V. B. Mišić, "Bridges of Bluetooth county: topologies, scheduling, and performance," *IEEE J. Select. Areas Commun. – Wireless Series* **21**(2):240–258, ©2004 IEEE. Reprinted with permission.)

- On the other hand, mean end-to-end delay for non-local traffic is lower in the scatternet topology with an MS bridge, due to the lower number of hops that the packets have to pass: three, compared to four in the case of the topology with an SS bridge.

- The scatternet with an MS bridge is more sensitive to the combination of high packet burst arrival rate or low traffic locality, and too short time interval between bridge exchanges. Here 'more sensitive' should be taken to mean that the relative increase in delays is higher, or more significant, in the region where these conditions apply. Again, these are extreme conditions that are not likely to be encountered in practice.

- The performance difference between the two topologies is not high, however, and considerations other than delay performance might be given priority when deciding on the topology of a scatternet.

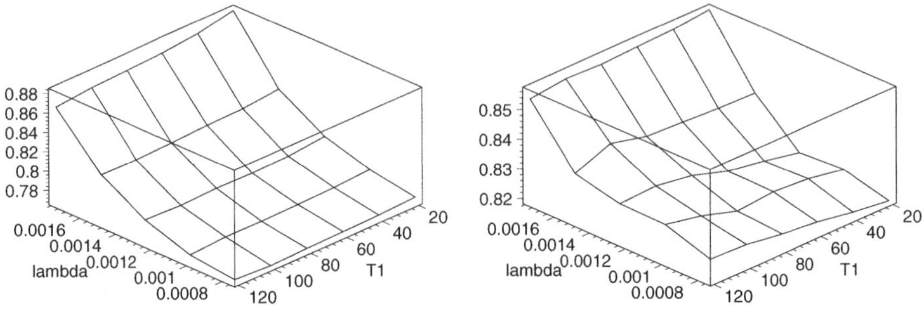

(a) Ratio of mean access delays.

(b) Ratio of end-to-end delays for non-local traffic.

FIGURE 11.12

SS bridge: ratio of delays in exhaustive vs. 1-limited polling as a function of burst arrival rate and time between bridge switches, T_1 – simulation results only. (From J. Mišić and V. B. Mišić, "Bridges of Bluetooth county: topologies, scheduling, and performance," *IEEE J. Select. Areas Commun. – Wireless Series* **21**(2):240–258, ©2004 IEEE. Reprinted with permission.)

Overall, the exhaustive polling scheme has been found to regularly outperform 1-limited polling, sometimes by as much as 30%, under a wide range of parameter values (burst arrival rate, time interval between bridge exchanges, and traffic locality). The difference gets smaller as burst arrival rate increases; at very high burst arrival rates 1-limited polling actually performs better than exhaustive polling. The actual mechanism of this behavior may be explained as follows. Under low loads, 1-limited polling will essentially 'waste' slots – i.e., there will be many empty POLL-NULL exchanges, or exchanges with a single data packet only and a POLL or NULL packet in the other direction. At the same time, exhaustive polling will 'pull out' all (or almost all) packets when a packet burst arrives. As the consequence, exhaustive service offers noticeably lower access delays as well as somewhat lower end-to-end delays. Under high loads, the proportion of 'wasted' slots decreases, and the performance of 1-limited polling improves relative to exhaustive polling. Exhaustive polling will still provide lower access delays, but the end-to-end delays will increase over those obtained with 1-limited polling. It should be noted that exhaustive polling does not guarantee fairness, unlike its 1-limited counterpart.

12

Adaptive bridge scheduling

Inter-piconet scheduling in a Bluetooth scatternet may be performed in an adaptive manner, so as to minimize end-to-end packet delays. In this chapter, we analyze the impact of different values of scatternet parameters on end-to-end delays and then investigate the possibility of minimization of the aforementioned delays. We show that such minimization is possible for scatternets with both Slave/Slave and Master/Slave bridges, and that it should be based on inter-piconet traffic. We also describe two simple algorithms that enable small scatternets to achieve near minimum delays in a wide range of traffic parameter values.

Using the previously derived expressions for end-to-end packet delays, in Section 12.1 we investigate whether those delays may be minimized through appropriate choice of some of scatternet parameters. Then, in Sections 12.2 and 12.3, we present two simple algorithms for adaptive inter-piconet scheduling in scatternets with an MS and SS bridge, respectively. The algorithms are referred to as Load Adaptive Master/slave Scheduling (LAMS) and Load Adaptive Slave/slave Scheduling (LASS) both of which dynamically adjust the time between bridge exchanges in order to minimize those delays and achieve near-optimal performance.

12.1 Minimization of delays

In Chapter 11 we have derived the expressions for end-to-end packet delays in simple scatternets with two piconets and a single bridge. Let us now investigate whether it would be possible to choose the values of different traffic and scatternet parameters so as to minimize one or the other end-to-end delay (or, preferably, both).

First, we note that the number of adjustable parameters is rather small. Some of them, such as packet burst arrival rate, burst length distribution, or traffic locality, are determined by the applications that communicate using Bluetooth, which is beyond control (at the MAC level). Others, such as mean burst length and packet length distribution, might be more amenable to control through appropriate segmentation and reassembly policies. These policies are handled at the higher layers of the Bluetooth protocol stack, and probably cannot be changed too frequently, let alone dynamically. Finally, the size of the scatternet may be controlled indirectly, through the scatternet formation algorithm(s). Once the scatternet is formed, the time to restructure it is long (sometimes even prohibitively long), and higher transmission delays might, in fact, be acceptable – although only as a lesser evil.

This leaves us with residence time T_1, or residence times T_1 and T_2 for the case of the SS bridge, as the only scatternet parameter that can be changed as often as necessary. Our objective is, then, to investigate (a) whether it is possible to modify the value of T_1 in order to reduce end-to-end packet delays, and (b) which of the delays we should strive to minimize – local or non-local, or maybe a combination of the two.

The answer to the first question is affirmative: the minimum of the local end-to-end delay may be found from $\partial \overline{W_{11}} / \partial T_1 = 0$, while keeping λ, $\overline{B}$, and P_l as parameters. Let us consider the scatternet with an SS bridge on account of its symmetry, and assume that $T_2 = T_1$. The loci of the solution of the resulting equation, together with the resulting delays, are shown in the left column of Fig. 12.1. Note that the orientation of λ and P_l axes in Figs. 12.1(a) and 12.1(e) have been changed for clarity.

If, on the other hand, we want to minimize the non-local end-to-end delay, we should solve the equation $\partial \overline{W_{12}} / \partial T_1 = 0$. The corresponding loci for T_1/T_2 and the resulting delays are shown in the right column of Fig. 12.1.

We note that similar results may be obtained for the MS bridge as well.

Two important observations can be made. First, values of T_1 in the low range of 10 to $40T$ generally minimize non-local delay, while values in the range of 50 to $400T$ lead to minimal local delays. The explanation is simple: local delay will be smaller if the local operation in the piconet goes on uninterrupted for as long as possible. In other words, local delay is minimized if bridge exchanges are performed as infrequently as possible. However, this causes very high delays for non-local traffic. Therefore, by minimizing the delay for local traffic we actually *penalize* non-local traffic – thus defeating the purpose with which the scatternet was formed in the first place.

On the other hand, shortening the time interval between bridge exchanges means that data packets with destinations in the other piconet are able to get through very quickly, thus reducing non-local delays. Frequent bridge exchanges do affect the performance of local traffic, as the piconet master has to stop servicing its own slaves. Yet the deterioration of delays is small, typically less than 10%, which should be perfectly acceptable in most cases.

The other important observation is that the optimum value for T_1/T_2 is not constant, but instead depends on packet arrival rate and traffic locality – in other words, the traffic load both within the piconets and through the bridge. No single value for T_1/T_2 can be found that works well in the entire range of burst arrival rates and traffic locality. This is the case even if some other performance measure, e.g., a weighted sum of mean local and non-local delays, were used as the basis for minimization [Mišić and Mišić, 2003*c*].

Similar conclusions may be made for the case of the MS bridge, except that there is only one bridge residence time (or, rather, the bridge absence time) to adjust. The resulting loci of T_1 that minimize the local and non-local delay are shown in Fig. 12.2; they do resemble the corresponding values obtained with the SS bridge, in

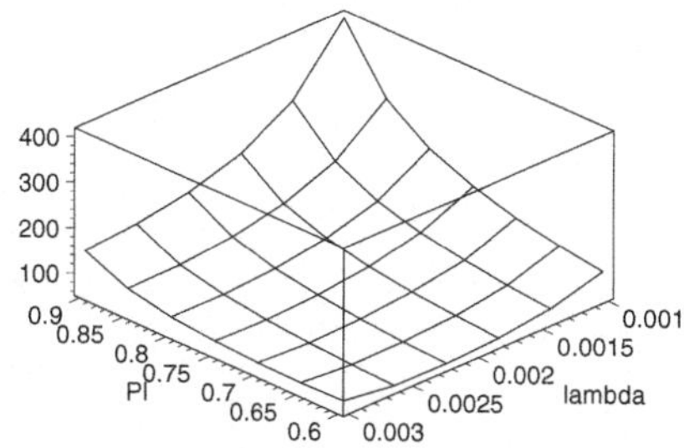

(a) Loci of T_1 values that minimize end-to-end delay for local traffic.

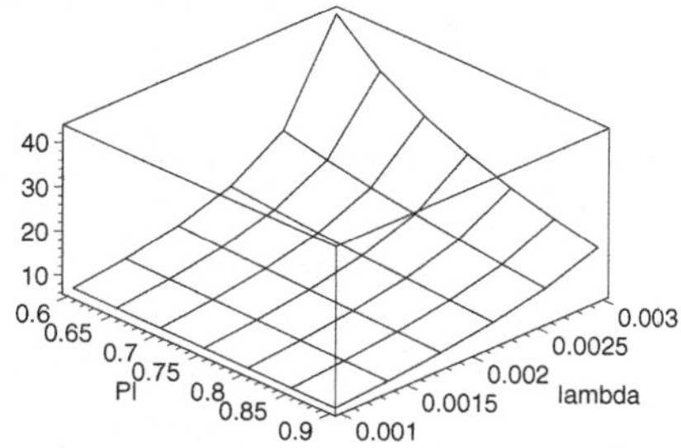

(b) Loci of T_1 values that minimize end-to-end delay for non-local traffic.

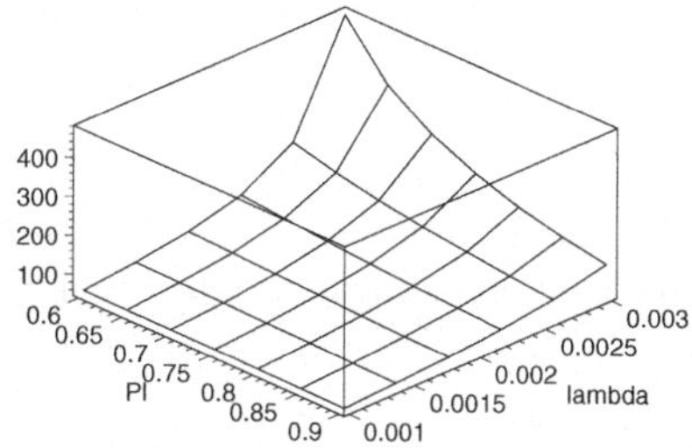

(c) Minimum end-to-end delay for local traffic.

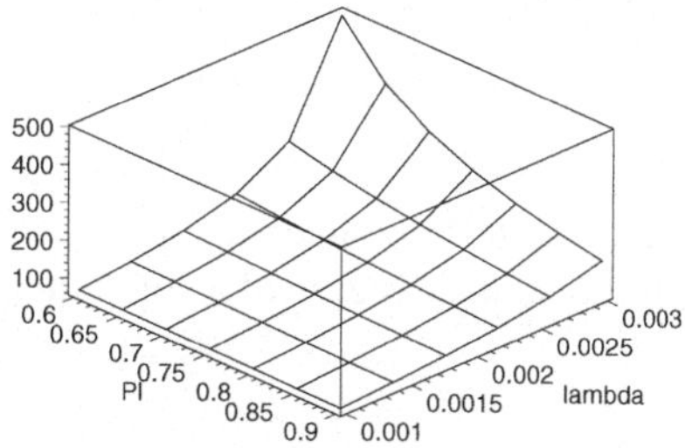

(d) End-to-end delay for local traffic when non-local delay is minimized.

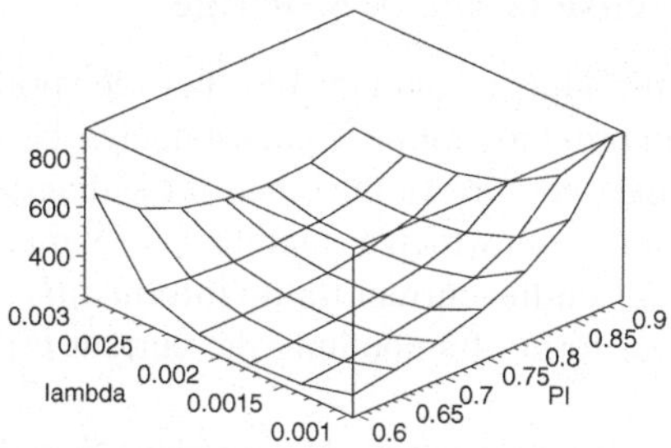

(e) End-to-end delay for non-local traffic when local delay is minimized.

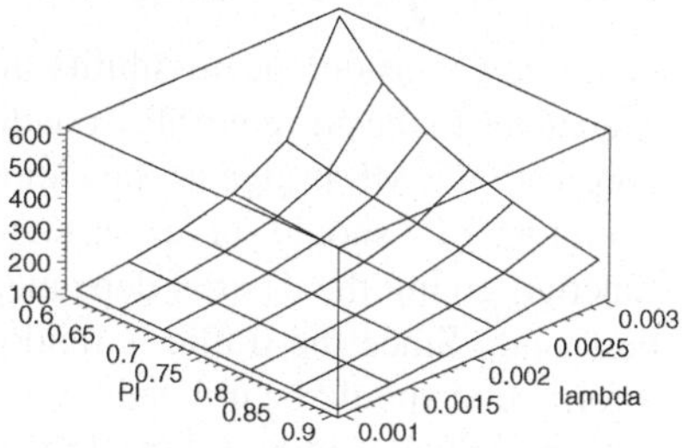

(f) Minimum end-to-end delay for non-local traffic.

FIGURE 12.1
Minimizing end-to-end delays in the scatternet with an SS bridge. (From V. B. Mišić and J. Mišić, "Adaptive inter-piconet scheduling in small scatternets," *ACM MC²R – Mobile Computing and Communications Review* **7**(2):45–58, © 2003 ACM. Reprinted with permission.)

qualitative terms at least. As with the SS bridge, no single value of T_1 works well in the entire range of usable packet burst arrival rate and traffic locality, local delays are minimized when bridge exchanges are performed as infrequently as possible,

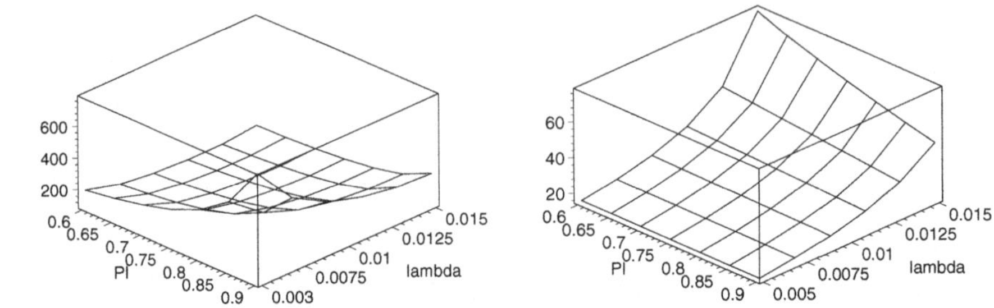

(a) Loci of T_1 values that minimize the delay for local traffic.

(b) Loci of T_1 values that minimize the delay for non-local traffic.

FIGURE 12.2
Loci of T_1 values that minimize end-to-end delay in the scatternet with an MS bridge.

and shorter time intervals between exchanges lead to improved delays for non-local traffic at the expense of a slight increase in the corresponding delays for local traffic.

12.2 Adaptive management: the case of the MS bridge

Let us now consider the possibility of dynamic adjustment of the time between bridge exchanges to the current traffic conditions in order to minimize packet delays, beginning with the MS bridge as the simpler of the two. One or the other of the diagrams from Fig. 12.2 should serve as a starting point, from which a suitable approximation function giving the desired dependency of T_1 on the current traffic information can be found. Since the traffic is random, some form of smoothing the current traffic information should be utilized.

One such algorithm, referred to as LAMS (Load Adaptive Master/slave Scheduling), has been described in [Mišić and Mišić, 2003*c*], and it operates as follows. In local operation, both piconet masters keep track of the instantaneous local traffic in their respective piconets; in addition, the bridge keeps track of its outgoing traffic.

When rendezvous comes, the bridge switches to the other piconet and the exchange begins. During the exchange, the bridge records the volume of incoming traffic.

When the exchange ends, the master of P2 informs the bridge of its local traffic. The bridge estimates the current traffic intensity and locality, finds the corresponding averages, and calculates the time to the next rendezvous using a suitable approximation function. The bridge and the master negotiate the HOLD mode using the calculated time interval as the hold timeout. The bridge then leaves the piconet P2 and returns to local operation, as does the master of P2.

This algorithm may be described with the pseudocode below.

```
// both piconets operate locally
bridge adds local packets to N_1
bridge adds non-local packets to N_out
master adds its local packets to N_2
// rendezvous time reached
// switch to exchange; record exchange data
add slots of incoming packets to N_in
...
// no more data, exchange ends
// master of P_2 sends N_1 to the bridge
// estimate current traffic intensity, locality
r = (N_2 + N_2 + N_out + N_in)/T_1
p = (N_1 + N_2)/(N_1 + N_in + N_2 + N_out)
ρ' = αr + (1 − α)ρ
π' = αp + (1 − α)π
// approximate time to next rendezvous
T_1 = a + b(ρ^2 − 1)/π
// request HOLD with hold time T_1
// master of P_2 acknowledges HOLD
// bridge switches back to local operation
reset slot counters
```

In the algorithm, the current traffic intensity, denoted with r, is obtained as the total number of data packets received during the local operation phase. The current traffic locality (denoted with p in order to distinguish it from P_l which is a long term mean value), measures the portion of the total traffic that passes through the bridge. Alternatively, data packet slots may be used, rather than simply packets, for a better approximation of time intervals in question.

In order to reduce the effects of traffic randomness, the algorithm uses Simple Exponential Smoothing [Delurgio, 1998] in which the average value ρ of traffic intensity r is obtained as $\rho' = \alpha r + (1-\alpha)\rho$, where ρ' is the new value of ρ, and α is the smoothing constant. The average traffic locality π is obtained (and periodically updated) in the same manner.

The dependency between the desired value of T_1 and the values of ρ' and π' has been obtained by approximating the loci from Fig. 12.2. The chosen function, $T_1 = a + b(\rho^2 - 1)/\pi$, is sufficiently simple so that even battery-limited devices should have no problem in evaluating it.

The performance of the LAMS algorithm has been verified through simulation; the corresponding diagrams are shown in Fig. 12.3. Approximation parameters were $\alpha = 0.1$ (the averaging does not need to remember too far back in the past), $a = 12T$ (this is also the minimum value for T_1, which is consistent with the upper bounds for synchronization delays outlined in Sec. 10.1), and $b = 24T$ (empirically obtained). The delays are quite close to the optimal values in a wide range of values of burst arrival rate λ, traffic locality P_l, and mean burst length $\overline{B}$.

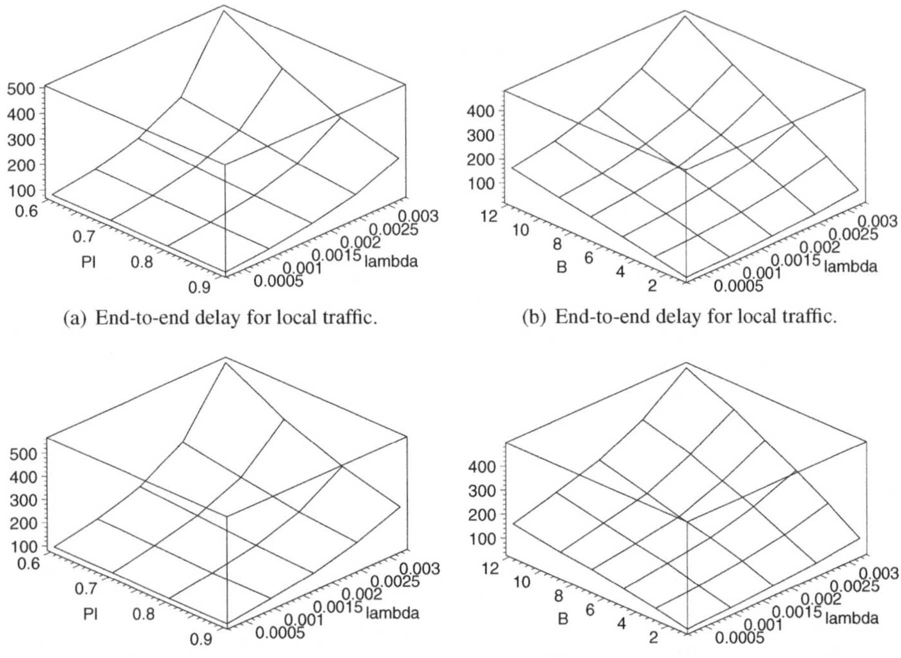

(a) End-to-end delay for local traffic.

(b) End-to-end delay for local traffic.

(c) End-to-end delay for non-local traffic.

(d) End-to-end delay for non-local traffic.

FIGURE 12.3

Adaptive minimization of non-local delays – simulation results for the MS bridge topology under LAMS scheduling. On the left, $\overline{B} = 5$, P_l parameter; on the right, $P_l = 0.9$, $\overline{B}$ is the parameter. (From V. B. Mišić and J. Mišić, "Adaptive inter-piconet scheduling in small scatternets," *ACM MC^{2}R – Mobile Computing and Communications Review* **7**(2):45–58, © 2003 ACM. Reprinted with permission.)

It may be interesting to note that mean non-local delays are almost the same as their non-local counterparts, which means that the algorithm provides equal treatment for local and non-local traffic.

The algorithm could be somewhat simplified if the participants were to skip the exchange of information about the total number of slots for local traffic. Either participant in the exchange would then calculate the next rendezvous on the basis of its local traffic, plus both the incoming and outgoing traffic; the missing information on local traffic in the other piconet would have to be estimated through some suitable algorithm. In this manner, the communication at the end of the exchange would be somewhat simplified. However, the savings of two or four slots per bridge cycle that would be obtained in this manner, are too small to justify the reduced accuracy. This is even more important in case there is wide asymmetry between the traffic in the two piconets.

12.3 Adaptive management: the case of the SS bridge

In the case of the SS bridge, however, the approach outlined above does not work well. There are two reasons for that, both of which are dictated by the scatternet topology.

First, in the MS bridge case, the bridge can collect traffic data from its own piconet, and the other master can simply send its data when the exchange ends. In this way, the information available to the bridge at the end of the exchange is always accurate; the error is caused by the estimation algorithm alone. In the SS bridge case, when the exchange with one piconet ends, the next exchange takes place in the *other* piconet. However, the decision on next rendezvous—which limits the duration of that next exchange—is made using the traffic estimate obtained from the *current* exchange. In other words, not only that we make an estimate – but we make it on the basis of insufficient information. In theory, the bridge might switch to the other piconet to get the traffic data, come back to make the appointment, and then switch again for the actual exchange. But switching from one piconet to another incurs additional synchronization delays, as shown in Chapter 10 and should not be done more often than is really necessary.

Second, in the MS bridge topology, the bridge is the master in one of the piconets, and usually can afford to wait until all queued packets are actually exchanged. However, in the SS bridge topology, the rendezvous appointment made between the bridge and one piconet master limits the exchange time with the other piconet. Since the traffic, both intra- and inter-piconet, is bursty, some of the exchanges will end before the rendezvous time, whereas others may take longer. In the latter case, some packets may still be waiting to be exchanged when the rendezvous time comes.

When this happens, the exchange might be allowed to proceed beyond the pending rendezvous, until all the queued packets are exchanged – but this solution is less than desirable from the performance standpoint. Namely, at the designated rendezvous time, the master of the other piconet will try to poll the bridge. The polling will continue until the bridge responds – and all packets queued at that master and its slaves will have to wait.

A better solution is to terminate the exchange per force, in which case some packets will be left to wait until the next exchange with that same piconet. Such packets will cause excessive delays, since they have to wait for at least T_1 to be forwarded to their destination. These delays will, of course, be reduced by minimizing the number of such packets, which may be accomplished by minimizing the number of exchange cycles that have to be force terminated.

The dependency of minimum end-to-end delays on packet burst arrival rate and traffic locality under such conditions is shown in Fig. 12.4. The values were obtained by running a series of simulation runs, each with a different value of T_1 (which was kept fixed during a single run), and hand-picking the value of T_1 that resulted in minimum end-to-end delay for non-local traffic. As can be seen, the delays are very close to the analytically obtained minima in Figs. 12.1(d) and 12.1(f).

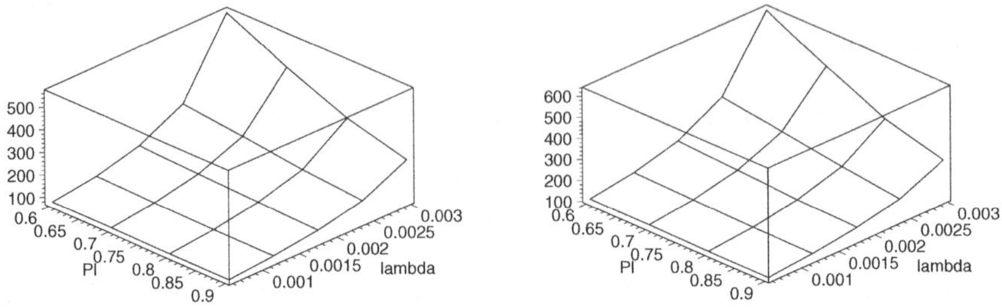

(a) Local delay variation. (b) Non-local delay variation.

FIGURE 12.4
End-to-end packet delays in the SS bridge topology under fixed T_1 scheduling: hand-picked minima. (From V. B. Mišić and J. Mišić, "Adaptive inter-piconet scheduling in small scatternets," *ACM MC^{2}R – Mobile Computing and Communications Review* 7(2):45–58, © 2003 ACM. Reprinted with permission.)

This observation, then, forms the basis for an adaptive bridge management algorithm for the SS bridge topology, which uses the manner in which the exchange terminates as a simple sensor variable. If all the queued packets have been exchanged before the next rendezvous, the allocated exchange time is able to handle the instantaneous volume of inter-piconet traffic and can be shortened. If the rendezvous time is reached and some packets are still waiting to be exchanged, the time allocated for exchange was too short and should be lengthened. As the next exchange takes place in the *other* piconet, the adjustment of the time interval should apply to the next exchange with the current piconet, i.e., the exchange *after* the immediately succeeding one. Therefore, the decision must be based on the outcome of *two* last exchanges, not just one, and we need four increment values (i.e., δ_{xy}) instead of just two. In our simulations, we found that the best results are obtained with $\delta_{00} = -2$, $\delta_{01} = 0$, $\delta_{10} = 1$, and $\delta_{11} = 3$.

The pseudocode of the algorithm, referred to as LASS (Load Adaptive Slave/slave Scheduling), is outlined below.

when rendezvous time reached **or** exchange complete
then if rendezvous time reached
    then if $f = 1$ **then** $\delta = \delta_{11}$ **else** $\delta = \delta_{10}$, $f' = 1$
    else // *exchange complete*
        if $f = 1$ **then** $\delta = \delta_{01}$ **else** $\delta = \delta_{00}$, $f' = 0$
        end if
        $T_1'(T_2') = T_1(T_2) + 2\delta$
        if $T_1'(T_2) < T_{min}$ **then** $T_1'(T_2') = T_{min}$
        // *request HOLD with hold time* T_1'
        // *current master acknowledges HOLD*
        // *switch to other piconet*
    end if

As in the case of LAMS, there should be a lower limit T_{min} for T_1/T_2: the efficiency drops sharply when T_1/T_2 approaches the sum of mean clock and frame synchronization times; the initial value we have used for simulations was $T_{min} = 12T$. No similar upper bound is necessary, because even after repeated bursts a period with no bridge traffic will eventually occur, and LASS is able to reduce T_1 quickly.

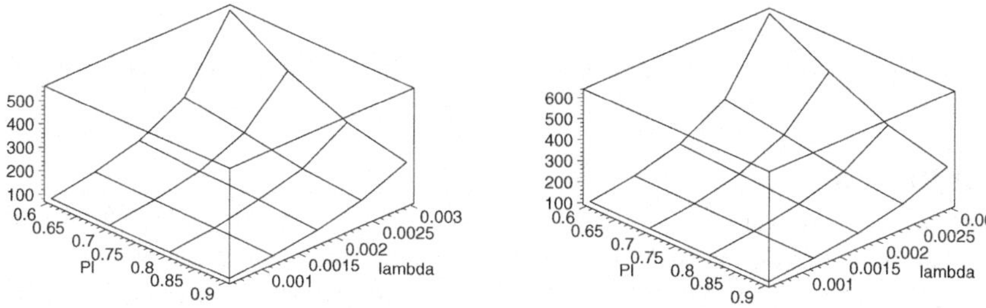

(a) End-to-end for local traffic.

(b) End-to-end for non-local traffic.

FIGURE 12.5
End-to-end packet delays in the SS bridge topology, as functions of burst arrival rate and traffic locality, under LASS scheduling. (From V. B. Mišić and J. Mišić, "Adaptive inter-piconet scheduling in small scatternets," *ACM MC²R – Mobile Computing and Communications Review* **7**(2):45–58, © 2003 ACM. Reprinted with permission.)

Fig. 12.5 shows the end-to-end delays for local and non-local traffic, respectively, as functions of packet burst arrival rate λ and traffic locality P_l. The mean delays are virtually equal to the hand-selected minima presented in Fig. 12.4 – but the adaptive algorithm requires no manual tuning in a wide range of traffic parameters. Furthermore, the delays obtained under LASS scheduling are very close to the analytically obtained minima from the right column of Fig. 12.1.

Finally, to demonstrate the reaction speed of the LASS algorithm, we have also varied the traffic load in the range 1:2 by varying the packet burst arrival rate for half of the slaves in each piconet. End-to-end delays for local and non-local traffic, averaged over an interval of 2.5s, are shown in Figs. 12.6(a) and 12.6(b), respectively. In all diagrams, dotted lines denote the number of packets generated during the averaging interval at that time, gray lines denote the average delays obtained for fixed $T_1 = 50T$, and solid black lines denote the average delays obtained with the LASS algorithm. Delays generally follow the instantaneous load, which is rather bursty. The local delay obtained with the LASS algorithm is indeed above the one obtained for a fixed T_1, but the difference is rather small. However, LASS provides substantially lower delay for non-local traffic in a wide range of piconet loads.

Similar behavior can be observed from Fig. 12.7, which shows large variations of bridge load in the range 1:5, obtained by varying the locality probability for some of

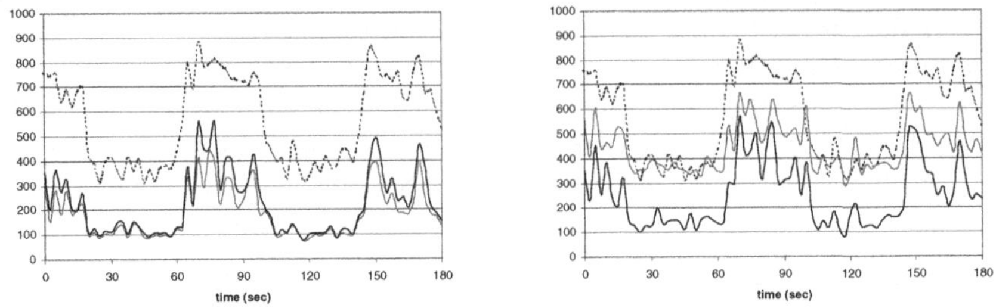

(a) Local delay under varying piconet load.

(b) Non-local delay under varying piconet load.

FIGURE 12.6
Large piconet load variations in the SS bridge topology: comparing LASS with fixed T_1 scheduling. (From V. B. Mišić and J. Mišić, "Adaptive inter-piconet scheduling in small scatternets," *ACM MC^2R – Mobile Computing and Communications Review* 7(2):45–58, © 2003 ACM. Reprinted with permission.)

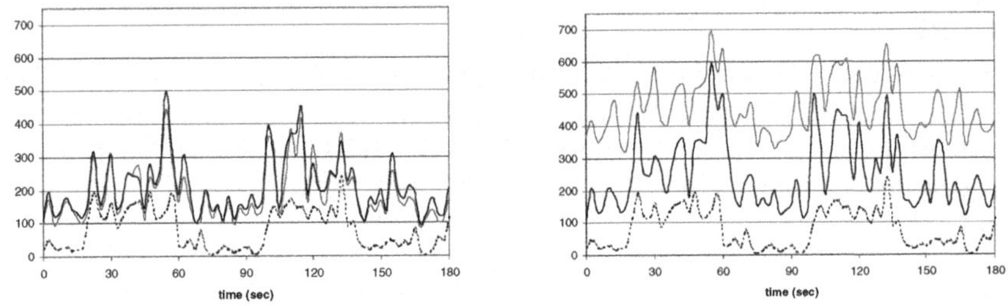

(a) Local delay.

(b) Non-local delay.

FIGURE 12.7
Large bridge load variations in the SS bridge topology: comparing LASS with fixed T_1 scheduling.

the slaves in the piconet, while total piconet load and packet burst arrival rate were kept constant. The dotted lines denote the number of packets generated during that period of time, gray lines denote the average delays obtained for fixed $T_1 = 50T$, and solid black lines denote the average delays obtained with the LASS algorithm.

Despite their excellent performance, the LASS algorithm and, to a lesser extent, the LAMS algorithm as well, suffer from a major drawback. Namely, they are both tailor made to the scatternet consisting of two piconets linked with a single bridge, and they cannot easily be scaled to more complex topologies. (These include scatternets with more than two piconets, but also scatternets in which a piconet may contain two or more bridges.) All the problems related to schedule clashes, which were mentioned in Chapter 10, will be present. The conclusion is that rendezvous based scheduling does not scale up well, and other solutions should be found.

13

Walk-in bridge scheduling

In this Chapter we discuss the third bridge scheduling mode, called walk-in bridge scheduling and present the analytical model and its performance evaluation. As we have shown in Chapter 10, the walk-in bridge scheduling mode operates without any schedule of rendezvous points, which makes it very attractive in practice. We also compare the performance and stability of the two operating modes, complete and limited exchange. In the complete exchange mode, the bridge may stay in each piconet for as long as there are more packets to exchange. This stay may take one or more piconet cycles. In the limited exchange mode, the bridge stays in each piconet for $K \geq 0$ piconet cycles (or $K+1$ exchanges) only and then leaves, even if there are still packets waiting to be delivered. Note that in the limiting case when $K \to \infty$, the limited exchange policy converges to the complete exchange one. Our analysis shows that the limited exchange mode performs better and has a wider stability region than the complete exchange mode. Simulations show that this scheduling approach offers good performance and excellent scalability as well.

The chapter is organized as follows. After describing our scheduling scheme and the associated queueing model in Section 13.1, we derive the probability distributions for packet service, vacation, and piconet cycle times in Section 13.2. Section 13.2 contains some observations related to service times and Section 13.3 contains derivations of delay components within the scatternet. Stability of the slave, master, and bridge queues is analyzed in Section 13.4, first analytically and then by simulations. Our final set of experiments in Section 13.5 demonstrates the scalability of the walk-in scheduling.

13.1 Scatternet model

Let us consider a scatternet with arbitrary topology in which a piconet may contain one or more bridges, and a bridge may link two or more scatternets, as shown in Fig. 13.1. The operation of the scatternet may be described with a queueing model shown in Fig. 10.7. For the time being, all queues are assumed to be of infinite size; the impact of finite buffer sizes will be discussed in Chapter 14. The traffic model has been described in Chapter 10. Let us focus on just two of the piconets, N and X, and the single bridge that links them. (This will not limit the generality of our

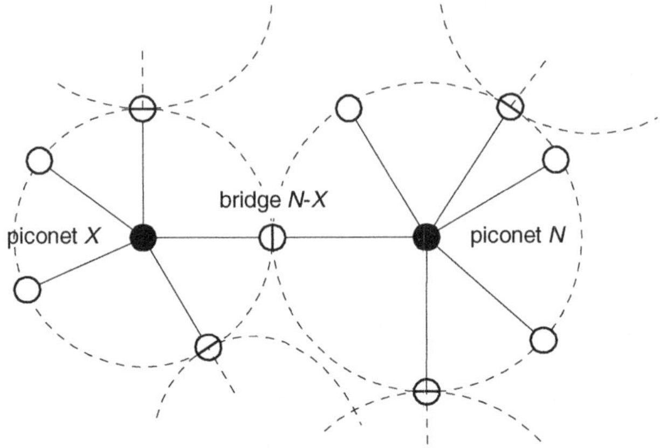

FIGURE 13.1
Portion of the scatternet under consideration, containing two piconets joined with a single bridge.

analysis, and it may easily be extended to more complex topologies.) The piconets have m_n and m_x members, respectively, which includes one master each and b_n (b_x) bridges. The bridge that links the piconets spends the time T_n (T_x) in piconet N (X); these times will be referred to as bridge residence times, and they are expressed in Bluetooth time slots $T = 625\mu s$.

We assume that both piconet masters poll their slaves in a round robin fashion using E-limited polling. Up to M_s data packets are exchanged during a single visit to a slave; we assume that all piconet masters use the same value for M_s, which may be chosen according to the criteria from Chapter 3.

The bridge switches between its piconets in a round robin fashion, without a predefined schedule. When present, the bridge is polled just like any other slave, except that the first packet to be sent to the bridge is always an empty POLL. The absence of a response to the initial POLL packet means that the bridge is not present, in which case the master simply moves on to the next slave. If the bridge is present, it responds with a NULL; the master then starts the exchange with actual data packets.

The bridge exchange lasts for M_b packets, or less if both queues are emptied. (Again, we assume that all piconet masters use the same value for M_b, but separate values for each piconet, or even each bridge, may be readily accommodated by our model.) The last condition may be explicitly signalized with a POLL-NULL frame, which complies with common Bluetooth practice. Thus, bridge exchanges last at most $M_b + 2$ frames, but the first and last frames always take only two slots each. This does result in wasted slots, as the first and last frame carry no data, but the impact on performance is minimal, as will be shown in subsequent analysis. Note that slots may be similarly wasted in virtually all other schemes, and some schemes even allow gaps between adjacent exchanges [Tan and Guttag, 2002].

Under the complete exchange policy, the bridge stays in the piconet for as many cycles as is necessary to allow all queued packets to be exchanged at once. Under the limited exchange policy, the bridge stays in the piconet for a limited time only; this time may be expressed in Bluetooth time slots or in piconet cycles. Since we are interested in queue stability and, ultimately, scatternet performance, we will assume that all the masters and all the bridges are aware of the actual policy used, including the time limit for exchanges, if any.

Piconet model

If the bridge burst arrival rates are known, each piconet can be analyzed in isolation using the analytical model for E-limited polling from Chapter 3. However, the results obtained there cannot simply be reused, since the packet burst arrival rate and the value of the polling parameter for an ordinary slave differ from those for the bridge. Therefore, we have to determine:

1. The joint probability distributions of the uplink/downlink queue lengths, frame service time, and channel service times for an ordinary slave (say, slave i);
2. The probability distribution of the channel service time for the bridge; and
3. The corresponding vacation times for ordinary slaves and the cycle time.

We note that there is a recursive relationship between those values since the probability distributions of the queue lengths depend on the vacation times, and vacation times depend on the channel service times (i.e., the probability distributions of the queue lengths of other piconet members). According to earlier analysis, an ordinary slave i is described with a total of $M_s - 1$ equations with unknowns $q^i_{0,0}$ and $\pi^{i,(\mu)}_{0,0}$, where $k = 1 \,..\, M_s - 1$, as follows:

$$\left(1 - \left(\frac{F^*_{is}(\lambda_{iu} - \lambda_{iu} G_b(z_j))}{z_j}\right)^{M_s}\right) q^i_{0,0} + \sum_{k=1}^{M_s-1} \left(1 - \left(\frac{F^*_{is}(\lambda_{iu} - \lambda_{iu} G_b(z_j))}{z_j}\right)^{M_s-k}\right) \pi^{i,(k)}_{0,0} = 0 \tag{13.1}$$

where

$$z_j = \sum_{n=1}^{\infty} \frac{e^{2\pi j n \sqrt{-1}/M_s}}{n!} \cdot \frac{d^{n-1}}{dz^{n-1}} \left(V^*_i(\lambda_{iu} - \lambda_{iu} G_b(z)) F^*_{is}(\lambda_{iu} - \lambda_{iu} G_b(z))\right)^{n/M_s} \Big|_{z=0} \tag{13.2}$$

and $F^*_{is}(s)$ represents the LST of the frame duration. This LST has already been calculated in Chapter 3, Equation (3.37), to be

$$\begin{aligned} F_{is}^{*}(s) &= PQ \cdot e^{-2s} \\ &+ \left(Q_i(0,1) + Q_i(1,0) + SP - 2 \cdot PQ \cdot G_p^{*}(s)\right) e^{-2s} \\ &+ (1 - Q_i(0,1) - Q_i(1,0) - SP + PQ)\,(G_p^{*}(s))^2 \end{aligned} \tag{13.3}$$

where

$$PQ = \sum_{\mu=1}^{M_s} \pi_{0,0}^{i,(\mu)} + q_{0,0}^{i}$$

and

$$SP = \sum_{\mu=1}^{M} \left(\Pi_{i,\mu}(0,1) + \Pi_{i,\mu}(1,0)\right)$$

The last, M_s-th equation is obtained from the condition $Q(1,1) + \sum_{\mu=1}^{M_s} \Pi_{i,\mu}(1,1) = 1$, and it reads:

$$M_s q_{0,0}^{i} + \sum_{k=1}^{M_s-1} (M_s - k)\pi_{0,0}^{i,(k)} = \frac{M_s(1 - \lambda_{iu}\overline{F_{is}B}) - \lambda_{iu}\overline{BV_i}}{1 - \lambda_{iu}\overline{F_{is}B} + \lambda_{iu}\overline{BV_i}} \tag{13.4}$$

where $V_i^{*}(s)$ represents the LST of the slave's i vacation time Bridge, when present in the piconet, is described with a similar system of equations with appropriate packet burst arrival rate and polling parameter M_b.

Since the PGF $V_i^{*}(\lambda_{iu} - \lambda_{iu}G_b(z))$ depends on the joint uplink/downlink queue length probability distributions at Markov imbedded points for all piconet members other than target slave i, each slave should be described with a separate set of equations similar to (13.1) and (13.4). Therefore, the target queue length distributions at embedded Markov points depend on the probability distributions of all Markov points at all the uplink and downlink queues. The exact expression for $V_i^{*}(\lambda_{iu} - \lambda_{iu}G_b(z))$ will be derived in the Section 13.2. However, initial derivation of the probability distribution of the number of packets in the slave i's uplink (downlink) queue after the vacation and after the uplink transmissions may be simplified if we assume that the PGF for the number of packet arrivals during the vacation period, $V_i^{*}(\lambda_{iu} - \lambda_{iu}G_b(z))$, depends only on values of $q_{0,0}^{i}$, and $\pi_{0,0}^{i,(\mu)}$ $\mu = 1..M_s$, but not on the other members of $Q_i(z, w)$ and $\Pi_{i,\mu}(z)$, $\mu = 1 \,..\, M_s$.

It is also possible to truncate the sum for calculation of z_j to the first few members only (actually, this number should be close to M_s for reasonable accuracy to be maintained). The solutions thus obtained may contain negligible imaginary part.

These simplifications allow us to describe the piconet N within the scatternet with $m_n - b_n - 1$ slaves with arbitrary arrival rates with $2M_s(m_n - b_n - 1) + 2M_b b_n$ equations. When the system is solved, the PGFs $\Pi_{i,\mu}(z, w)/\Pi_{i,\mu}(1,1)$, $\mu = 1..M_s$ and $Q_i(z, w)/Q_i(1,1)$ can be determined. Again, a relatively simple solution may be obtained under the assumption that the vacation time does not depend recursively on the $Q_i(z, w)$ (distributions of the number of packets upon the return from the vacation). Unfortunately, this assumption is not quite correct, and the system will have to be solved in an iterative manner; a detailed discussion can be found in Chapter 3.

One such system of equations is needed for each piconet in the scatternet.

13.2 Service, vacation, and cycle times

We are now able to determine the service times for the uplink and downlink queues for ordinary slave and the bridges, as well as the vacation time for the pair of queues corresponding to one slave or bridge.

The LST for the duration of channel service time for the ordinary slave with NULL packets included, expressed in slots, has been derived in Chapter 3 to be

$$S_i^*(s) = \sum_{k=0}^{M_s-1} P_{f,k} \prod_{\mu=1}^{k} (F^{*\mu}(s)) e^{-2s} + P_{f,M_s} \prod_{\mu=1}^{M_s} F^{*\mu}(s) \tag{13.5}$$

where

$$\begin{aligned}
P_{f,0} &= \frac{q_{0,0}^i}{Q_i(1,1)} \\
P_{f,1} &= \frac{Q_i(1,1) - q_{0,0}^i}{Q_i(1,1)} \cdot \frac{\pi_{0,0}^{i,(1)}}{\Pi_{i,1}(1,1)} \\
P_{f,k} &= \frac{Q_i(1,1) - q_{0,0}^i}{Q_i(1,1)} \cdot \prod_{\mu=1}^{k-1} \frac{\Pi_{i,\mu}(1,1) - \pi_{0,0}^{i,(\mu)}}{\Pi_{i,\mu}(1,1)} \cdot \frac{\pi_{0,0}^{i,(k)}}{\Pi_{i,k}(1,1)}, \\
&\qquad k = 2 \mathinner{..} M_s - 1 \\
P_{f,M_s} &= 1 - \sum_{k=0}^{M_s-1} P_{f,k} \\
F^{*1}(s) &= \frac{Q_i(1,0) + Q_i(0,1) - 2q_{0,0}^i}{Q_i(1,1)} G_p^*(s) e^{-s} \\
&+ \frac{Q_i(1,1) - Q_i(1,0) - Q_i(0,1) + q_{0,0}^i}{Q_i(1,1)} G_p^*(s)^2 \\
F^{*\mu}(s) &= \frac{\Pi_{i,\mu-1}(1,0) + \Pi_{i,\mu-1}(0,1) - 2\pi_{0,0}^{i,(\mu-1)}}{\Pi_{i,\mu-1}(1,1)} G_p^*(s) e^{-s} \\
&+ \frac{\Pi_{i,\mu-1}(1,1) - \Pi_{i,\mu-1}(1,0) - \Pi_{i,\mu-1}(0,1) + \pi_{0,0}^{i,(\mu-1)}}{\Pi_{i,\mu-1}(1,1)} G_p^*(s)^2, \\
&\qquad \mu = 2 \mathinner{..} M_s
\end{aligned} \tag{13.6}$$

Mean value of the service time is equal to $\overline{S_i} = -S_i^{*'}(0)$.

In the presence of b_n bridges in the piconet N, and assuming that bridges have equal traffic load, the downlink bridge arrival rate toward the piconet x is

$$\lambda_{b_{n,x}} = \frac{(1-P_l)}{b_n} \sum_{\substack{i=2 \\ \neq j}}^{m_n} \lambda_{iu} \tag{13.7}$$

where the master is the piconet member with the index 0, and the bridge has the index j.

Consider again the bridge between piconets N and X, shown in Fig. 13.1, where the residence times for the bridge are T_n and T_x, respectively, and the corresponding PGFs are $T_n(z)$ and $T_x(z)$. When the bridge switches into a piconet, its master will not be aware of it until its first attempt to poll the bridge. The time between joining the piconet and first poll by the master may be described with the PGFs $T_{syn,n,x}(z)$ and $T_{syn,x,n}(z)$. These synchronization times may be derived as follows.

Under walk-in bridge scheduling, the piconet cycle time affects the bridge cycle time in piconet N and vice versa, and the latter cannot be determined independently of the former. Therefore, we must start by finding the PGF for the piconet cycle time in N without bridge interruptions, using an equation similar to (11.18). Note that for zero length bridge transfers, $C_n^t(z)|0 = C_n^p(z)|0$. The only difference from (11.18) would be the multiplication with z^2 in order to model the polling of bridge while it is absent from the piconet: $C_n^p(s)|0 = (S_n^p(z)|0)^{m_n-2}z^2$. Then, we need to find the probability distribution of the bridge synchronization time in N, which is the interval from the time slot in which the bridge switches to the piconet N until the first time slot in which the master polls the bridge.

Under walk-in scheduling, the master will poll all of the slaves, including the bridge, in every cycle. If the bridge is absent, the master will spend two slots (POLL packet, followed by a one-slot silence) and continue with polling other slaves [Bluetooth SIG, 2001*b*], as shown in Fig. 13.2 (a). For convenience, we have assumed that the piconet cycle starts by polling the bridge.

The bridge may join piconet N (i.e., synchronize its clock to the clock of N) at any time; the position of this time instant will be uniformly distributed over the span of the piconet cycle of N – which has started by polling the bridge that was *not* present at the time. The time interval from the moment in which the bridge has joined the piconet, to the moment in which the master polls the bridge, is the bridge synchronization time, as shown in Fig. 13.2 (b).

The probability distribution for the bridge synchronization time, $T_{syn2,x,n}$, may be derived by using the results of the theory of continuous-time and discrete-time renewal processes [Kleinrock, 1972; Cooper, 1990]. $T_{syn,x,n}$ is equal to the backward recurrence time in the discrete-time renewal process, where the interval between two successive renewal points is given by the piconet cycle time when the bridge is absent, $C_n^p|0$. The probability that the piconet cycle time when the bridge is absent lasts exactly j time slots is given by $r_j^{(0)} = \frac{1}{j!}\frac{d^j}{dz^j}C_n^p(0)|0$. Now, the probability

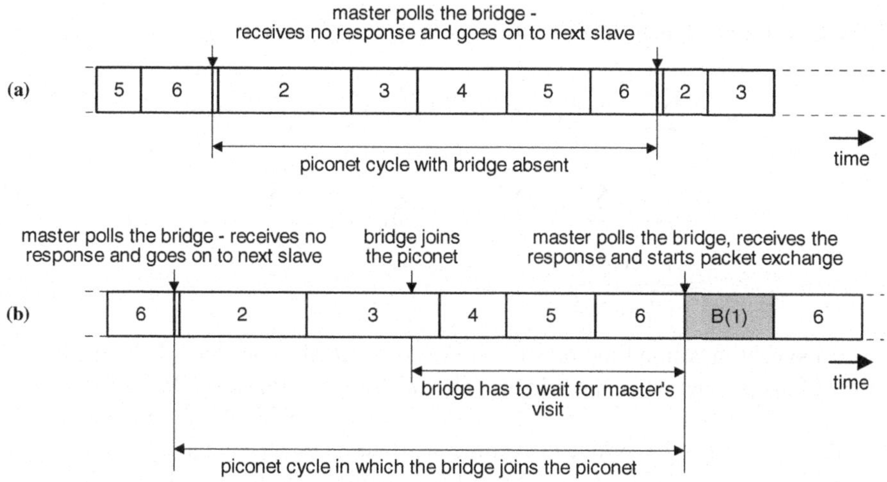

FIGURE 13.2
Pertaining to the bridge synchronization upon joining the piconet with a total of six slaves, and bridge is slave no. 1. (From J. Mišić, V. B. Mišić, and K. L. Chan, "Performance of Bluetooth bridge scheduling algorithms," *Computer Communications* **27**(12):1143-1151, © 2004 Elsevier B.V.)

that the piconet cycle in which the bridge joins piconet N will last exactly j time slots is

$$\frac{j}{\overline{\overline{C_n^P|0}}} \frac{1}{j!} \frac{d^j}{dz^j} C_n^P(0)|0 = \frac{j r_j^{(0)}}{\overline{\overline{C_n^P|0}}}, \qquad j = 1, 2, \ldots \tag{13.8}$$

The probability that the bridge joins this piconet cycle exactly k slots before it will be polled by the master is

$$\frac{1}{j} \frac{j r_j^{(0)}}{\overline{\overline{C_n^P|0}}} = \frac{r_j^{(0)}}{\overline{\overline{C_n^P|0}}}, \qquad k = 0 \ldots j-1 \tag{13.9}$$

Then, the probability that the bridge joins the piconet cycle of arbitrary length (described with the PGF $C_n^P(z)|0$), exactly k slots before it will be polled by the master, is

$$\text{Prob}[T_{syn,x,n} = k] = \sum_{j=k+1}^{\infty} \frac{r_j^{(0)}}{\overline{\overline{C_n^P|0}}} \tag{13.10}$$

The PGF for the synchronization time is

$$\begin{aligned}
T_{syn,x,n}(z) &= \sum_{k=0}^{\infty} z^k \mathrm{Prob}[T_{syn,x,n} = k] = \sum_{k=0}^{\infty} z^k \sum_{j=k+1}^{\infty} \frac{r_j^{(n)}}{\overline{\overline{C_n^p|0}}} \\
&= \frac{1}{\overline{\overline{C_n^p|0}}} \sum_{j=1}^{\infty} r_j^{(0)} \sum_{k=0}^{j-1} z^k = \frac{1}{\overline{\overline{C_n^p|0}}(1-z)} \sum_{j=1}^{\infty} r_j^{(0)} (1-z^j) \\
&= \frac{1 - C_n^p(z)|0}{\overline{\overline{C_n^p|0}}(1-z)}
\end{aligned} \tag{13.11}$$

Mean synchronization time may be shown to be equal to one half of the mean cycle time of the piconet when the bridge is absent. Note that this result holds for both MS and SS bridges: after all, in both cases the bridge will be only a slave in piconet N.

The moments of the bridge cycle time distribution may be obtained from the equation $T_{cyc,n}(z) = z^{T_1} T_r(z) T_{syn,x,n}(z)$. This equation is similar to (11.10) which holds in the case of rendezvous bridge scheduling, but with the addition of $T_{syn,x,n}(z)$. However, $T_{syn,x,n}(z)$ depends only on the traffic parameters in N and does not introduce any new variables. Therefore, once the PGF for the bridge cycle is known, we can determine the complete expressions for $C_n^t(z)$ and $C_X^t(z)$.

The packets are sent from the master's outgoing queue toward the bridge much in the same manner as they are sent from the 'normal' downlink queues toward corresponding ordinary slaves. Hence, we may describe uplink/downlink queue for one slave with Equations (13.1) and (13.4), but using M_b in place of M_s; actually, we will use M_{bx} in piconet X and M_{bn} in piconet N. Note that the master-bridge link will experience longer vacation times than the ordinary slave-master links, due to bridge absence from the piconet. However, this analogy helps us estimate the outgoing service time without having to solve the entire system.

As noted above, there are two different policies for the bridge management under E-limited intra-piconet polling. Let us now investigate the performance of these approaches in more detail.

Complete exchange policy

As a matter of convenience, let us assume that the bridge carries more traffic from piconet N to piconet X than in the opposite direction. Consider the time interval between two successive bridge departures from piconet N which we will refer to as the *bridge cycle*. The PGF for the bridge cycle has the form:

$$T_{cyc}(z) = T_{syn,n,x}(z) T_x(z) T_{syn,x,n}(z) T_n(z) \tag{13.12}$$

The LST for the bridge cycle $T^*_{cyc}(s)$ can be obtained by substituting variable z with e^{-s}. During the bridge cycle, the outgoing queue of the piconet N master was filled at the rate $\lambda_{b_{n,x}}$. Similarly to the approach taken in Chapters 3 and 11, we obtain the PGF for the number of packet arrivals in the bridge queue during the bridge cycle as

$$N_{n,x}(z) = T^*_{cyc}(\lambda_{b_{n,x}} - \lambda_{b_{n,x}} G_b(z)) \tag{13.13}$$

The PGF for the number of bridge cycles needed to empty this queue (i.e., deliver all of its packets to the bridge) is

$$Ns^n_{n,x}(z) = \left\lceil N_{n,x}(z^{\frac{1}{M_{bn}}}) \right\rceil \tag{13.14}$$

where $\lceil \ldots \rceil$ denote that every exponent in the PGF should be rounded to the next larger integer. Then the PGF for the corresponding number of slots is

$$Nc^n_{n,x}(z) = Ns^n_{n,x}(C^+_n(z)) \tag{13.15}$$

where $C^+_n(z)$ denotes the PGF for the duration of the cycle time in piconet N with the bridge present. Mean number of slots needed to empty the master's outgoing queue is

$$\overline{Nc^n_{n,x}} = \left\lceil \frac{\lambda_{b_{n,x}} \overline{BT_{cyc}}}{M_{bn}} \right\rceil \overline{C^+_n} - \overline{C^-_n} \tag{13.16}$$

where $\overline{C^-_n}$ denotes the mean cycle time in piconet N when the bridge is not present. The last expression may be approximated with:

$$\overline{Nc^n_{n,x}} = \frac{\lambda_{b_{n,x}} \overline{BT_{cyc}}}{M_{bn}} \overline{C_n} \tag{13.17}$$

where $\overline{C_n}$ denotes the mean cycle time averaged over a long time (i.e., many cycles with and without the bridge present).

After switching to piconet X, the bridge has to deliver all queued packets to the master. Note that the bridge queue cannot be refilled as long as the bridge is with piconet X. However, the bridge uplink service time may be calculated in a different way. The PGF for the number of service periods in piconet X needed to empty the bridge is similar to one given by (13.14). The PGF for the equivalent number of slots is

$$Nc^x_{n,x}(z) = Ns^x_{n,x}(C^+_x(z)) \tag{13.18}$$

where $C^+_x(z)$ denotes the PGF for the duration of the cycle time in piconet X when bridge is present in the piconet. The mean number of slots may be approximated with

$$\overline{Nc^x_{n,x}} = \frac{\lambda_{b_{n,x}} \overline{B}(\overline{T_n} + \overline{T_x} + \overline{T_{syn,n,x}} + \overline{T_{syn,x,n}})}{M_{bx}} \overline{C_x} \tag{13.19}$$

In this case, the bridge residence times are described with

$$\begin{aligned} Nc^n_{n,x}(z) &= T_n(z) \\ Nc^x_{n,x}(z) &= T_x(z) \end{aligned} \tag{13.20}$$

and the respective equalities hold for mean values as well. The residence time in piconet N may easily be obtained as

$$\overline{T_n} = \frac{(\overline{T_{syn,n,x}} + \overline{T_{syn,x,n}}) \dfrac{M_{bx}}{\lambda_{b_{n,x}} \overline{BC_x}}}{\dfrac{M_{bn}}{\lambda_{b_{n,x}} \overline{BC_n}} \dfrac{M_{bx}}{\lambda_{b_{n,x}} \overline{BC_x}} - \dfrac{M_{bn}}{\lambda_{b_{n,x}} \overline{BC_n}} - \dfrac{M_{bx}}{\lambda_{b_{n,x}} \overline{BC_x}}} \tag{13.21}$$

It is easy to show that $\overline{T_x} = \overline{T_n}\left(\frac{\overline{C_x}}{M_{bx}}\right) / \left(\frac{\overline{C_n}}{M_{bn}}\right)$.

Under the complete exchange policy, the master's outgoing queue in piconet N will be emptied at a rate of M_{bn} packets per cycle, and possibly refilled by packets picked up from the slaves between two visits to the bridge. The exchange lasts until there are no more packets to exchange (which is detected via a POLL-NULL sequence), and the bridge then leaves the piconet. As long as the bridge is absent, it takes only two time slots (i.e., one POLL packet and one empty slot) for the master to detect this absence. The probability that the bridge will be present in the piconet N when the master polls it (hit probability) is

$$H_n = \frac{\lceil \overline{T_n}/\overline{C_n} \rceil}{(\overline{T_n} + \overline{T_x} + \overline{T_{syn,n,x}} + \overline{T_{syn,x,n}})/\overline{C_n}} \approx \frac{\lambda_{b_{n,x}} \overline{B}\,\overline{C_n}}{M_{bn}} \tag{13.22}$$

Previously calculated PGFs of bridge residence times, bridge synchronization times and hit probability directly lead to the calculation of the vacation time for the bridge when we consider it as a piconet member; we will calculate it in Section 13.2. However, the PGF for the bridge downlink service time which assumes that bridge will exchange M_{bn} packets in every cycle (except, possibly, the last one) can be approximated with

$$Sb_{n,x,d}(z) = H_n (G_p(z))^{M_{bn}} + (1 - H_n)\, z + z \tag{13.23}$$

Mean bridge service time for the complete exchange is

$$\overline{Sb_{n,x,d}} = \lambda_{b_{n,x}} \overline{B}\,\overline{C_n} \left(\overline{L} - \frac{1}{M_{bn}}\right) + 2 \tag{13.24}$$

Finally, the PGF of uplink service time in piconet X is

$$Sb_{n,x,u}(z) = H_x \cdot (G_p(z))^{M_{bx}} + (1 - H_x)\, z + z \tag{13.25}$$

As noted above, we assume that $\lambda_{b_{n,x}} \geq \lambda_{b_{x,n}}$, hence the service time in the direction with the smaller bridge traffic is

$$Sb_{x,n,d}(z) = H_n \frac{\overline{N_{x,n}}}{\overline{N_{n,x}}} (G_p(z))^{M_{bx}} + \left(1 - H_n \frac{\overline{N_{x,n}}}{\overline{N_{n,x}}}\right) z + z \tag{13.26}$$

and the corresponding mean bridge service time is

$$\overline{Sb_{x,n,d}} = \lambda_{b_{x,n}} \overline{B}\,\overline{C_n}\, M_{bx} \left(\overline{L} - \frac{1}{M_{bx}}\right) + 2 \tag{13.27}$$

Limited exchange policy

Let us assume that the bridge residence time limit is expressed in the number of exchanges with the master. In other words, let the bridge stay in piconet N for at most $K+1$ (where $K = 0, 1, 2 \ldots$) exchanges of up to M_{bn} packets with the master, interleaved with K piconet cycles. Of course, the bridge may leave earlier if there are no more packets to exchange. (Note that when $K \to \infty$, the bridge will stay in the piconet for as many exchange cycles as is necessary to empty both master and bridge queues – which actually corresponds to the complete exchange policy.) Under this policy, the PGF for the maximum bridge residence time in piconet N is

$$T_n^{max}(z) = (G_p(z))^{2M_{bn}}(C_n(z))^K \tag{13.28}$$

By the same token, the maximum bridge residence time in piconet x is

$$T_x^{max}(z) = (G_p(z))^{2M_{bx}}(C_x(z))^K \tag{13.29}$$

Let us consider the time interval between two successive departures of the bridge from piconet N, i.e., the bridge cycle. The PGF for the maximum bridge cycle is

$$T_{cyc}^{max}(z) = T_{syn,n,x}(z)T_x^{max}(z)T_{syn,x,n}(z)T_n^{max}(z) \tag{13.30}$$

The number of packets that have entered the master's outgoing queue in that time period has the PGF:

$$N_{n,x}(z) = T_{cyc}^{*max}(\lambda_{b_{n,x}} - \lambda_{b_{n,x}}G_b(z)) \tag{13.31}$$

The master's outgoing queue in piconet N behaves much in the same manner as under the complete exchange policy, except that the duration of bridge residence is limited and we do not wait for this queue to clear up. We will focus on the case of $K = 0$ since it promises lowest delays for inter-piconet traffic. The probability that the bridge will be present in the piconet when the master attempts to poll it, is

$$H_n = \frac{\overline{T_n}}{\overline{T_n} + \overline{T_x} + \overline{T_{syn,n,x}} + \overline{T_{syn,x,n}}} \tag{13.32}$$

The PGF for the bridge downlink service time may be roughly approximated by

$$Sb_{n,x,d}(z) = H_n(G_p(z))^{\overline{N_{n,x}}} + (1 - H_n)\,z + z \tag{13.33}$$

with the mean value of

$$\overline{Sb_{n,x,d}} = \lambda_{b_{n,x}}\overline{B}\,\overline{C_n}\,\overline{L} - H_n + 2. \tag{13.34}$$

The PGF for the uplink service period of the bridge while present in the piconet X is similar to (13.33), but with H_x and $\overline{C_x}$ as parameters. Again, previous discussions referred to the bridge direction with the larger bridge traffic; the PGF for the service time in the direction with the smaller traffic is

$$Sb_{x,n,d}(z) = H_n(G_p(z))^{\overline{N_{x,n}}} + (1 - H_n)\,z + z \tag{13.35}$$

with the mean value of

$$\overline{Sb_{x,n,d}} = \lambda_{b_{x,n}}\overline{B}\,\overline{C_n}\,\overline{L} - H_n + 2. \tag{13.36}$$

Piconet cycle time

The LST of service time for the pair of uplink and downlink queues that correspond to the ordinary slave i is $S_i^*(s)$, which is given by (13.5). The approximate LST for service time for the bridge (slave j) is

$$S_{bj}^*(s) = Sb_{n,x,d}^{*j}(s)Sb_{n,x,u}^{*j}(s) \tag{13.37}$$

Given that piconet N has $8 \geq m_n \geq 2$ members, one of which is the master and b_n of which are bridges, the PGF for the vacation time observed by an ordinary slave is

$$V_i^*(s) = \prod_{\substack{j=2 \\ j \neq i}}^{m_n} S_j^*(s) \tag{13.38}$$

The vacation time for the bridge in piconet N under complete exchange policy corresponds to the ordinary vacation time for E-limited service when bridge is present in the network while it corresponds to the bridge absence time otherwise. The PGF for this vacation time is

$$V_i^*(s) = H_n \prod_{\substack{j=2 \\ j \neq i}}^{m_n} S_j^*(s) + (1 - H_n)T_{syn,n,x}^*(s)T_x^*(s)T_{syn,x,n}^*(s) \tag{13.39}$$

For limited exchange policy with $K = 0$, it has the form

$$V_i^*(s) = T_{syn,n,x}^*(s)T_x^*(s)T_{syn,x,n}^*(s) \tag{13.40}$$

When the vacation time for the bridge is known, Equations (13.1) and (13.4) can be set, and the exact service time $Sb_{n,x}^j$, for bridge j between piconets N and X, can be calculated using Equation (13.5).

The cycle time of the piconet N with the bridge present, then becomes

$$C_n^{*+}(s) = \prod_{i=2}^{m_n} S_i^*(s) \tag{13.41}$$

and its mean value will be:

$$\overline{C_n} = \sum_{\substack{i=2 \\ i \neq j}}^{m_n} \overline{S_i} + \sum_{j=2}^{b_n+1} \overline{Sb_{n,x}^j} \tag{13.42}$$

By taking into account the approximate mean bridge service times from (13.24) and (13.27), the cycle time for the case of complete exchange becomes

$$\overline{C_n^+} = \frac{\sum_{\substack{i=2 \\ i \neq j}}^{m_n} \overline{S_i} + 2b_n}{1 - \overline{B}\left(\overline{L} - \frac{1}{M_{bn}}\right) \sum_{j=2}^{b_n+1} \left(\lambda_{b_{n,x_j}} + \lambda_{b_{x_j,n}}\right)} \tag{13.43}$$

The mean cycle time for the limited exchange has a similar form.

As the distribution of the number of packets in uplink and downlink queues are calculated via vacation times and vice versa, there is a recursion here, and the system of equations may be solved iteratively as in Chapter 3.

Observations related to the service times

We note that (13.5) and (13.43) still depend on parameters $\sum_{k_d=0}^{\infty} q^i_{ku,kd}/Q_i(1,1)$ and $\sum_{k_u=0}^{\infty} q^i_{ku,kd}/Q_i(1,1)$, where $k_u, k_d = 0 \ldots M_s - 1$. Notice, however, that there is interaction between the slaves in one piconet: the service time for slave i will depend on the service times of all other slaves, since burst arrivals are accumulated in the uplink queue of slave i while other slaves are being serviced (i.e., during the vacation). If the service times for other slaves are smaller, the service time for slave i will also be smaller since less packets will be accumulated in the buffer, and vice versa. Of course, the service time for the bridge indirectly depends on the service times for slave through the piconet cycle time. Due to this coupling among the slaves, the mean cycle time actually depends on the sum of arrival rates in all the slaves, rather than on their individual arrival rates, provided that differences in arrival rates are not high (say, within 80% of each other). A similar observation has been made for a single piconet in Chapter 3, page 43.

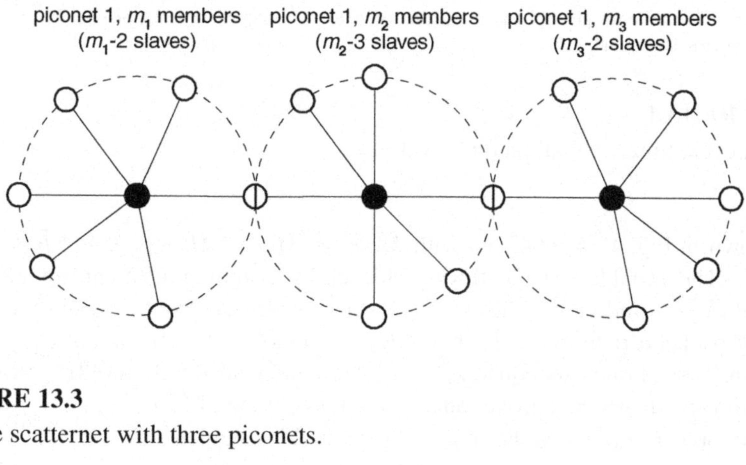

FIGURE 13.3
Simple scatternet with three piconets.

Consider a simple array of three piconets as shown in Fig. 13.3, and observe the dependency of piconet cycle time in 'end' piconets (i.e., piconets 1 and 20 on the total piconet load, variation of load among the slaves, and locality probability, which

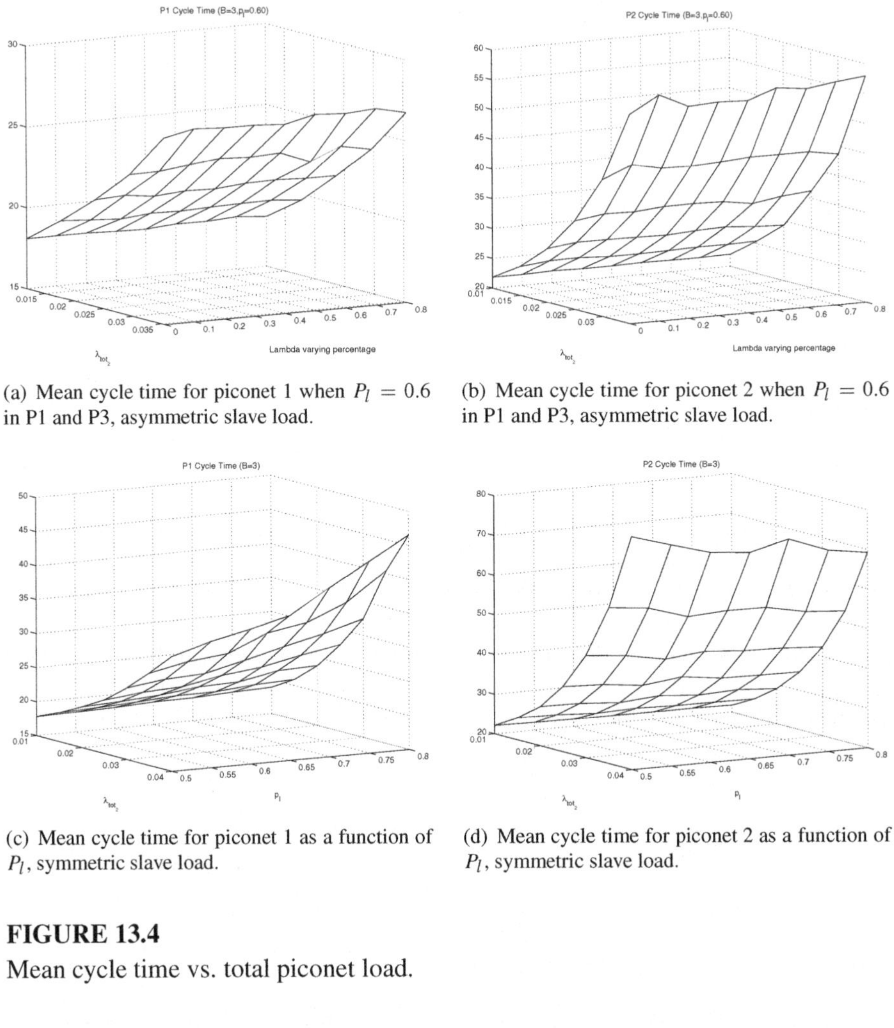

(a) Mean cycle time for piconet 1 when $P_l = 0.6$ in P1 and P3, asymmetric slave load.

(b) Mean cycle time for piconet 2 when $P_l = 0.6$ in P1 and P3, asymmetric slave load.

(c) Mean cycle time for piconet 1 as a function of P_l, symmetric slave load.

(d) Mean cycle time for piconet 2 as a function of P_l, symmetric slave load.

FIGURE 13.4
Mean cycle time vs. total piconet load.

is shown in Fig. 13.4 (obtained with $M_s = 3$, $M_{b1} = M_{b2} = 9$, and $\overline{B} = 3$ in all cases). The variable slave load was achieved by assigning different (uplink) packet arrival rates to individual slaves. Let λ stand for the arithmetic mean of all slaves' uplink packet arrival rates. Then, the load variation of 0 corresponds to symmetric load (all arrival rates are equal, $\lambda_i = \lambda$), while the load variation of 0.5 corresponds to uniformly distributed arrival rates in the range $(0.5\lambda, 1.5\lambda)$.

The total arrival rate in the 'middle' piconet 2 is calculated as

$$\lambda_{tot} = \lambda(m_1 + m_3 - 4)(1 - P_l) + \lambda(m_2 - 3)$$

where λ denotes mean packet burst arrival rate per ordinary slave in the whole scatternet. As we see, the cycle time is not sensitive to the slave heterogeneity, but is

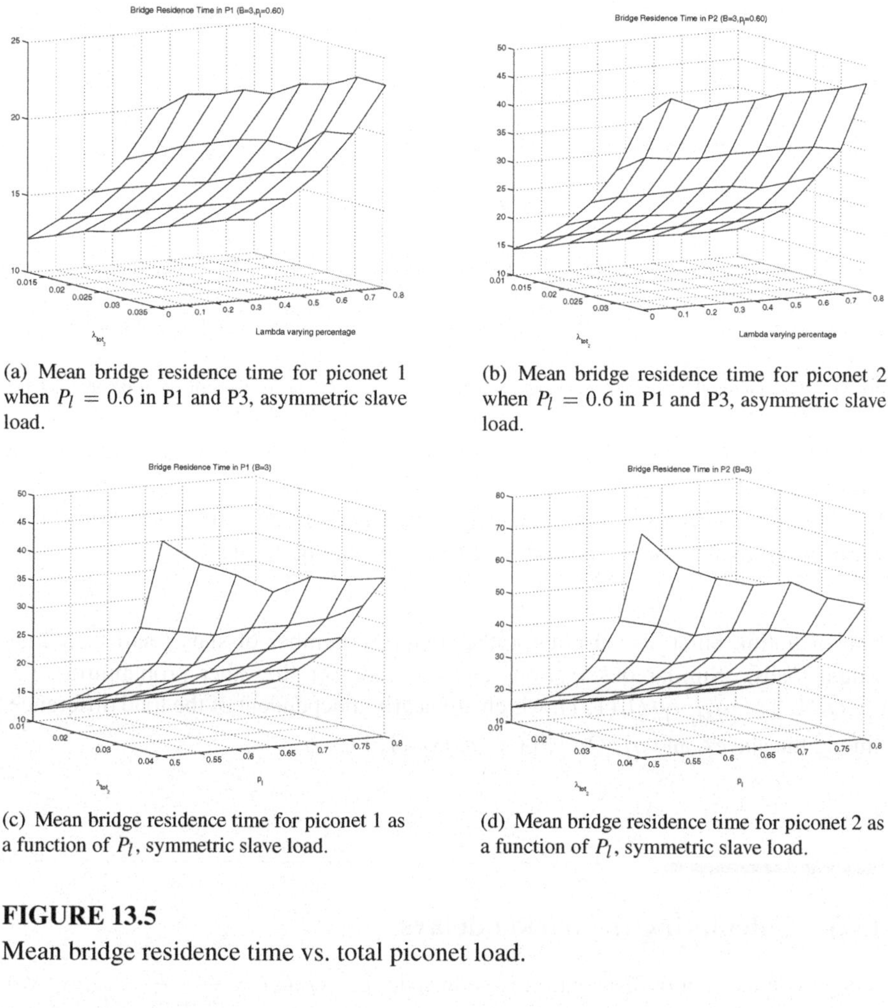

(a) Mean bridge residence time for piconet 1 when $P_l = 0.6$ in P1 and P3, asymmetric slave load.

(b) Mean bridge residence time for piconet 2 when $P_l = 0.6$ in P1 and P3, asymmetric slave load.

(c) Mean bridge residence time for piconet 1 as a function of P_l, symmetric slave load.

(d) Mean bridge residence time for piconet 2 as a function of P_l, symmetric slave load.

FIGURE 13.5
Mean bridge residence time vs. total piconet load.

sensitive to the amount of bridge traffic; this dependency becomes more pronounced at higher total load.

Bridge residence time, shown in Fig. 13.5, exhibits similar behavior.

Similar observation holds for the marginal probability that the slave's queue is empty upon the return from the vacation $q_0^i/Q_i(1,1) = \sum_{jd=0}^{\infty} q_{0,jd}^i/Q_i(1,1)$. (The corresponding dependencies are shown in Fig. 13.6.) Due to the coupling among the uplink queues of the slaves and bridges, the values of $q_0^i/Q_i(1,1)$ at both slaves and bridges are insensitive to the variability of arrival rates among the slaves. This phenomenon has been explained in Chapter 3, in the context of simple piconets operating under E-limited scheduling.

By extension, the value of the probability $q_0^i/Q_i(1,1)$ primarily depends on the

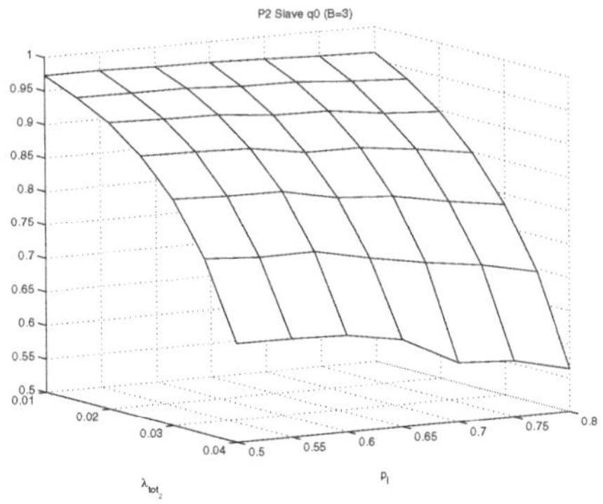

(a) $q_0^i/Q_i(1,1)$ for the slave in piconet 2 when λ_{tot} and P_l are varying.

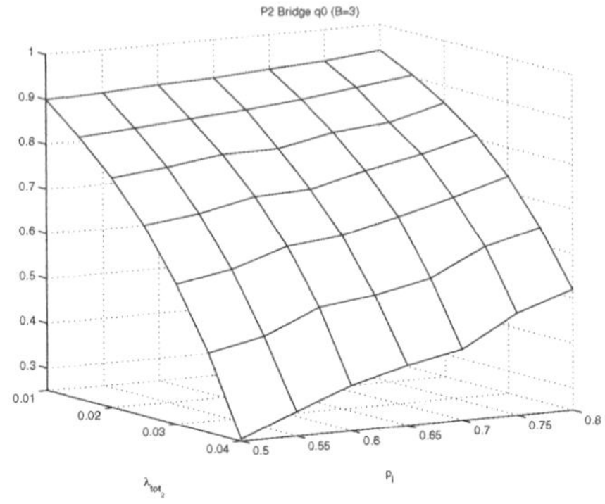

(b) $q_0^i/Q_i(1,1)$ for the bridge in piconet 2 when λ_{tot} and P_l are varying.

FIGURE 13.6
Probabilities that the slave uplink queue contains no packets upon return from the vacation vs. total piconet load and load variation among the slaves.

sum of uplink burst arrival rates, rather than the number of slaves and their individual loads. Similar observations can be made for the other probabilities, i.e., $q_1^i/Q_i(1,1)\,..\,q_{M_s-1}^i/Q_i(1,1)$, which are nearly independent of the load differences among the slaves, and $q_1^i/Q_i(1,1) \ll q_0^i/Q_i(1,1)$.

13.3 Calculating the packet delays

Since the burst length distribution is geometric, i.e., $G_b(z) = \dfrac{z}{\overline{B} + z - z\overline{B}}$, we will introduce the substitution $s = \lambda_{iu} - \lambda_{iu} z/(\overline{B} + z - z\overline{B})$ in the expression for $Q_i(z, w)$. By using the decomposition principle [Takagi, 1991], the LST for the packet access delay at the slave uplink queue becomes:

$$W_{ai}^*(s) = \frac{s(1 - \lambda_{iu}\overline{F_{is}B})}{s - \lambda_{iu} + \lambda_{iu} G_b(F_{is}^*(s))} \cdot \frac{1 - G_b(F_{is}^*(s))}{\overline{B}(1 - F_{is}^*(s))} \cdot \frac{1 - V_i^*(s)}{s\overline{V_i}} \cdot \frac{Q_i\left(1 - \dfrac{s}{\lambda_{iu}\overline{B} - s\overline{B} + s}, 1\right)}{Q_i(1,1)V_i^*(s)} \tag{13.44}$$

The first term in this expression corresponds to the time needed to serve the first packet in the burst in the $M^{[x]}/G/1$ system. The second term corresponds to the time needed to serve the given target packet in the burst. The third term corresponds to the time needed to serve packets which arrive during the vacation, but before the target burst. Finally, the last term corresponds to time needed to serve packets which were already in the uplink queue when the vacation had started.

Mean value of the access delay is obtained as $\overline{W_{ai}} = -W_{ai}^{*'}(0)$, which amounts to

$$\overline{W_{ai}} = \frac{\lambda_{iu}\overline{B}(\overline{F_{is}^2})}{(1-\lambda_{iu}\overline{F_{is}}\overline{B})} + \frac{\overline{B^{(2)}}\overline{2}}{2\overline{B}(1-\lambda_{iu}\overline{F_{is}}\overline{B})} + \frac{\overline{V_i^2}}{2\overline{V_i}} - \overline{V_i} + \frac{Q_i'(1,1)}{\lambda_{iu}\overline{B}Q_i(1,1)} \tag{13.45}$$

where $\overline{V_i^2} = V_i^{*''}(0)$.

Two properties of the mean access delay may be deduced from the last expression:

1. Under constant offered load $\rho_i = 2\lambda_{iu}\overline{LB}$, and under fixed value of M_s, the access delay will increase with the mean burst size $\overline{B}$. This increase is due to the increase of the second factorial moment of the burst (second term) and the increased number of packets in the uplink queue at the beginning of the vacation (last term).

2. Under constant offered load and fixed mean burst size $\overline{B}$, the mean access delay decreases when the value of M increases. This is due to the decreased number of packets in the uplink queue at the beginning of vacation (fourth and fifth terms in last expression).

Simulation results confirm these observations, as can be seen from Fig. 13.7.

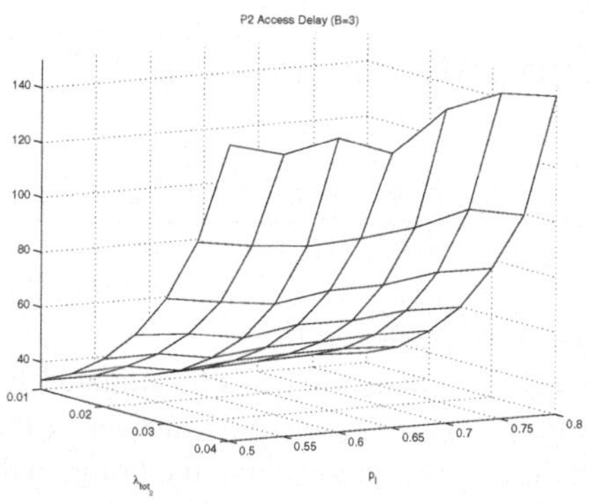

(a) Access delay in piconet 2 when $P_l = 0.6$ with asymmetric slaves.

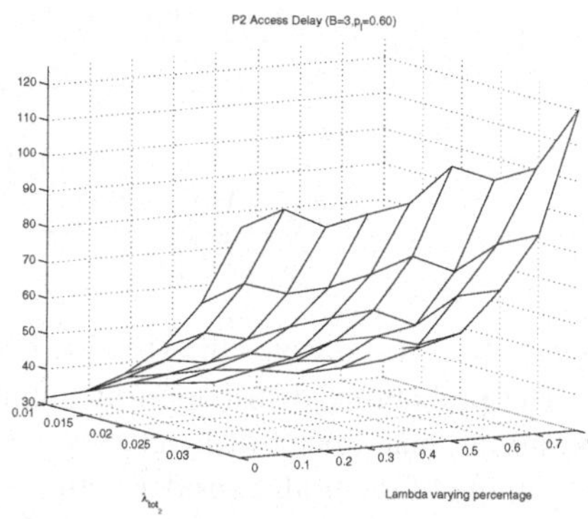

(b) Access delay in piconet 2 when P_l and total arrival rate are varying.

FIGURE 13.7

Mean access delay in piconet 2 for $M_{b1} = M_{b2} = 6$, $M_s = 3$, and $\overline{B} = 3$.

The calculation of downlink service and vacation time, and subsequently of the downlink queueing delay, proceeds along a similar path as that for the corresponding variables in the uplink, except that the modified values of λ_{id} and the PGF $G_{b,i,d}(z)$ should be taken into account. The complete calculation of modified burst size distribution and downlink packet burst arrival rate is similar to the one given in Chapter 3; the important conclusion is that the original burst size is almost preserved if the polling parameter M_s is equal or larger than the average burst size. Let us introduce the substitution $u = \lambda_{id} - \lambda_{id} w/(\overline{B} + w - w\overline{B})$ in the expression for $Q_i(z, w)$. The LST for the downlink delay then becomes

$$W_{di}^*(u) = \frac{u(1 - \lambda_{id}\overline{F_{is}}\,\overline{B})}{u - \lambda_{id} + \lambda_{id} G_{b,i,d}(F_{is}^*(u))} \cdot \frac{1 - G_{b,i,d}(F_{is}^*(u))}{\overline{B}(1 - F_{is}^*(u))} \cdot \frac{1 - V_i^*(u)}{u\overline{V_i}} \cdot \frac{Q_i\left(1, 1 - \dfrac{u}{\lambda_{i,d}\overline{B} - u\overline{B} + u}\right)}{Q_i(1,1)V_i^*(u)} \tag{13.46}$$

and the mean downlink delay is

$$\overline{W_{di}} = \frac{\lambda_{id}\overline{B_{i,d}}\,\overline{L^2}}{(1 - \lambda_{id}\overline{F_{is}}\,\overline{B_{i,d}})} + \frac{\overline{B_{i,d}^{(2)}}\,\overline{F_{is}}}{2\overline{B_{i,d}}(1 - 2\lambda_{id}\overline{F_{is}}\,\overline{B_{i,d}})} + \frac{\overline{V_i^2}}{2\overline{V_i}} - \overline{V_i} + \frac{Q_i'(1,1)}{\lambda_{id}\overline{B_{i,d}}Q_i(1,1)} \tag{13.47}$$

The downlink bridge delay in piconet n has the same form as the downlink delay toward the destination slave.

$$Wb_{n,x,d}^*(s) = \frac{s(1 - 2\lambda_{b_{n,x}}\overline{L}\,\overline{B_d})}{s - \lambda_{b_{n,x}} + \lambda_{b_{n,x}} G_{b,d,i}\left((G_p^*(s))^2\right)} \cdot \frac{1 - G_{b,d,i}\left((G_p^*(s))^2\right)}{\overline{B_d}\left(1 - \left(G_p^*(s))^2\right)\right)} \cdot \frac{1 - Vb_{n,x}^*(s)}{s\overline{Vb_{n,x}}} \cdot \frac{Q_i\left(1, 1 - \dfrac{s}{\lambda_{b_{n,x}}\overline{B_{i,d}} - s\overline{B_{i,d}} + s}\right)}{Q_i(1,1)Vb_{n,x}^*(s)} \tag{13.48}$$

where $Vb_{n,x}^*(s)$ denotes the vacation time observed by the bridge.

The probability distribution of uplink delays in piconet x can be estimated by observing that a burst of length given with PGF $N_{n,x}(z)$ from 13.31, arrives at the beginning of the bridge residence time. After the arrival of the burst, the bridge will be polled in the E-limited mode, i.e., M_{bn} packets will be transmitted in one piconet cycle. Therefore, the LST for the uplink bridge's delay will be:

$$Wb_{n,x,u}^*(s) = \frac{1 - N_{n,x}(C_x^*(\frac{s}{M_{bx}}))}{\overline{N_{n,x}}(1 - C_x^*(\frac{s}{M_{bx}}))} \tag{13.49}$$

13.4 Stability considerations

The mechanism of scatternet operation affects the probability distribution of service times for each device, and therefore affects the stability of all the queues. Scatternet stability has two aspects: first, the queues involved in the bridge exchange must be stable; second, the uplink and downlink queues that correspond to ordinary slaves must be stable. As might have been expected, both aspects of stability are critically dependent on the bridge management policy.

Bridge stability condition means that the mean number of packets to arrive in the master's outgoing queue must be less than the mean number of packets delivered to the bridge. From the denominator of (13.21), we obtain the condition

$$\frac{M_{bn}}{\lambda_{b_{n,x}}\overline{B}\,\overline{C_n}}\frac{M_{bx}}{\lambda_{b_{n,x}}\overline{B}\,\overline{C_x}} - \frac{M_b}{\lambda_{b_{n,x}}\overline{B}\,\overline{C_n}} - \frac{M_{bx}}{\lambda_{b_{n,x}}\overline{B}\,\overline{C_x}} > 0 \tag{13.50}$$

Since the mean cycle time is a monotonically increasing function of the total load, and $\lambda_{b_{n,x}}$ is the fraction of the total load, the upper bound for bridge packet arrival rate is attained when the cycle time reaches its maximum:

$$M_{bn}M_{bx} - M_b\lambda_{b_{n,x}}\overline{B}\,\overline{C_x^{max}} - M_{bx}\lambda_{b_{n,x}}\overline{B}\,\overline{C_n^{max}} > 0 \tag{13.51}$$

where the maximum value of mean cycle time $\overline{C_n^{max}}$ may be determined from (13.43):

$$\overline{C_n^{max}} = \min\left(\left(2(m_n - b_n - 1)M_s\overline{L} + 2b_nM_b\overline{L}\right), \frac{2(m_n - b_n - 1)M_s + 2b_n}{1 - \overline{B}\left(\overline{L} - \frac{1}{M_b}\right)\sum_{j=2}^{b_n+1}\left(\lambda_{b_{n,x_j}} + \lambda_{b_{x_j,n}}\right)} \right) \tag{13.52}$$

The value for $\overline{C_x^{max}}$ may be determined in similar way.

By solving (13.51) for $\lambda_{b_{n,x}}$, we may obtain the limit on maximum total load in the piconet since $\lambda_{b_{n,x}} = (1 - P_l)\sum_{\substack{i=2 \\ i\neq j}}^{m_n} \lambda_{iu}$ where λ_{iu} denotes uplink arrival rate for slave i in piconet n. By solving for λ_{iu}, the maximum total load for stable operation of bridge can be determined.

On the other hand, the slave uplink queue will be stable if the average number of packets that arrive during the average cycle time may be serviced in that time, $\lambda_{iu}\overline{B}\,\overline{C_n^{max}} < M_s$, from which a different λ_{iu} may be found.

Both stability conditions must hold simultaneously, which means that the stable region is always determined by the stricter condition of the two. Of course, all piconets must be stable for the scatternet to be stable.

In the case of limited exchange policy, the bridge residence time is limited to $K+1$ exchange cycles interleaved with K piconet cycles:

$$\lambda_{b_{n,x}}\overline{B}\left(\overline{T_{syn,n,x}} + \overline{T_{syn,x,n}} + K\overline{C_n} + K\overline{C_x} + 2\overline{L}M_b + 2\overline{L}M_b\right) < (K+1)M_b \tag{13.53}$$

and the bridge stability condition becomes

$$\sum_{\substack{i=2 \\ i\neq j}}^{m_n} \lambda_{iu} < \frac{\dfrac{b_n}{(1-P_l)\overline{B}}(K+1)M_b}{\left(\overline{T_{syn,n,x}} + \overline{T_{syn,x,n}} + K\overline{C_n^{max}} + K\overline{C_x^{max}} + 2\overline{L}M_b + 2\overline{L}M_b\right)} \tag{13.54}$$

As before, the slave stability condition dictates that

$$\lambda_{iu}\overline{B}\,\overline{C_n^{max}} < M_s \tag{13.55}$$

and the stability of the piconet is determined by the stricter condition of the two.

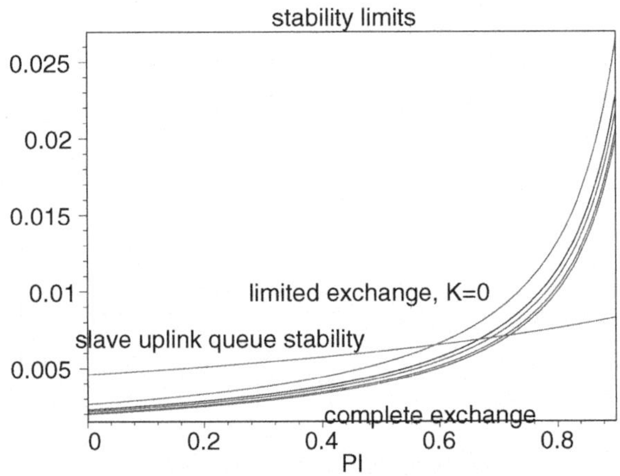

FIGURE 13.8
Stability conditions as functions of traffic locality P_l, with $M_b = 12$. (From V. B. Mišić, J. Mišić, and K. L. Chan, "Walk-in bridge scheduling in Bluetooth scatternets," *Cluster Computing* **8**(2/3):197–210, © 2005 Springer Science + Business Media, Inc. With kind permission of Springer Science and Business Media.)

The maximum allowable packet burst arrival rates for the scatternet with two piconets, for a fixed $M_b = 12$, are shown in Fig. 13.8. The nearly horizontal curve denotes the slave stability condition, which does not depend on the value of K. The family of faster rising curves represent the bridge stability conditions for $K = 0, 1, 2, 3$ and $K \to \infty$ (which corresponds to complete exchange policy). The stable region is below the curve for uplink queue stability *and* the curve for the chosen value of K.

As can be seen, the highest allowable packet burst arrival rate is obtained for $K = 0$, when the bridge stays in each piconet for a single exchange cycle only.

Note also that the limited exchange policy outperforms the complete exchange one whenever inter-piconet traffic is a significant portion of the total traffic. For scatternets with predominantly local traffic, P_l is close to one, and slave stability limit—which does not depend on the bridge management policy—takes precedence.

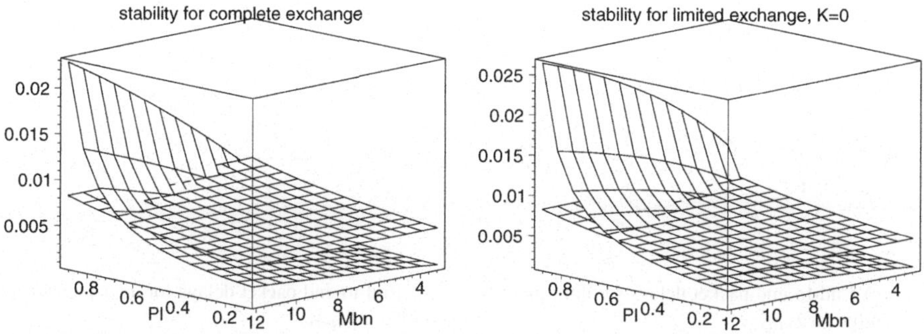

(a) Complete exchange policy.

(b) Limited exchange policy with $K = 0$.

FIGURE 13.9

Stability conditions as the function of traffic locality P_l and the value of M_b. (From V. B. Mišić, J. Mišić, and K. L. Chan, "Walk-in bridge scheduling in Bluetooth scatternets," *Cluster Computing* **8**(2/3):197–210, © 2005 Springer Science + Business Media, Inc. With kind permission of Springer Science and Business Media.)

In Fig. 13.9, maximum allowable burst arrival rates are shown as functions of traffic locality and the number of packets to be exchanged in a single visit, M_b, for $\overline{B} = 3$. The slightly slanted surfaces correspond to the slave stability limit, while the curved surfaces correspond to the bridge stability limits; the stable region is the space below one *and* the other. Again, the slave stability limits the traffic in cases where local traffic is dominant; for higher portion of non-local traffic, bridge stability becomes a more restrictive factor. The limited exchange policy offers a slightly wider stability region than the complete exchange one.

These results were confirmed through delay measurements in the scatternet topology of Fig. 13.3. The scatternet had three piconets operating under limited exchange

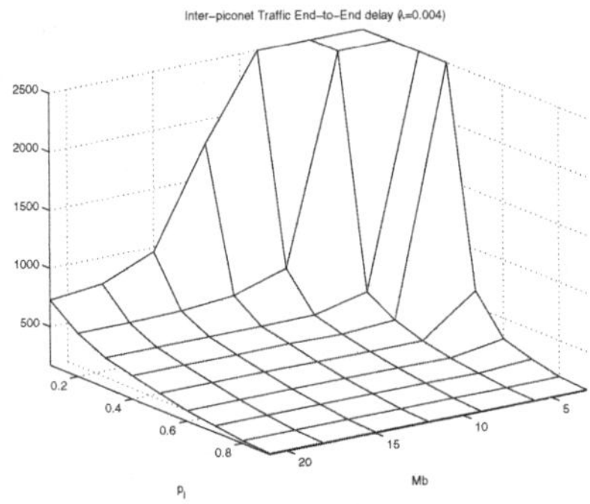

(a) End-to-end delay for inter-piconet traffic, under constant packet burst arrival rate.

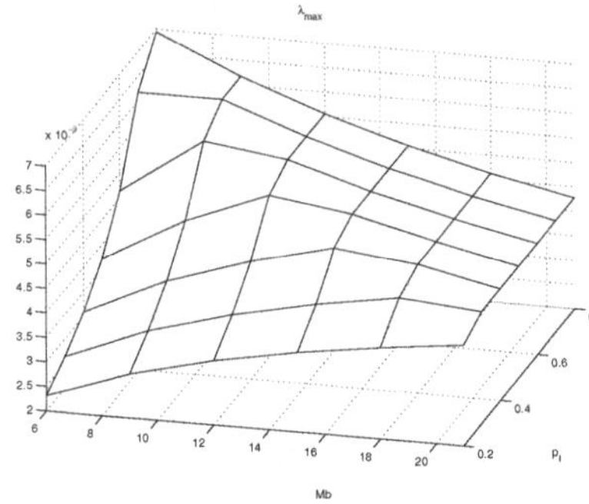

(b) Maximum packet burst arrival rate, λ_{max}, that does not lead to instability.

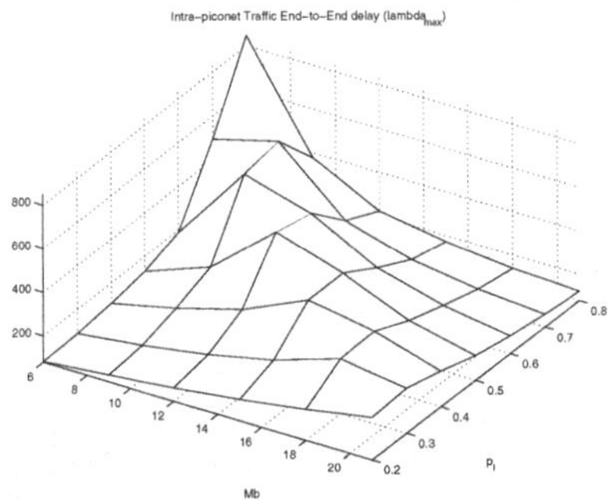

(c) End-to-end packet delays for intra-piconet traffic, at λ_{max}.

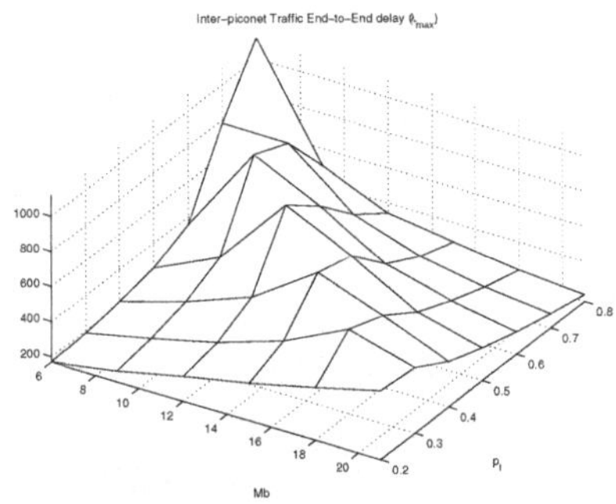

(d) End-to-end packet delays for inter-piconet traffic, at λ_{max}.

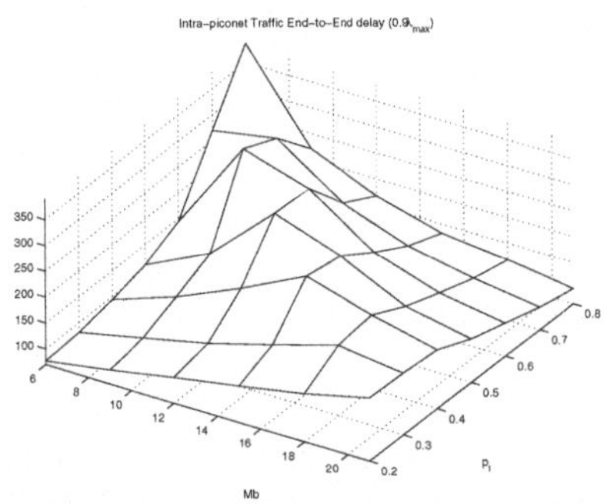

(e) End-to-end packet delays for intra-piconet traffic, at $0.9\lambda_{max}$.

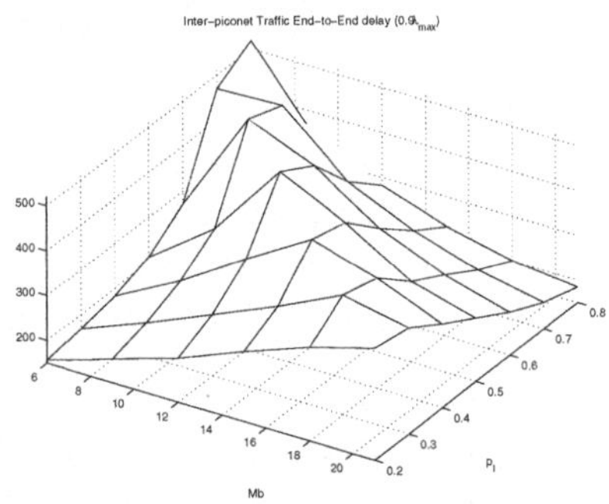

(f) End-to-end packet delays for inter-piconet traffic, at $0.9\lambda_{max}$.

FIGURE 13.10

Pertaining to the stability of the scatternet under walk-in scheduling. (From V. B. Mišić, J. Mišić, and K. L. Chan, "Walk-in bridge scheduling in Bluetooth scatternets," *Cluster Computing* **8**(2/3):197–210, © 2005 Springer Science + Business Media, Inc. With kind permission of Springer Science and Business Media.)

policy with $K = 0$, $\overline{B} = 3$, and $M_s = 3$.

First, Fig. 13.10(a) shows the end-to-end packet delays for inter-piconet traffic under constant packet burst arrival rate of $\lambda = 0.004$ per Bluetooth time slot; this value exceeds the stability limit for some combinations of traffic locality P_l and M_b. As can be seen, the delays rise steeply when the stability condition is exceeded. Second, Fig. 13.10(b) shows the maximum packet burst arrival rates that do not lead to instability (detected through excessive end-to-end packet delays), which closely corresponds to the stability region of Fig. 13.9(b).

End-to-end packet delays for intra- and inter-piconet traffic at the edge of the stability region are shown in Figs. 13.10(c) and 13.10(d), respectively. The delays for inter-piconet traffic are slightly higher, due to the increased number of hops that such traffic has to pass through – four, as opposed to only two for intra-piconet traffic. Note that absolute delays can reach values of $800T = 0.5s$ and above. However, if the arrival rates are reduced to 90% of the maximum values, the delays drop sharply, as shown in Figs. 13.10(e) and 13.10(f).

In general, the simulation results vividly confirm the validity of analytical conclusions presented above.

13.5 Scalability

In order to assess the scalability of the walk-in scheduling approach, we have measured intra- and inter-piconet delays under variable scatternet topology, using limited exchange policy with $K = 0$. In all cases, local traffic has exactly two hops to go – from the source slave to the master, and from the master to the destination slave. Non-local traffic is always targeted toward the adjacent piconet(s), so that it has exactly four hops to go. This choice simplifies routing, which is yet undefined in Bluetooth specification [Bluetooth SIG, 2001*b*], and makes comparisons easier. The packet burst arrival rate was fixed at $\lambda = 0.0045$ per Bluetooth time slot per slave. Traffic locality values were $P_l = 0.6$ in piconets with one bridge only, and $P_l = 0.2$ in piconets with two and three bridges.

Topology 1 is similar to a linear chain in which each piconet (except for the first and the last) has two bridges, and each bridge connects two piconets, as shown in Fig. 13.11(a). Each piconet has five ordinary slaves. Measured values of the throughput for intra- and inter-piconet traffic, as well as their sum, are shown in Fig. 13.11(b), while the mean packet delays in piconets 1 through 9 are shown in Fig. 13.11(c).

As can be seen, the delays in piconets 1 and 2 are essentially independent of the total number of piconets. Since piconet 1 has only one bridge and only one adjacent piconet, whereas piconet 2 has two of each, the total load in piconet 1 is smaller, and its delays are lower.

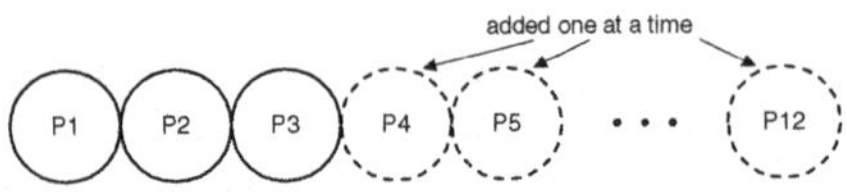

(a) Scatternet topology.

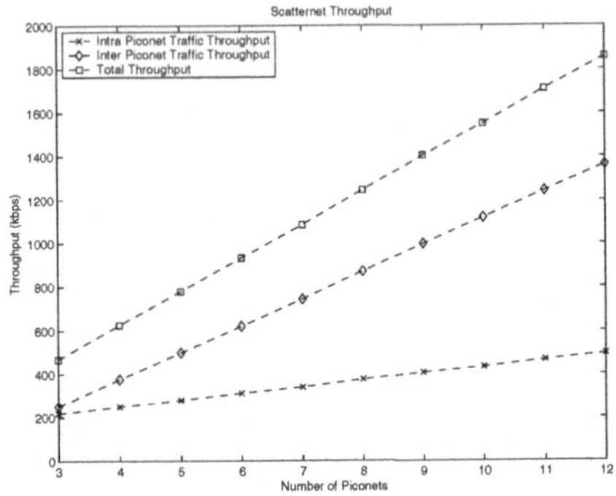

(b) Scatternet throughput.

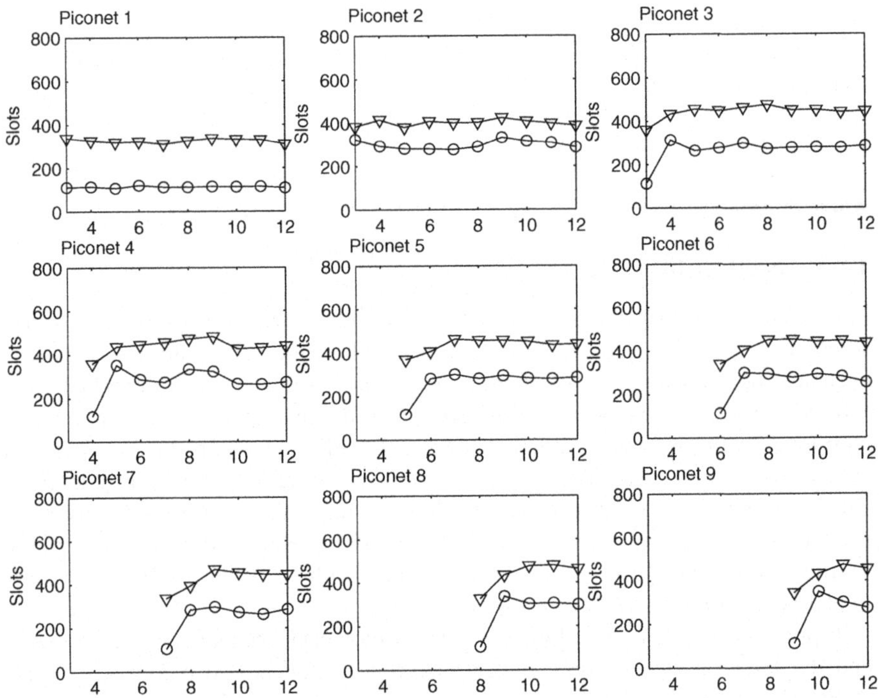

(c) Mean packet delays in piconets 1 through 9: circles denote local delays, triangles denote inter-piconet delays.

FIGURE 13.11
Pertaining to scalability of walk-in scheduling: topology 1. (From V. B. Mišić, J. Mišić, and K. L. Chan, "Walk-in bridge scheduling in Bluetooth scatternets," *Cluster Computing* **8**(2/3):197–210, © 2005 Springer Science + Business Media, Inc. With kind permission of Springer Science and Business Media.)

From the diagrams of delays in piconets 3 to 9, a repetitive pattern may be observed. Each of these piconets first appears at the edge of the scatternet, and initially contains one bridge only; the load in such a piconet is equal to that of piconet 1, $\rho_1 = 0.567$, and so are the delays. In the next step, another piconet is added, together with another bridge, and the edge piconet becomes a 'middle' one. Its total load rises to the level of piconet 2, $\rho_2 = 0.729$, and so do the delays. As more piconets are added, the loads and delays in the 'middle' piconets remain essentially the same.

The overall throughput is a linear function of the number of piconets, and so are its local and non-local components.

Topology 2 is a modified chain shown in Fig. 13.12(a). It has four piconets initially, with two new piconets added at a time, as shown in the diagram. In this case, each piconet has four ordinary slaves. Some piconets (1, 3, 5, ...) have a single bridge in every configuration, piconet 2 has three bridges in every configuration, while the others (4, 6, ...) have one bridge at first, and three bridges thereafter. The total throughput, together with its local and non-local components, is shown in Fig. 13.12(b); again, all of these are linear functions of the number of piconets.

Mean packet delays are shown in Fig. 13.12(c), from which a repetitive pattern may again be observed. The load in piconets 1, 3, and 4 is small at first, $\rho_1 = 0.41$, since each of them has a single bridge only. Piconet 2 has three bridges and much higher load of $\rho_{2a} = 0.71$; consequently, its delays are noticeably higher. When piconets 5 and 6 are added, their load is again small, $\rho_1 = 0.41$, while the piconet 4 is augmented with two more bridges, and its load rises to $\rho_{2b} = 0.67$; the delays rise. At the same time, some of the traffic that went from piconet 4 to piconet 2 is diverted toward piconets 5 and 6. Hence, the total load of piconet 2 is reduced to $\rho_{2b} = 0.67$, and its delays decrease.

As part of the load of piconet 2 comes from/to piconet 3, the reduced delays in piconet 3 will affect the inter-piconet delays in piconet 2: these decrease as well.

Adding piconets 7 and 8 produces similar effects in piconets 6, and (indirectly) in piconets 4 and 5 as well. However, piconet 2 still has three bridges; therefore, its load does not change, and neither do its delays.

From these experiments, two important conclusions may be drawn. First, the total throughput is a linear function of the number of piconets and it is virtually unaffected by the bridge scheduling algorithm. Note that the relative difference in throughput between Figs. 13.11(b) and 13.12(b) is caused by the fact that piconets of topology 1 have five packet-generating slaves each, while those of topology 2 have only four each.

Second, both local and non-local delays in the scatternet primarily depend on the total load of the piconet itself, and—to some extent—on the number of bridges and the bridge load. (Note that the load is affected by the exact position of the piconet in the scatternet topology.) However, delays *do not* depend in any noticeable way on the total number of piconets in the scatternet.

Therefore, we may conclude that the walk-in bridge scheduling provides both excellent performance and scalability, without requiring a schedule of rendezvous points like the other approaches.

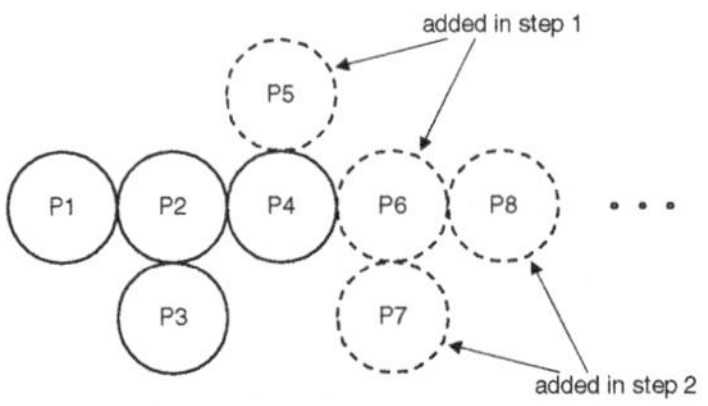

(a) Scatternet topology.

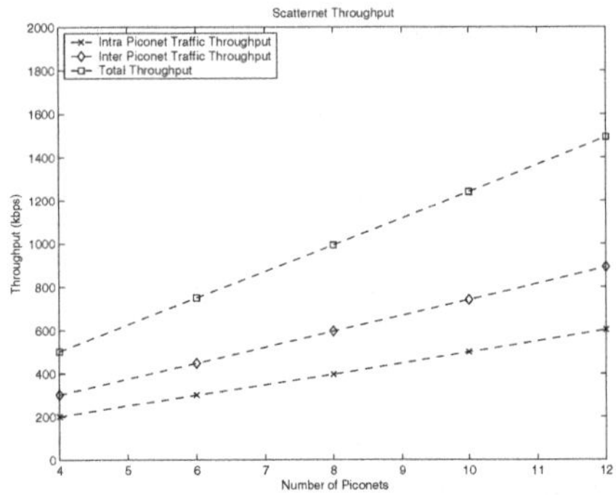

(b) Scatternet throughput.

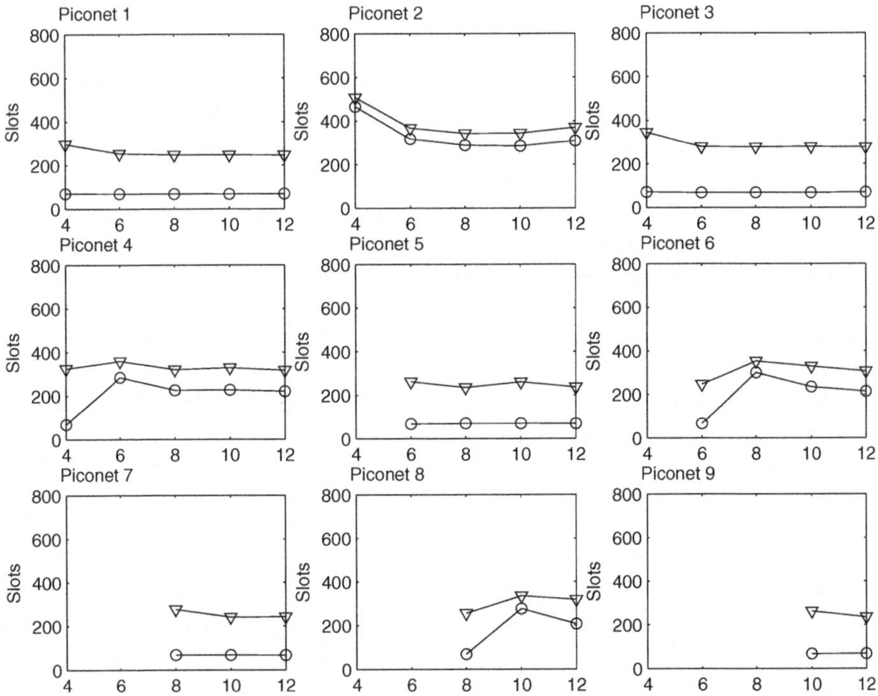

(c) Mean packet delays in piconets 1 through 9: circles denote local delays, triangles denote inter-piconet delays.

FIGURE 13.12

Pertaining to scalability of walk-in scheduling: topology 2. (From V. B. Mišić, J. Mišić, and K. L. Chan, "Walk-in bridge scheduling in Bluetooth scatternets," *Cluster Computing* **8**(2/3):197–210, © 2005 Springer Science + Business Media, Inc. With kind permission of Springer Science and Business Media.)

14

Scatternet with finite buffers

Performance analyses of scatternets operating under different bridge scheduling algorithms have been done under the assumption that buffers with infinite capacity are available to implement different queues in Bluetooth devices. While useful as the first approximation, this assumption is simply unrealistic. Namely, not only that buffers will be finite, but most of them will in fact be rather small in size, in particular those on mobile and/or battery-operated devices.

This chapter presents the performance analysis of the scatternet operating under walk-in bridge scheduling to scatternets where all devices have finite buffers, and where all piconets operate under E-limited polling. After describing our scheduling scheme and the associated queueing model for the joint uplink/downlink queue length at vacation termination time and packet departure times in Section 14.1, we derive the probability distributions for the packet service times (for ordinary slaves, piconet masters, and bridges), vacation times, and piconet cycle times in Section 14.3. Section 14.4 presents the derivation of blocking probabilities and components of end-to-end packet delays within the scatternet. The queueing theoretic analysis is confirmed through simulations, the results of which are presented in Section 14.5. We also present some important conclusions regarding the choice of scheduling parameters and buffer sizes for large scatternets.

14.1 Scatternet model with finite buffers

Let us consider an arbitrary topology scatternet in which a piconet may contain one or more bridges, and a bridge may link two or more scatternets. Our analysis will focus on just two of the piconets, N and X, and the single bridge that links them, as shown in Fig. 14.1. (This will not limit the generality of our analysis, and it may easily be extended to more complex topologies.) The piconets have m_n and m_x members, respectively, which includes one master each and b_n (b_x) bridges. The bridge that links the piconets spends the time T_n (T_x) in piconet N (X); these times will be referred to as bridge residence times, and they are expressed in Bluetooth time slots $T = 625\mu s$.

The operation of each of the piconets in the scatternet may be described with a queueing model shown in Fig. 14.2. The main difference from the queueing model

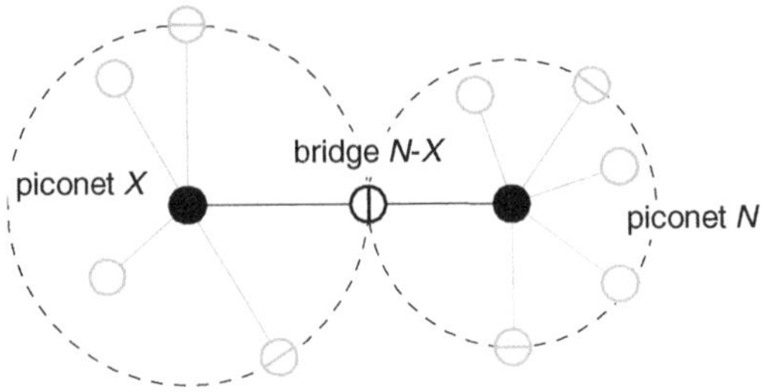

FIGURE 14.1
Part of the scatternet: two piconets linked through a single bridge.

described in Chapter 10 and subsequently used as the foundation for analysis in Chapters 11 to 13, lies in the fact that the bridges have finite sizes: K_u, K_d, and K_b, for slaves, master, and bridge queues, respectively.

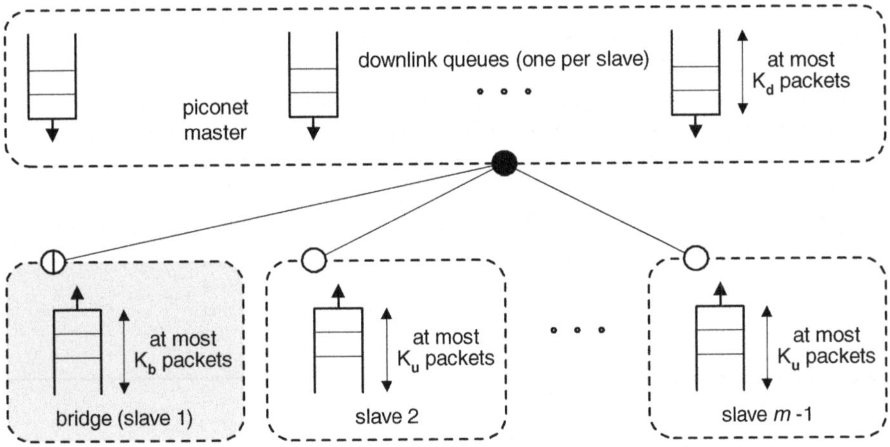

FIGURE 14.2
Pertaining to Bluetooth scatternet operation

We assume that both piconet masters poll their slaves using E-limited polling. Up to M_s data packets are exchanged between the master and an ordinary slave during a single master's visit to a slave. We assume that all piconet masters use the same value for M_s. All devices are assumed to use the same segmentation/reassembly protocol; therefore, the mean burst size has the same value for all devices. In this case, the value of $M_s \approx \overline{B}$ can preserve the bursts from uplink to downlink queues, and lead to minimal end-to-end packet delays, as shown in Chapter 3.

According to the walk-in bridge scheduling approach, the bridge switches between the piconets in a round robin fashion without a predefined schedule. When present, the bridge is polled just like any other slave, except that the first packet to be sent to the bridge is always an empty POLL. The absence of a response to the initial POLL packet means that the bridge is not present, in which case the master simply moves on to the next slave. If the bridge is present, it responds with a NULL; the master then starts the exchange with actual data packets.

The exchange lasts for M_b packets, or less if both queues are emptied. Both piconet masters use the same value for M_b, but separate values for each piconet, or even each bridge, may be readily accommodated by our model. If the emptying of both queues is explicitly signalized with a POLL/NULL frame, as is common in Bluetooth, the bridge exchanges last at most $M_b + 2$ frames, with the first and last frames taking only two slots each.

Under the complete exchange policy, the bridge stays in the piconet for as many cycles as is necessary to allow all queued packets to be exchanged at once. Under the limited exchange policy, the bridge stays in the piconet for a limited time only; this time is expressed in piconet cycles. In the analysis that follows, we will consider the performance of both policies.

If the bridge burst arrival rates are known, each piconet can be analyzed separately using the analytical model for E-limited polling given in Chapter 4. (Of course, the different values for the polling parameter M_b and packet burst arrival rate have to be taken into account.) Therefore, we have to determine the joint probability distributions of the uplink/downlink queue lengths at the moments of uplink packet departure. This in turn requires the solution of the Markov chain containing imbedded Markov points that correspond to the vacation termination times and uplink packet departure times. There are $M_s + 1$ imbedded Markov points for an ordinary slave, and $M_b + 1$ such points for the bridge – when it is present in the piconet. The resulting probability distribution of the uplink/downlink queue length in Markov points for slave i is represented with $q^i_{k_u,k_d}$ (where $k_u = 0, 1 \,..\, K_u$, $k_d = 0, 1 \,..\, K_d$), while $\pi^{i,(\mu)}_{k_u,k_d}$ (where $\mu = 1 \,..\, M_s$) denote queue lengths at the return from vacation and queue length after uplink packet departures, respectively. The total number of equations that describe the uplink and downlink queue probability distribution in Markov points for a single slave is, therefore, $(K_u + 1)(K_d + 1)(M_s + 1)$.

Using these distributions, we can determine the other distributions that are of interest, i.e., the probability distributions of frame service time and channel service times for ordinary slave i (fortunately these times depend only on the queue length probability distributions in Markov points), channel service time for bridges, the corresponding vacation times for ordinary slave and the cycle time, and probability distribution of the queue length at arbitrary time which is needed to calculate the blocking probability and access delay.

As in the case of infinite buffers, there is a recursive relationship between those values since the probability distributions of the queue lengths depend on the vacation times. Vacation times in turn depend on the channel service times and, by extension, on the probability distributions of the queue lengths of other piconet members.

14.2 Uplink/downlink queue length distribution in Markov points

Given that bridge arrival rates are known, and vacation times for individual slaves can be found, each slave can be described with the set of equations which model slave's uplink/downlink queue length at the return from the vacation and after each uplink transmission.

Let $f_{is}(x)$ and $v_i(x)$ stand for the pdfs of the frame transmission time and vacation time, respectively, at the uplink queue of slave i; the corresponding LST transforms will be $F_{is}^*(s)$ and $V_i^*(s)$, respectively. We will also make use of the following probabilities:

- The probability of k_u packet arrivals at the slave i's uplink queue during the frame time, denoted with a_{k_u};
- The probability of k_d packet arrivals at the master's downlink queue during the frame time, denoted with a_{k_d};
- The probability of k_u packet arrivals at the slave i's uplink queue during the vacation time (i.e., while the master is servicing other slaves), denoted with f_{k_u};
- The probability of k_d packet arrivals in the master's downlink during the vacation time, denoted with f_{k_d}.

These probabilities can be calculated as

$$
\begin{aligned}
a_{k_u} &= \sum_{l=0}^{\infty} \frac{1}{k_u!} \frac{d^{k_u}}{dz^{k_u}} (G_b(z))^l \Big|_{z=0} \int_0^{\infty} \frac{(\lambda_{iu}x)^l}{l!} e^{-\lambda_{iu}x} f_{is}(x)dx \\
&= \frac{1}{k_u!} \frac{d^{k_u}}{dz^{k_u}} \left(F_{is}^*(\lambda_{iu} - \lambda_{iu}G_b(z))\right)\Big|_{z=0} \\
a_{k_d} &= \sum_{l=0}^{\infty} \frac{1}{k_d!} \frac{d^{k_d}}{dz^{k_d}} (G_b(z))^l \Big|_{z=0} \int_0^{\infty} \frac{(\lambda_{id}x)^l}{l!} e^{-\lambda_{id}x} f_{is}(x)dx \\
&= \frac{1}{k_d!} \frac{d^{k_d}}{dz^{k_d}} \left(F_{is}^*(\lambda_{id} - \lambda_{id}G_b(z))\right)\Big|_{z=0} \\
f_{k_u} &= \sum_{l=0}^{\infty} \frac{1}{k_u!} \frac{d^{k_u}}{dz^{k_u}} (G_b(z))^l \Big|_{z=0} \int_0^{\infty} \frac{(\lambda_{iu}x)^l}{l!} e^{-\lambda_{iu}x} v_i(x)dx \\
&= \frac{1}{k_u!} \frac{d^{k_u}}{dz^{k_u}} V_i^*(\lambda_{iu} - \lambda_{iu}G_b(z))\Big|_{z=0} \\
f_{k_d} &= \sum_{l=0}^{\infty} \frac{1}{k_d!} \frac{d^{k_d}}{dz^{k_d}} (G_b(z))^l \Big|_{z=0} \int_0^{\infty} \frac{(\lambda_{iu}x)^l}{l!} e^{-\lambda_{iu}x} v_i(x)dx \\
&= \frac{1}{k_d!} \frac{d^{k_d}}{dz^{k_d}} V_i^*(\lambda_{iu} - \lambda_{iu}G_b(z))\Big|_{z=0}
\end{aligned}
\tag{14.1}
$$

Similar expressions hold for the number of arrivals in the downlink queue. The probabilities that the uplink queue of the (ordinary) slave i contains k_u packets and that the downlink queue toward that slave contains k_d packets in imbedded Markov points, satisfy the following equations:

$$
\begin{aligned}
\pi^{i,(1)}_{k_u,k_d} &= \sum_{j_u=1}^{k_u+1}\sum_{j_d=1}^{k_d+1} q^i_{j_u,j_d} a_{k_u-j_u+1} a_{k_d-j_d+1} + \sum_{j_d=1}^{k_d+1} q^i_{0,j_d} a_{k_d-j_d+1} a_{k_u} \\
&+ \sum_{j_u=1}^{k_u+1} q^i_{j_u,0} a_{k_u-j_u+1} a_{k_d}, \\
&\qquad 0 \le k_u \le K_u-2, 0 \le k_d \le K_d-2 \\
\pi^{i,(1)}_{K_u-1,k_d} &= \sum_{j_u=1}^{K_u}\sum_{j_d=1}^{k_d+1} q^i_{j_u,j_d} \left(\sum_{k_u=K_u-j_u}^{\infty} a_{k_u}\right) a_{k_d-j_d+1} \\
&+ \sum_{j_d=1}^{k_d+1} q^i_{0,j_d} a_{k_d-j_d+1} \sum_{k_u=K_u}^{\infty} a_{k_u} + \sum_{j_u=1}^{k_u+1} q^i_{j_u,0} a_{k_d} \sum_{k_u=K_u-j_u}^{\infty} a_{k_u}, \\
&\qquad 0 \le k_d \le K_d-2 \\
\pi^{i,(1)}_{k_u,K_d-1} &= \sum_{j_u=1}^{k_u+1}\sum_{j_d=1}^{K_d} q^i_{j_u,j_d} a_{k_u-j_u+1} \sum_{k_d=K_d-j_d}^{\infty} a_{k_d} \\
&+ \sum_{j_d=1}^{K_d} q^i_{0,j_d} a_{k_u} \sum_{k_d=K_d-j_d}^{\infty} a_{k_d} + \sum_{j_u=1}^{k_u+1} q^i_{0,j_d} a_{k_u-j_u+1} \sum_{k_d=K_d}^{\infty} a_{k_d}, \qquad (14.2) \\
&\qquad 0 \le k_u \le K_u-2 \\
\pi^{i,(\mu)}_{k_u,k_d} &= \sum_{j_u=1}^{k_u+1}\sum_{j_d=1}^{k_d+1} \pi^{i,(\mu-1)}_{j_u,j_d} a_{k_u-j_u+1} a_{k_d-j_d+1} \\
&+ \sum_{j_d=1}^{k_d+1} \pi^{i,(\mu-1)}_{0,j_d} a_{k_d-j_d+1} a_{k_u} + \sum_{j_u=1}^{k_u+1} \pi^{i,(\mu-1)}_{j_u,0} a_{k_u-j_u+1} a_{k_d}, \\
&\qquad 0 \le k_u \le K_u-2, 0 \le k_d \le K_d-2,\ \mu = 2\,..\,M_s \\
\pi^{i,(\mu)}_{K_u-1,k_d} &= \sum_{j_u=1}^{K_u}\sum_{j_d=1}^{k_d+1} \pi^{i,(\mu-1)}_{j_u,j_d} \left(\sum_{k_u=K_u-j_u}^{\infty} a_{k_u-j_u+1}\right) a_{k_d-j_d+1} \\
&+ \sum_{j_d=1}^{k_d+1} \pi^{i,(\mu-1)}_{0,j_d} a_{k_d-j_d+1} \sum_{k_u=K_u}^{\infty} a_{k_u} \\
&+ \sum_{j_u=1}^{k_u+1} \pi^{i,(\mu-1)}_{j_u,0} a_{k_d} \sum_{k_u=K_u-j_u}^{\infty} a_{k_u-j_u+1}, \\
&\qquad 0 \le k_d \le K_d-2,\ \mu = 2\,..\,M_s
\end{aligned}
$$

continued on next page . . .

... continued from previous page

$$
\begin{aligned}
\pi^{i,(\mu)}_{k_u,K_d-1} &= \sum_{j_u=1}^{k_u+1}\sum_{j_d=1}^{K_d}\pi^{i,(\mu-1)}_{j_u,j_d}a_{k_u-j_u+1}\sum_{k_d=K_d-j_d}^{\infty}a_{k_d-j_d+1} \\
&\quad+\sum_{j_d=1}^{K_d}\pi^{i,(\mu-1)}_{0,j_d}a_{k_u}\sum_{k_d=K_d-j_d}^{\infty}a_{k_d} \\
&\quad+\sum_{j_u=1}^{k_u+1}\pi^{i,(\mu-1)}_{j_u,0}a_{k_u-j_u+1}\sum_{k_d=K_d}^{\infty}a_{k_d}, \\
&\qquad 0 \le k_u \le K_u-2,\ \mu = 2\,..\,M_s \\
q^i_{k_u,k_d} &= \left(\sum_{m=1}^{M-1}\pi^{i,(\mu)}_{0,0}+q^i_{0,0}\right)f_{k_u}f_{k_d}+\sum_{j_u=0}^{k_u}\sum_{j_d=0}^{k_d}\pi^{i,(\mu)}_{j_u,j_d}f_{k_u-j_u}f_{k_d-j_d}, \\
&\qquad 0 \le k_u \le K_u-2,\ 0 \le k_d \le K_d-2 \qquad (14.2) \\
q^i_{K_u-1,k_d} &= \left(\sum_{\mu=1}^{M-1}\pi^{i,(\mu)}_{0,0}+q^i_{0,0}\right)f_{k_d}\sum_{k_u=K_u}^{\infty}f_{k_u} \\
&\quad+\sum_{j_u=0}^{k_u}\sum_{j_d=0}^{k_d}\pi^{i,(\mu)}_{j_u,j_d}f_{k_d-j_d}\sum_{k_u=K_u-j_u}^{\infty}f_{k_u} \\
&\qquad 0 \le k_d \le K_d-2 \\
q^i_{k_u,K_d-1} &= \left(\sum_{\mu=1}^{M-1}\pi^{i,(\mu)}_{0,0}+q^i_{0,0}\right)f_{k_u}\sum_{k_d=K_d}^{\infty}f_{k_d} \\
&\quad+\sum_{j_u=0}^{k_u}\sum_{j_d=0}^{k_d}\pi^{i,(\mu)}_{j_u,j_d}f_{k_u-j_u}\sum_{k_d=K_d-j_d}^{\infty}f_{k_d}, \\
&\qquad 0 \le k_u \le K_u-2
\end{aligned}
$$

As always, the sum of all probabilities must be equal to 1, hence

$$
\sum_{k_u=0}^{K_u}\sum_{k_d=0}^{K_d}q^i_{k_u,k_d}+\sum_{\mu=1}^{M_s}\sum_{k_u=0}^{K_u-1}\sum_{k_d=0}^{K_d-1}\pi^{i,(\mu)}_{k_u,k_d}=1 \qquad (14.3)
$$

Additional M_s+1 more equations should also be set for (K_u, K_d) cases (this is left as an exercise to the reader). Solving this system of equations gives the probability distribution of uplink and downlink queue lengths in Markov points. As mentioned above, the corresponding probability distributions for bridge queue lengths, when the bridge is present in the piconet, require the system with M_b+1 Markov points. However, before solving it, the probability distributions of the service time and vacation time have to be found as functions of queue length distributions.

14.3 Service, vacation, and cycle times

The channel service time is defined as time for a single visit to the given slave, from the first poll packet of the master until either a total of M_s (or M_b) frames have been exchanged or an empty frame has been encountered. The LST for the channel service time, including empty packets, may be found using the expression derived in Chapter 4 as

$$S_i^*(s) = \sum_{k=0}^{M_s-1} P_{f,k} \prod_{\mu=1}^{k} (F^{*\mu}(s))e^{-2s} + P_{f,M_s} \prod_{\mu=1}^{M_s} F^{*\mu}(s) \qquad (14.4)$$

where $P_{f,k}$ denote the probabilities that the channel service time will take exactly k data frames:

$$\begin{aligned}
P_{f,0} &= \frac{q^i_{0,0}}{\sum_{k_u=0}^{K_u-1}\sum_{k_d=0}^{K_d-1} q^i_{k_u,k_d}} \\
P_{f,1} &= \left(1 - \frac{q^i_{0,0}}{\sum_{k_u=0}^{K_u-1}\sum_{k_d=0}^{K_d-1} q^i_{k_u,k_d}}\right) \frac{\pi^{i,(1)}_{0,0}}{\sum_{k_u=0}^{K_u-1}\sum_{k_d=0}^{K_d-1} \pi^{i,(1)}_{k_u,k_d}} \\
P_{f,k} &= \left(1 - \frac{q^i_{0,0}}{\sum_{k_u=0}^{K_u-1}\sum_{k_d=0}^{K_d-1} q^i_{k_u,k_d}}\right) \\
&\cdot \prod_{\mu=1}^{k-1}\left(1 - \frac{\pi^{i,(\mu)}_{0,0}}{\sum_{k_u=0}^{K_u-1}\sum_{k_d=0}^{K_d-1} \pi^{i,(\mu)}_{k_u,k_d}}\right) \cdot \frac{\pi^{i,(k)}_{0,0}}{\sum_{k_u=0}^{K_u-1}\sum_{k_d=0}^{K_d-1} \pi^{i,(k)}_{k_u,k_d}}, \\
&\qquad k = 2\,..\,M_s - 1 \\
P_{f,M_s} &= 1 - \sum_{k=0}^{M_s-1} P_{f,k}
\end{aligned} \qquad (14.5)$$

The LST for the duration of $\mu - th$ frame ($\mu = 1\,..\,M_s$) is

$$F^{*1}(s) = \frac{\sum_{k_u=1}^{K_u-1} q^i_{k_u,0} + \sum_{k_d=1}^{K_d-1} q^i_{0,k_d}}{\sum_{k_u=0}^{K_u-1}\sum_{k_d=0}^{K_d-1} q^i_{k_u,k_d}} G_p^*(s)e^{-s} + \frac{\sum_{k_u=1}^{K_u-1}\sum_{k_d=1}^{K_d-1} q^i_{k_u,k_d}}{\sum_{k_u=0}^{K_u-1}\sum_{k_d=0}^{K_d-1} q^i_{k_u,k_d}} (G_p^*(s))^2$$

$$F^{*\mu}(s) = \frac{\sum_{k_u=1}^{K_u-1} \pi^{i,(\mu)}_{k_u,0} + \sum_{k_d=1}^{K_d-1} \pi^{i,(\mu)}_{0,k_d}}{\sum_{k_u=0}^{K_u-1}\sum_{k_d=0}^{K_d-1} \pi^{i,(\mu)}_{k_u,k_d}} G_p^*(s)e^{-s} + \frac{\sum_{k_u=1}^{K_u-1}\sum_{k_d=1}^{K_d-1} \pi^{i,(\mu)}_{k_u,k_d}}{\sum_{k_u=0}^{K_u-1}\sum_{k_d=0}^{K_d-1} \pi^{i,(\mu)}_{k_u,k_d}} G_p^*(s)^2,$$

$$\mu = 2\,..\,M_s \tag{14.6}$$

Mean value of the service time is equal to $\overline{S_i} = -S_i^{*'}(0)$.

Service times for the bridge

In the presence of b_n bridges in the piconet N, and assuming that bridges carry equal traffic, the downlink bridge arrival rate toward the piconet x is

$$\lambda_{b_{n,x}} = \frac{(1-P_l)}{b_n} \sum_{\substack{i=2 \\ i \neq j}}^{m_n} \lambda_{iu} \tag{14.7}$$

where the bridge is the piconet member with the index j.

Let us consider just one bridge, e.g., the one between piconets N and X shown in Fig. 14.1. If the residence times for the bridge are T_n and T_x, respectively, the corresponding PGFs will be $T_n(z)$ and $T_x(z)$. As shown in Chapter 13, upon joining a piconet, the bridge will have to wait for some time before the master polls it. These synchronization times, described with the PGFs $T_{syn,n,x}(z)$ and $T_{syn,x,n}(z)$, may be derived from the results of theory of continuous-time and discrete-time renewal processes [Kleinrock, 1972; Cooper, 1990]. Here we just repeat the main result obtained in Chapter 13, page 239, i.e., that the PGF of synchronization time $T_{syn,x,n}$ is

$$T_{syn,x,n}(z) = \sum_{k=0}^{\infty} z^k \mathcal{P}[T_{syn,x,n} = k] = \frac{1 - C_n^-(z)}{\overline{C_n^-}(1-z)} \tag{14.8}$$

where $\mathcal{P}(x)$ denotes the probability of event x, and $C_n^-(z)$ denotes the PGF of the duration of piconet cycle time when the bridge is absent. Mean synchronization time is equal to one half of the mean cycle time of the piconet when the bridge is absent.

The master's outgoing queue toward the bridge operates in the same manner as the other downlink queues servicing ordinary slaves. Hence, we may describe the uplink/downlink queue for the bridge in Markov points with a system of equations similar to that for an ordinary slave, but using M_b in place of M_s. A total of $(M_b+1)(K_b+1)^2$ equations are needed to fully describe the behavior of the uplink/downlink queue of a bridge. Note, however, that the bridge link will experience larger vacation times than the slave links, due to bridge absences.

We have seen in Chapter 10 and, subsequently, in Chapter 13, that bridge exchanges can be operated under two different policies, referred to as complete and limited exchange. Let us first investigate the performance of the limited exchange policy. As a matter of convenience, let us assume that the bridge carries more traffic from piconet N to piconet X than in the opposite direction. Let us consider the bridge cycle – i.e., the time interval between two successive bridge departures from piconet N – which has the PGF of the form

$$T_{cyc}(z) = T_{syn,n,x}(z)T_x(z)T_{syn,x,n}(z)T_n(z) \tag{14.9}$$

During the bridge cycle, the outgoing queue of the piconet N master was refilled at the rate $\lambda_{b_{n,x}}$.

The probability that the bridge will be present in the piconet when the master attempts to poll it, is

$$H_n = \frac{\overline{T_n}}{\overline{T_n} + \overline{T_x} + \overline{T_{syn,n,x}} + \overline{T_{syn,x,n}}} \tag{14.10}$$

Given that the piconet N has $8 \geq m_n \geq 2$ members (including the master and b_n bridges), the LST for the vacation time observed by an ordinary slave is

$$V_i^*(s) = \prod_{\substack{j=2 \\ j\neq i}}^{m_n} S_j^*(s) \tag{14.11}$$

where $S_i^*(s)$ denotes the LST of the service time for the uplink/downlink queue pair that corresponds to slave i, calculated in (14.4).

The vacation time for the bridge in piconet N corresponds to the ordinary vacation time under E-limited polling, when the bridge is present in the piconet, and to the bridge absence time otherwise; the corresponding LST is

$$V_i^*(s) = H_n \prod_{\substack{j=2 \\ j\neq i}}^{m_n} S_j^*(s) + (1-H_n)T_{syn,n,x}^*(s)T_x^*(s)T_{syn,x,n}^*(s) \tag{14.12}$$

When the vacation time for the bridge is known, equations for queue length in imbedded Markov points can be set for the bridge, together with other bridges and ordinary slaves, and the LST for the service time $Sb_{n,x}^{*j}(s)$ for the bridge j can be calculated from (14.4).

The cycle time of the piconet N with the bridge present can be described with

$$C_n^{*+}(s) = \prod_{i=2}^{m_n} S_i^*(s) \tag{14.13}$$

Knowing the LSTs for the bridge service time and piconet cycle time, we can calculate the bridge residence time in piconet N in the case of limited exchange

$$T_n^*(s) = Sb_{n,x}^{*j}(s)(C_n^{*+}(s))^K \tag{14.14}$$

where $K = 0, 1, 2 \ldots$. The probability that the bridge is present in the piconet may be obtained by substituting mean bridge residence times in both piconets and the corresponding mean synchronization times into (14.10).

Under the complete exchange policy, we begin by determining the number of packet arrivals to the bridge queue during the bridge cycle as

$$N_{n,x}(z) = T_{cyc}^*(\lambda_{b_{n,x}} - \lambda_{b_{n,x}} G_b(z)) \tag{14.15}$$

The PGF for the number of bridge exchanges cycles needed to empty this queue (i.e., deliver those packets to the bridge) is

$$Ns_{n,x}^n(z) = \left\lceil N_{n,x}(z^{\frac{1}{M_b}}) \right\rceil \tag{14.16}$$

where $\lceil \ldots \rceil$ denote that every exponent in the PGF should be rounded to the next larger integer. Then the PGF for the corresponding number of slots, i.e., for the bridge residence time in piconet N is

$$T_n(z) = Ns_{n,x}^n(C_n^+(z)) \tag{14.17}$$

where $C_n^+(z)$ denotes the PGF for the duration of the cycle time in piconet N when the bridge is present.

14.4 Blocking probability and packet delays

In order to calculate the buffer blocking probabilities at slaves and bridges, we have to find probability distribution of uplink/downlink queue length at arbitrary time. This analysis is based on the prior derivation of uplink/downlink queue length distribution at Markov points. Let us first note that LST for the frame duration time for the ordinary slave has the form

$$\begin{aligned} F_{is}^*(s) = &\left(q_{0,0}^i + \sum_{\mu=1}^{M_s} \pi_{0,0}^{i,(\mu)} \right) e^{-2s} \\ &+ \left(\sum_{k_u=1}^{K_u-1} (q_{k_u,0}^i + \sum_{\mu=1}^{M_s} \pi_{k_u,0}^{i,(\mu)}) + \sum_{k_d=1}^{K_d-1} (q_{0,k_d}^i + \sum_{\mu=1}^{M_s} \pi_{0,k_d}^{i,(\mu)}) \right) G_p^*(s) e^{-s} \\ &+ \sum_{k_u=1}^{K_u-1} \sum_{k_d=1}^{K_d-1} \left(q_{k_u,k_d}^i + \sum_{\mu=1}^{M_s} \pi_{k_u,k_d}^{i,(\mu)} \right) (G_p^*(s))^2 \end{aligned} \tag{14.18}$$

Also, let us denote the probability that a vacation starts after the uplink transmission as $h_i = \sum_{\mu=1}^{M-1} \pi_{0,0}^{i,(\mu)} + \sum_{k_u=0}^{K_u-1} \sum_{k_d=0}^{K_d-1} \pi_{k_u,k_d}^{i,(M)}$. The probability that a vacation will start after an arbitrary Markov point is $q_{0,0}^i + h_i$. Then, the average time interval between two consecutive Markov points at slave i is

$$\eta_i = (q_{0,0}^i + h_i)\overline{V_i} + (1 - q_{0,0}^i - h_i)\overline{F_{is}} \tag{14.19}$$

By using the probability distribution of the uplink queue length in Markov points, we will derive the probability distribution of the uplink queue length at arbitrary time between two Markov points, together with the PDF of the remaining vacation time (if the previous Markov point was the start of vacation), or the PDF of the remaining frame service time (if the previous Markov point was the start of the packet service). We will introduce the following variables:

- The pdf of the vacation time $v_i(x)$, and its PDF $V_i(x)$,
- Uplink and downlink queue length at arbitrary time $L_{q,i}$,
- Elapsed vacation time $V_{\_,i}$,
- Remaining vacation time as $V_{+,i}$,
- The number of packet arrivals (results of packet burst arrivals) in the slave's uplink queue during the elapsed vacation time as $A_u(V_\_)$, and corresponding number of arrivals in the downlink queue as $A_d(V_\_)$,
- The pdf of the frame service time $f_{is}(x)$ and its PDF $F_{is}(x)$,
- Elapsed frame service time as $X_{\_,i}$,
- Remaining frame service time as $X_{+,i}$, and
- The number of packet arrivals (results of packet burst arrivals) in the slave's uplink queue during the elapsed frame service time as $A_u(X_\_)$, and corresponding number of arrivals in the downlink queue as $A_d(X_\_)$.

According to the results from the renewal theory [Kleinrock, 1972; Takagi, 1991], in the case of frame service time, the pdf of the elapsed vacation time is $\dfrac{1 - V_i(x)}{\overline{V_i}}$ and pdf of the remaining vacation time is $\dfrac{v_i(x)}{1 - V_i(x)}$.

For the time between the start of the vacation and end of vacation we define the joint probability of the uplink/downlink queue lengths and remaining vacation time as:

$$\Omega^*_{k_u,k_d,i}(s) = \int_0^\infty e^{-sy} \mathcal{P}[L_{q,i} = (k_u, k_d), y < V_{+,i} < y + dy], \qquad 0 \le k_u \le K_u, 0 \le k_d \le K_d \tag{14.20}$$

For the time between the start and end of the frame service for phase $1 \leq \mu \leq M_s$ we define joint probability of the uplink/downlink queue lengths and remaining frame service time as:

$$\Pi^*_{k_u,k_d,\mu,i}(s) = \int_0^\infty e^{-sy}\mathcal{P}[L_{q,i} = (k_u, k_d), y < X_{+,i} < y + dy], \qquad (14.21)$$
$$1 \leq k_u \leq K_u, 1 \leq k_d \leq K_d, 1 \leq \mu \leq M_s$$

Then, by using the probabilities of the uplink queue state in the previous Markov point, we obtain

$$\begin{aligned}
&\Omega^*_{k_u,k_d,i}(s) = \\
&\quad \frac{\overline{V_i}}{\eta_i}\left(q^i_{0,0} + \sum_{\mu=1}^{M_s-1}\pi^{i,(\mu)}_{0,0}\right) E[e^{-sV_{+,i,u}}|A_u(V_{_,i}) = k_u, A_d(V_{_,i}) = k_d] \\
&\qquad \cdot\mathcal{P}[A_u(V_{_,i}) = k_u]\mathcal{P}[A_d(V_{_,i}) = k_d] \\
&\quad + \frac{\overline{V_i}}{\eta_i}\sum_{j_u=0}^{k_u}\sum_{j_d=0}^{k_d}\pi^{i,(M_s)}_{j_u,j_d} E[e^{-sV_{+,i,u}}|A_u(V_{_,i}) = k_u - j_u, A_d(V_{_,i}) = k_d - j_d] \\
&\qquad \cdot\mathcal{P}[A_u(V_{_,i}) = k_u - j_u]\mathcal{P}[A_d(V_{_,i}) = k_d - j_d], \\
&\qquad 0 \leq k_u \leq K_u - 1, 0 \leq k_d \leq K_d - 1 \\
&\Omega^*_{K_u,k_d,i}(s) = \\
&\quad \frac{\overline{V_i}}{\eta_i}\left(q^i_{0,0} + \sum_{\mu=1}^{M_s-1}\pi^{i,(\mu)}_{0,0}\right)\sum_{k_u=K_u}^{\infty} E[e^{-sV_{+,i}}|A_u(V_{_,i}) = k_u, A_d(V_{_,i}) = k_d] \\
&\qquad \cdot\mathcal{P}[A_u(V_{_,i}) = k_u]\mathcal{P}[A_d(V_{_,i}) = k_d] \\
&\quad + \frac{\overline{V_i}}{\eta_i}\sum_{j_u=0}^{k_u}\sum_{j_d=0}^{k_d}\pi^{i,(M)}_{j_u,j_d}\sum_{j_u=0}^{k_u}\sum_{k_u=K_u-j_u}^{\infty} E[e^{-sV_{+,i}}|A_u(V_{_,i}) = k_u, A_d(V_{_,i}) = k_d - j_d] \\
&\qquad \cdot\mathcal{P}[A_u(V_{_,i}) = k_u]\mathcal{P}[A_d(V_{_,i}) = k_d - j_d] \\
&\qquad 0 \leq k_d \leq K_d - 1 \\
&\Omega^*_{k_u,K_d,i}(s) = \\
&\quad \frac{\overline{V_i}}{\eta_i}\left(q^i_{0,0} + \sum_{\mu=1}^{M_s-1}\pi^{i,(\mu)}_{0,0}\right)\sum_{k_d=K_d}^{\infty} E[e^{-sV_{+,i}}|A_u(V_{_,i}) = k_u, A_d(V_{_,i}) = k_d] \\
&\qquad \cdot\mathcal{P}[A_u(V_{_,i}) = k_u]\mathcal{P}[A_d(V_{_,i}) = k_d] \\
&\quad + \frac{\overline{V_i}}{\eta_i}\sum_{j_u=0}^{k_u}\sum_{j_d=0}^{k_d}\pi^{i,(M)}_{j_u,j_d}\sum_{k_d=K_d-j_d}^{\infty} E[e^{-sV_{+,i}}|A_u(V_{_,i}) = k_u - j_u, A_d(V_{_,i}) = k_d] \\
&\qquad \cdot\mathcal{P}[A_u(V_{_,i}) = k_u - j_u]\mathcal{P}[A_d(V_{_,i}) = k_d], \\
&\qquad 0 \leq k_u \leq K_u - 1
\end{aligned}$$

continued on next page ...

(14.22)

... continued from previous page

$$\Pi^*_{k_u,k_d,1,i}(s) = \frac{F_{is}}{\eta_i}\sum_{j_u=1}^{k_u}\sum_{j_d=1}^{k_d} q^i_{j_u,j_d} E[e^{-sX_{+,i}}|A_u(X_{-,i}) = k_u - j_u, A_d(X_{-,i}) = k_d - j_d], \cdot\mathcal{P}[A_u(X_{-,i}) = k_u - j_u]\mathcal{P}[A_d(X_{-,i}) = k_d - j_d] \quad 1 \le k_u \le K_u - 1, 1 \le k_d \le K_d - 1$$

$$\Pi^*_{K_u,k_d,1,i}(s) = \frac{F_{is}}{\eta_i}\sum_{j_u=1}^{K_u}\sum_{j_d=1}^{k_d} q^i_{j_u,j_d}\sum_{k_u=K_u-j_u}^{\infty} E[e^{-sX_{+,i}}|A_u(X_{-,i}) = k_u, A_d(X_{-,i}) = k_d - j_d], \cdot\mathcal{P}[A_u(X_{-,i}) = k_u]\mathcal{P}[A_d(X_{-,i}) = k_d - j_d] \quad 1 \le k_d \le K_d - 1$$

$$\Pi^*_{k_u,K_d,1,i}(s) = \frac{F_{is}}{\eta_i}\sum_{j_u=1}^{k_u}\sum_{j_d=1}^{K_d} q^i_{j_u,j_d}\sum_{k_d=K_d-j_d}^{\infty} E[e^{-sX_{+,i}}|A_u(X_{-,i}) = k_u - j_u, A_d(X_{-,i}) = k_d] \cdot\mathcal{P}[A_u(X_{-,i}) = k_u - j_u]\mathcal{P}[A_d(X_{-,i}) = k_d], \quad 1 \le k_u \le K_u - 1$$

$$\Pi^*_{k_u,k_d,\mu,i}(s) = \frac{F_{is}}{\eta_i}\sum_{j_u=1}^{k_u}\sum_{j_d=1}^{k_d} \pi^{i,(\mu)}_{j_u,j_d} E[e^{-sX_{+,i}}|A_u(X_{-,i}) = k_u - j_u, A_d(X_{-,i}) = k_d - j_d] \cdot\mathcal{P}[A_u(X_{-,i}) = k_u - j_u]\mathcal{P}[A_d(X_{-,i}) = k_d - j_d], \quad 1 \le k_u \le K_u - 1, 1 \le k_d \le K_d - 1, 2 \le \mu \le M_s$$

$$\Pi^*_{K_u,k_d,\mu,i}(s) = \frac{F_{is}}{\eta_i}\sum_{j_u=1}^{K_u}\sum_{j_d=1}^{k_d} \pi^{i,(\mu)}_{j_u,j_d}\sum_{k_u=K_u-j_u}^{\infty} E[e^{-sX_{+,i}}|A_u(X_{-,i}) = k_u, A_d(X_{-,i}) = k_d - j_d,] \cdot\mathcal{P}[A_u(X_{-,i}) = k_u]\mathcal{P}[A_d(X_{-,i}) = k_d - j_d], \quad 2 \le \mu \le M_s, 1 \le k_d \le K_d - 1$$

$$\Pi^*_{k_u,K_d,\mu,i}(s) = \frac{F_{is}}{\eta_i}\sum_{j_u=1}^{k_u}\sum_{j_d=1}^{K_d} \pi^{i,(\mu)}_{j_u,j_d}\sum_{k_d=K_d-j_d}^{\infty} E[e^{-sX_{+,i}}|A_u(X_{-,i}) = k_u - j_u, A_d(X_{-,i}) = k_d,] \cdot\mathcal{P}[A_u(X_{-,i}) = k_u - j_u]\mathcal{P}[A_d(X_{-,i}) = k_d], \quad 2 \le \mu \le M_s, 1 \le k_u \le K_d \tag{14.22}$$

In an analogous fashion, additional $M_s + 1$ equations should also be set for (K_u, K_d) cases, but we omit them for brevity.

Equations (14.22) can be simplified as follows. For clarity, let us distinguish between the PGFs for uplink and downlink burst sizes as $G_{b_u}(z)$ and $G_{b_d}(z)$, respectively, even though we have initially assumed that they will be the same, provided M_s and M_b are larger than $\overline{B}$.

$$
\begin{aligned}
\phi^*_{k_u,k_d}(s) &= E[e^{-sV_{+,i}}|A_u(V_{-,i}) = k_u, A_d(V_{-,i}) = k_d] \\
&\cdot \mathcal{P}[A_u(V_{-,i}) = k_u]\mathcal{P}[A_d(V_{-,i}) = k_d] \\
&= \sum_{l_u=0}^{\infty} \frac{1}{k_u!} \frac{d^{k_u}}{dz_u^{k_u}} \left(G_{b_u}(z_u)\right)^{l_u}\Big|_{z_u=0} \sum_{l_d=0}^{\infty} \frac{1}{k_d!} \frac{d^{k_d}}{dz_d^{k_d}} \left(G_{b_d}(z_d)\right)^{l_d}\Big|_{z_d=0} \\
&\cdot \int_0^{\infty} \frac{(\lambda_{iu}x)^{l_u}}{l_u!} e^{-\lambda_{iu}x} \frac{(\lambda_{id}x)^{l_d}}{l_d!} e^{-\lambda_{id}x} \frac{1 - V_i(x)}{\overline{V_i}} dx \\
&\int_0^{\infty} e^{-sy} \frac{v_i(x+y)}{1 - V_i(x)} dy \\
\psi^*_{k_u,k_d}(s) &= E[e^{-sX_{+,i}}|A_u(X_{-,i}) = k_u, A_d(X_{-,i}) = k_d] \\
&\cdot \mathcal{P}[A_u(X_{-,i}) = k_u]\mathcal{P}[A_d(X_{-,i}) = k_d] \\
&= \sum_{l_u=0}^{\infty} \frac{1}{k_u!} \frac{d^{k_u}}{dz_u^{k_u}} \left(G_{bu}(z_u)\right)^{l_u}\Big|_{z_u=0} \sum_{l_d=0}^{\infty} \frac{1}{k_d!} \frac{d^{k_d}}{dz_d^{k_d}} \left(G_{bd}(z_d)\right)^{l_d}\Big|_{z_d=0} \\
&\cdot \int_0^{\infty} \frac{(\lambda_{iu}x)^{l_u}}{l_u!} e^{-\lambda_{iu}x} \frac{(\lambda_{id}x)^{l_d}}{l_d!} e^{-\lambda_{id}x} \frac{1 - F_{is}(x)}{\overline{F_{is}}} dx \\
&\cdot \int_0^{\infty} e^{-sy} \frac{f_s(x+y)}{1 - F_{is}(x)}
\end{aligned}
\tag{14.23}
$$

By introducing the substitution $u = x + y$, and by changing the order and limits of integration of variables x and u, we obtain

$$
\begin{aligned}
\phi^*_{k_u,k_d}(s) &= \frac{1}{\overline{V_i} k_u! k_d!} \\
&\cdot \frac{d^{k_u}}{dz_u^{k_u}} \frac{d^{k_d}}{dz_d^{k_d}} \frac{V_i^*(-\lambda_{iu}G_{bu}(z_u) + \lambda_{iu} - \lambda_{id}G_{bd}(z_d) + \lambda_{id}) - V_i^*(s)}{\lambda_{iu}G_{bu}(z_u) + \lambda_{id}G_{bd}(z_d) - \lambda_{iu} - \lambda_{id} + s}\Bigg|_{z_u=0, z_d=0} \\
\psi^*_{k_u,k_d}(s) &= \frac{1}{\overline{F_{is}} k_u! k_d!} \\
&\cdot \frac{d^{k_u}}{dz_u^{k_u}} \frac{d^{k_d}}{dz_d^{k_d}} \frac{F_{is}^*(-\lambda_{iu}G_{bu}(z_u) + \lambda_{iu} - \lambda_{id}G_{bd}(z_d) + \lambda_{id}) - F_{is}^*(s)}{\lambda_{iu}G_{bu}(z_u) + \lambda_{id}G_{bd}(z_d) - \lambda_{iu} - \lambda_{id} + s}\Bigg|_{z_u=0, z_d=0}
\end{aligned}
\tag{14.24}
$$

Then, the system (14.22) can be transformed to

$$
\begin{aligned}
\Omega^*_{k_u,k_d,i}(s) &= \frac{\overline{V_i}}{\eta_i} \left(q^i_{0,0} + \sum_{\mu=1}^{M_s-1} \pi^{i,(\mu)}_{0,0} \right) \phi^*_{k_u,k_d}(s) \\
&+ \frac{\overline{V_i}}{\eta_i} \sum_{j_u=0}^{k_u} \sum_{j_d=0}^{k_d} \pi^{i,(M_s)}_{j_u,j_d} \phi^*_{k_u-j_u,k_d-j_d} \\
&\qquad 1 \le k_u \le K_u - 1, 1 \le k_d \le K_d - 1
\end{aligned}
\tag{14.25}
$$

continued on next page ...

... continued from previous page

$$\Omega^*_{K_u,k_d,i}(s) = \frac{\overline{V_i}}{\eta_i}\left(q^i_{0,0} + \sum_{\mu=1}^{M_s-1}\pi^{i,(\mu)}_{0,0}\right)\sum_{k_u=K_u}^{\infty}\phi^*_{k_u,k_d}(s)$$
$$+ \frac{\overline{V_i}}{\eta_i}\sum_{j_u=0}^{K_u}\sum_{j_d=0}^{k_d}\pi^{i,(M_s)}_{0,0}\sum_{k_u=K_u-j_u}^{\infty}\phi^*_{k_u,k_d-j_d}(s),$$
$$1 \le k_d \le K_d - 1$$

Note: a similar expression holds for $\Omega^*_{k_u,K_d}(s)$.

$$\Omega^*_{K_u,K_d,i}(s) = \frac{\overline{V_i}}{\eta_i}\left(q^i_{0,0} + \sum_{\mu=1}^{M_s-1}\pi^{i,(\mu)}_{0,0}\right)\sum_{k_u=K_u}^{\infty}\sum_{k_d=K_d}^{\infty}\phi^*_{k_u,k_d}(s)$$
$$+ \frac{\overline{V_i}}{\eta_i}\sum_{j_u=0}^{K_u}\sum_{j_d=0}^{K_d}\pi^{i,(M)}_{0,0}\sum_{k_u=K_u-j_u}^{\infty}\sum_{k_d=K_d-j_d}^{\infty}\phi^*_{k_u,k_d}(s)$$

$$\Pi^*_{k_u,k_d,1,i}(s) = \frac{\overline{F_{is}}}{\eta_i}\sum_{j_u=1}^{k_u}\sum_{j_d=1}^{k_d}q^i_{j_u,j_d}\psi^*_{k_u-j_u,k_d-j_d}(s),$$
$$1 \le k_u \le K_u - 1, 1 \le k_d \le K_d - 1$$

$$\Pi^*_{K_u,k_d,1,i}(s) = \frac{\overline{F_{is}}}{\eta_i}\sum_{j_u=1}^{K_u}\sum_{j_d=1}^{k_d}q^i_{j_u,j_d}\sum_{k_u=K_u-j_u}^{\infty}\psi^*_{k_u,k_d-j_d}(s), \tag{14.25}$$
$$1 \le k_d \le K_d - 1$$

Note: a similar expression holds for $\Pi^*_{k_u,K_d,1,i}(s)$.

$$\Pi^*_{K_u,K_d,1,i}(s) = \frac{\overline{F_{is}}}{\eta_i}\sum_{j_u=1}^{K_u}\sum_{j_d=1}^{K_d}q^i_{j_u,j_d}\sum_{k_u=K_u-j_u}^{\infty}\sum_{k_d=K_d-j_d}^{\infty}\psi^*_{k_u,k_d}(s)$$

$$\Pi^*_{k_u,k_d,\mu,i}(s) = \frac{\overline{F_{is}}}{\eta_i}\sum_{j_u=1}^{k_u}\sum_{j_d=1}^{k_d}\pi^{i,(\mu)}_{j_u,j_d}\psi^*_{k_u-j_u,k_d-j_d}(s),$$
$$1 \le k_u \le K_u - 1, 1 \le k_d \le K_d - 1, 2 \le \mu \le M$$

$$\Pi^*_{K_u,k_d,\mu,i}(s) = \frac{\overline{F_{is}}}{\eta_i}\sum_{j_u=1}^{K_u}\sum_{j_d=1}^{k_d}\pi^{i,(\mu)}_{j_u,j_d}\sum_{k_u=K_u-j_u}^{\infty}\psi^*_{k_u,k_d-j_d}(s),$$
$$1 \le k_d \le K_d - 1, 2 \le \mu \le M_s$$

Note: a similar expression holds for $\Pi^*_{k_u,K_d,\mu,i}(s)$.

$$\Pi^*_{K_u,K_d,\mu,i}(s) = \frac{\overline{F_{is}}}{\eta_i}\sum_{j_u=1}^{K_u}\sum_{j_d=1}^{K_d}\pi^{i,(\mu)}_{j_u,j_d}\sum_{k_u=K_u-j_u}^{\infty}\sum_{k_d=K_d-j_u}^{\infty}\psi^*_{k_u,k_d}(s),$$
$$2 \le \mu \le M_s$$

The joint distribution of the uplink/downlink queue size for a single slave at arbitrary time is given by

$$
\begin{aligned}
\mathcal{P}[L_{q,i} = (0,0)] &= \Omega^*_{0,0,i}(0) \\
\mathcal{P}[L_{q,i} = (0,k_d)] &= \Omega^*_{0,k_d,i}(0) + \sum_{\mu=1}^{M_s} \Pi^*_{0,k_d,\mu,i}(0), \qquad 1 \le k_d \le K_d \\
\mathcal{P}[L_{q,i} = (k_u,0)] &= \Omega^*_{k_u,0,i}(0) + \sum_{\mu=1}^{M_s} \Pi^*_{k_u,0\mu,i}(0), \qquad 1 \le k_u \le K_u \\
\mathcal{P}[L_{q,i} = (k_u,k_d)] &= \Omega^*_{k_u,k_d,i}(0) + \sum_{\mu=1}^{M_s} \Pi^*_{k_u,k_d,\mu,i}(0), \\
&\qquad 1 \le k_u \le K_u - 1, 1 \le k_d \le K_d - 1 \\
\mathcal{P}[L_{q,i} = (K_u,k_d)] &= \Omega^*_{K_u,k_d,i}(0) + \sum_{\mu=1}^{M_s} \Pi^*_{K_u,k_d,\mu,i}(0), \ 1 \le k_d \le K_d \\
\mathcal{P}[L_{q,i} = (k_u,K_d)] &= \Omega^*_{k_u,K_d,i}(0) + \sum_{\mu=1}^{M_s} \Pi^*_{k_u,K_d,\mu,i}(0), \ 1 \le k_u \le K_u
\end{aligned}
\tag{14.26}
$$

Then, we can calculate the burst blocking probability in the uplink queue of the slave i under the total rejection policy. Total rejection, in this context, means that the entire packet burst will be rejected if the available buffer space is insufficient to hold all the packets in the burst.

$$
\begin{aligned}
P_{B,i,u} &= \sum_{k_u=0}^{K_u} \sum_{k_d=0}^{K_d} \mathcal{P}[L_{q,i} = (k_u,k_d)]\mathcal{P}[\text{burst} > K_u - k_u] \\
&= \sum_{k_u=0}^{K_u} \sum_{k_d=0}^{K_d} \mathcal{P}[L_{q,i} = (k_u,k_d)] \sum_{l=K_u-k_u}^{\infty} g_l
\end{aligned}
\tag{14.27}
$$

where $g_l = \frac{1}{l!}\frac{d^l}{dz^l} G_{bu}(z)|_{z=0}$ denotes the mass probability of burst size being equal to l packets.

Finally we are able to determine the LST for the access delay for the first packet in the burst since joint queue length distributions contain information about remaining service/vacation time. Let us denote blocking probability for the first packet of the burst in slave's uplink queue as $P^g_{B,i,u} = \sum_{k_d=0}^{K_d} \left(\Omega^*_{K_u,k_d,u}(0) + \sum_{\mu=1}^{M_s} \Pi^*_{K_u,k_d,\mu,i}(0) \right)$.
Then the access delay of the first packet in the burst becomes

$$W^*_{g,ai}(s) = \frac{\sum_{k_d=0}^{K_d}\left(\sum_{k_u=0}^{K_u-1} \Omega^*_{k_u,k_d,i}(s)F^*_{is}(s)^{k_u}V^*_i(s)^{\lfloor k/M_s \rfloor}\right)}{1-P^g_{B,i,u}} + \frac{\sum_{k_d=0}^{K_d}\left(\sum_{k_u=1}^{K_u-1}\sum_{\mu=1}^{M_s} \Pi^*_{k_u,k_d,\mu,i}(s)F^*_{is}(s)^{(k_u-1)}V^*_i(s)^{\lfloor (k_u+\mu-1)/M_s \rfloor}\right)}{1-P^g_{B,i,u}} \tag{14.28}$$

The mean access delay for the first packet in the burst is determined as

$$\begin{aligned}\overline{W_{g,ai}} &= -\frac{d}{ds}W^*_{g,ai}(s)|_{s=0} \\ &= \frac{\sum_{k_d=0}^{K_d}\sum_{k_u=1}^{K_u-1} k_u\left(\sum_{\mu=1}^{M_s}\pi^{i,(\mu)}_{k_u,k_d}\right) + \frac{K_u}{\lambda_{iu}}\frac{P^g_{B,i,u}}{1-P^g_{B,i,u}} - \overline{F_{is}}}{\lambda^2_{iu}(1-P^g_{B,i,u})\eta_i}\end{aligned} \tag{14.29}$$

In order to calculate the delay of an arbitrary packet in the burst, we need the distribution of the distance of that arbitrary packet from the first packet in the burst. The probability that there are exactly k packets before an arbitrary packet in the burst is $g^-_k = \frac{1}{\overline{B}}\sum_{j=k+1}^{\infty} g_j$. Then, the LST for the delay of the arbitrary packet in the accepted burst becomes

$$W^*_{ai}(s) = \frac{\sum_{k_d=0}^{K_d}\sum_{k_u=0}^{K_u-1}\Omega^*_{k_u,k_d,i}(s)\sum_{j=1}^{K_u-k_u} g_j \sum_{w=1}^{j-1} g^-_w F^*_{is}(s)^{(k_u+w)}V^*_i(s)^{\lfloor (k_u+w)/M_s \rfloor}}{1-P_{B,i,u}} + \frac{\sum_{k_d=0}^{K_d}\sum_{k_u=1}^{K_u-1}\sum_{\mu=1}^{M_s}\Pi^*_{k_u,k_d,\mu,i}(s)\sum_{j=1}^{K_u-k_u} g_j \sum_{w=1}^{j-1} g^-_w F^*_{is}(s)^{(k_u-1)}V^*_i(s)^{\lfloor (k_u+\mu+w-1)/M_s \rfloor}}{1-P_{B,i,u}} \tag{14.30}$$

LST for the downlink delay can be calculated in a similar manner. The delay through the bridge can be also calculated in analogous way with use of polling parameter M_b.

14.5 Simulation results

In order to verify the analytical results presented above, we have modified our Bluetooth scatternet simulator to include finite buffers. The scatternet had six piconets operating under E-limited polling, and connected as shown in Fig. 14.3. The bridges use the walk-in scheduling with the limited exchange policy. Each slave generates traffic which targets other slaves in the same piconet with the probability of P_l, and slaves in any other piconet with the probability $(1 - P_l)/5$. In case of nonadjacent piconets, the inter-piconet traffic is routed through the shortest path. When two such paths exist (e.g., traffic from piconet P1 to piconet P6 may go through P2 or P4), traffic is split evenly between those paths.

As can be observed, the 'exterior' piconets P1, P3, and P5 will behave in an identical manner, thanks to the symmetry of the topology, as is the case with the 'interior' piconets P2, P4, and P6; the difference is highlighted by shading.

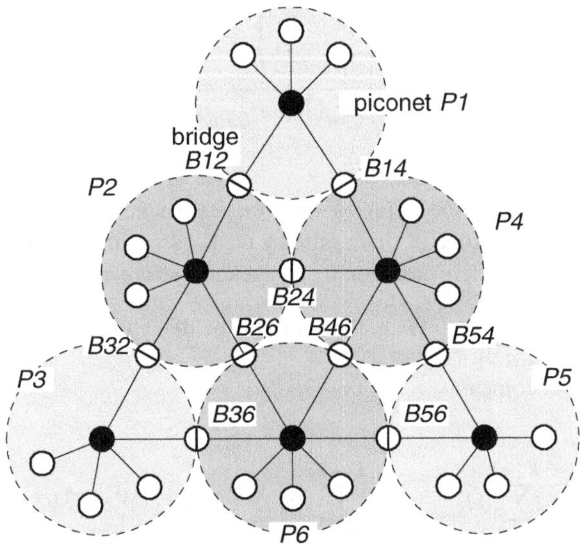

FIGURE 14.3
Topology of the scatternet under consideration.

Default parameter values are as follows: mean burst size $\overline{B} = 3$ with five slot packets used throughout the simulation, poling parameter values $M_s = 3$ and $M_b = 15$, device buffer sizes $K_d = 100$ and $K_b = 40$ (both expressed in baseband packets), and traffic locality $P_l = 0.6$.

Blocking probabilities and access delays at slaves

The first set of simulation results shows slave buffer blocking probabilities for exterior and interior piconets, respectively. In both cases, the size of slave buffers was varied between 10 and 30 baseband packets, and the packet burst arrival rate was varied between 0.001 and 0.006 (burst arrivals per Bluetooth slot). The bridges were present in each piconet for K cycles, or $K + 1$ exchanges. Note that $K = 0$ means that the bridge will be present only until its exchange is over and then leave, while $K > 1$ means that bridge will be polled $K + 1$ times and spend $K - 1$ full piconet cycles before leaving. The polling parameter for the bridges was fixed at $M_b = 15$.

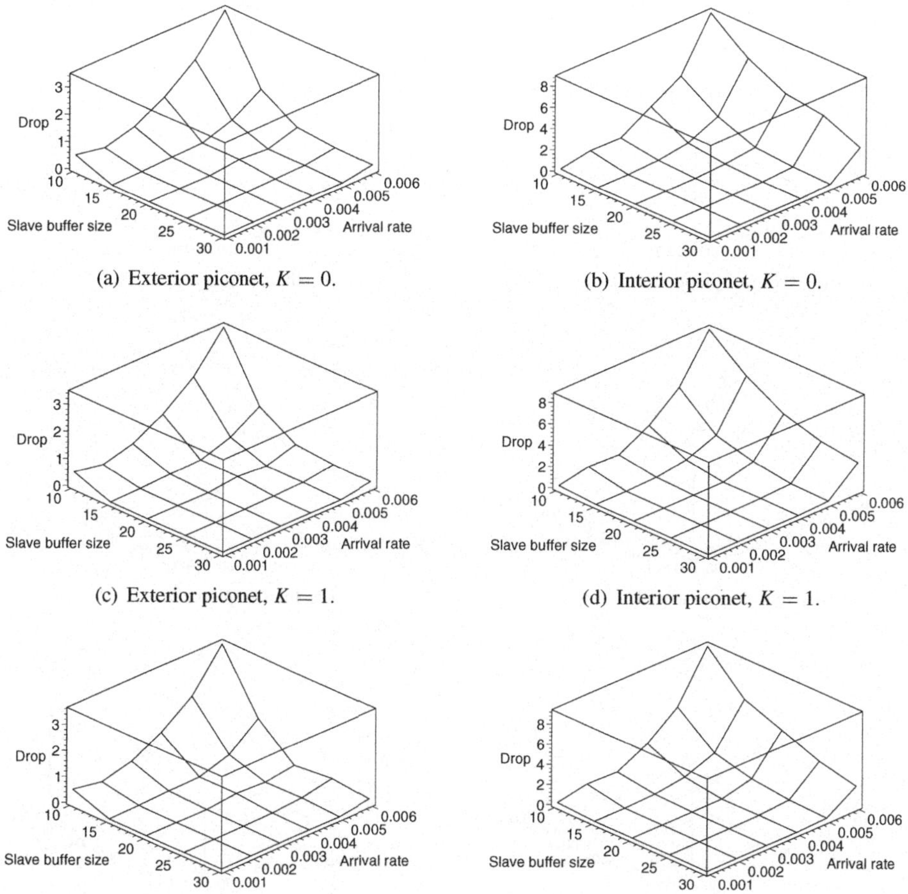

(a) Exterior piconet, $K = 0$.

(b) Interior piconet, $K = 0$.

(c) Exterior piconet, $K = 1$.

(d) Interior piconet, $K = 1$.

(e) Exterior piconet, $K = 2$.

(f) Interior piconet, $K = 2$.

FIGURE 14.4
Slave buffer blocking rate with fixed $M_s = 5$.

The diagrams in Fig. 14.4 show the blocking probability, expressed in percents, for two different piconets, one in the exterior group and another in the interior group, respectively. They were obtained by fixing the polling parameter for the slaves to $M_s = 5$, and varying the number of cycles between $K = 0$ and 2. As can be seen, higher values of K do lead to a slight reduction in the blocking probability, although the effects are rather minuscule. Note, however, that the interior piconet has experienced more than twice the blocking probability of the exterior one. This is due to the longer cycle and vacation times caused by higher load – the interior piconets have four bridges each, while the exterior ones have only two bridges each.

The diagrams in Fig. 14.5 show the effects of varying M_s between 2 and 8 while

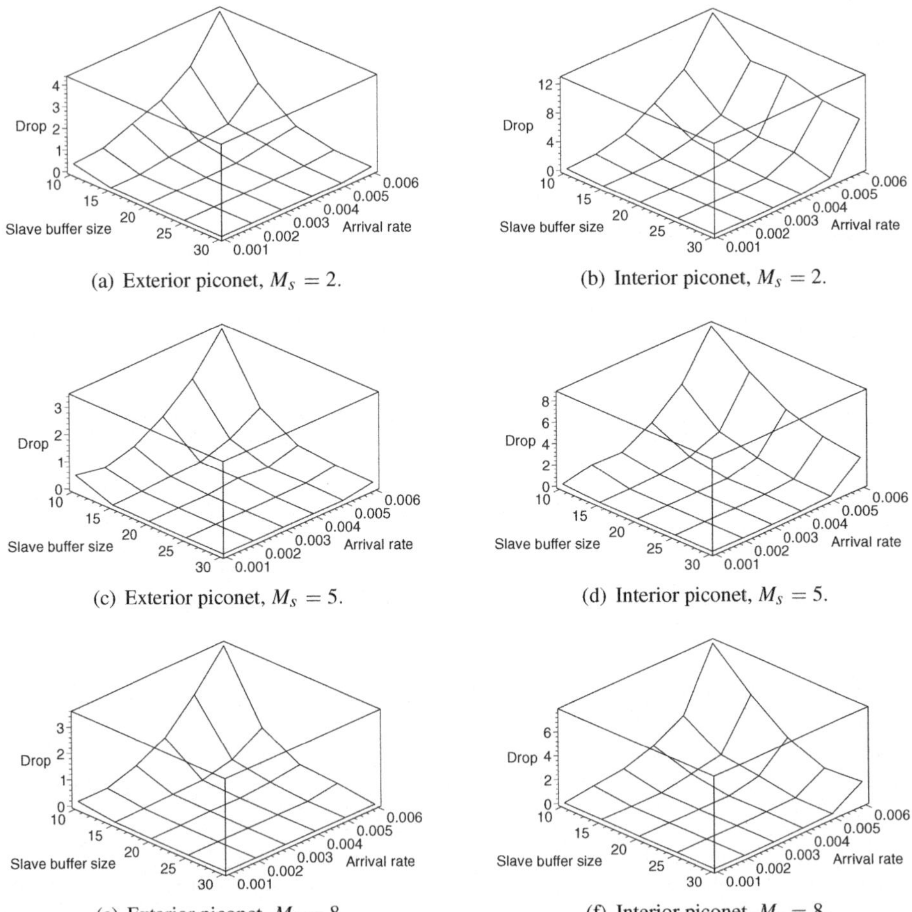

(a) Exterior piconet, $M_s = 2$.

(b) Interior piconet, $M_s = 2$.

(c) Exterior piconet, $M_s = 5$.

(d) Interior piconet, $M_s = 5$.

(e) Exterior piconet, $M_s = 8$.

(f) Interior piconet, $M_s = 8$.

FIGURE 14.5
Slave buffer blocking rate with fixed $K = 1$.

keeping a fixed $K = 1$. Higher values of the bridge polling parameter do lead to a reduction in blocking probabilities, as could be expected from our earlier discussions. This effect, however, is more pronounced in interior piconets, again on account of their higher traffic load; increasing the number of packets to be exchanged with the bridge helps reduce this load and, indirectly, the blocking probability as well.

Blocking probabilities at the bridges

The second set of experiments investigates the bridge blocking probability under varying K and M_s. In this case, we have also varied the bridge buffer size from $K_b = 10$ to 20 baseband packets, as well as the locality probability from $P_l = 0.2$

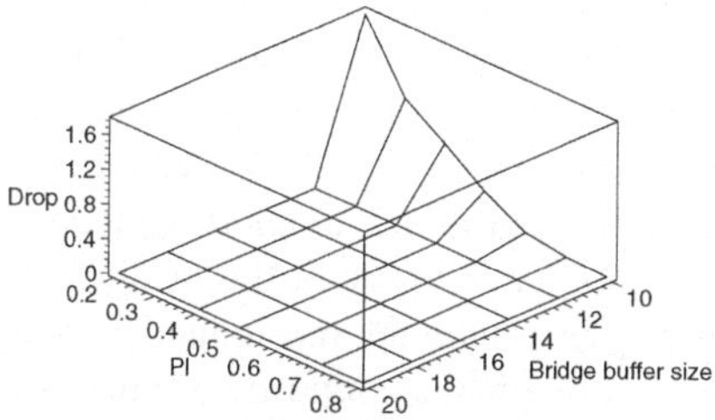

(a) Bridge from P1 to P2, $K = 0$.

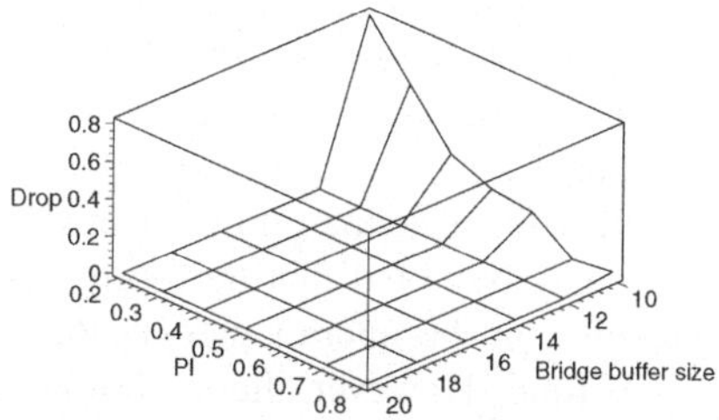

(b) Bridge from P2 to P1, $K = 0$.

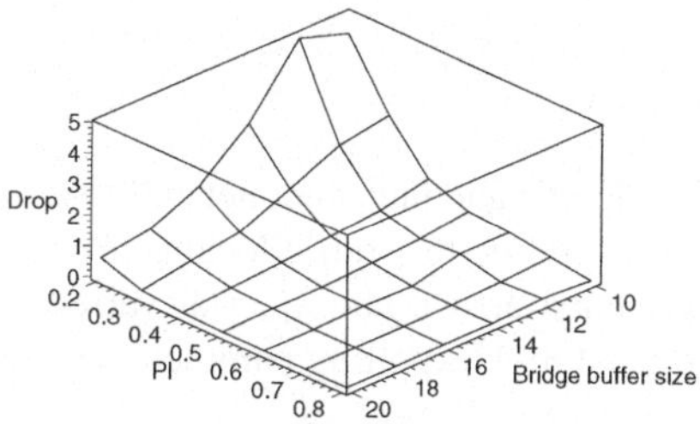

(c) Bridge from P1 to P2, $K = 1$.

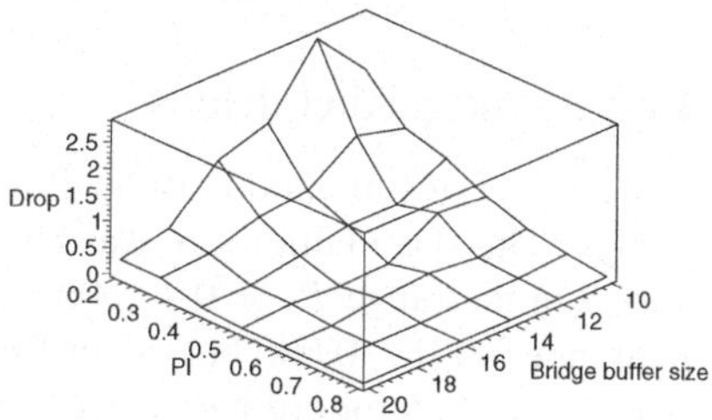

(d) Bridge from P2 to P1, $K = 1$.

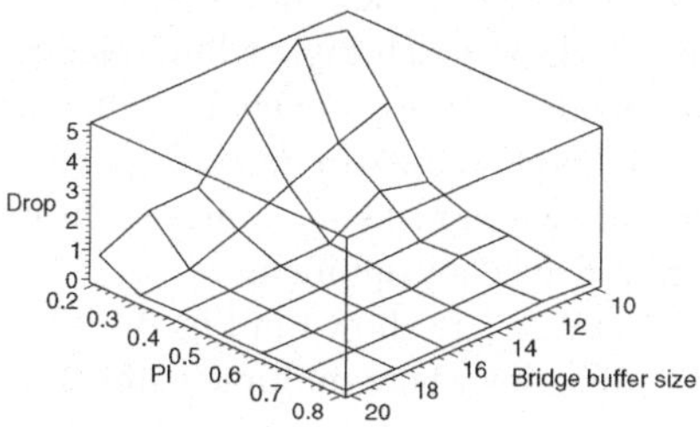

(e) Bridge from P1 to P2, $K = 2$.

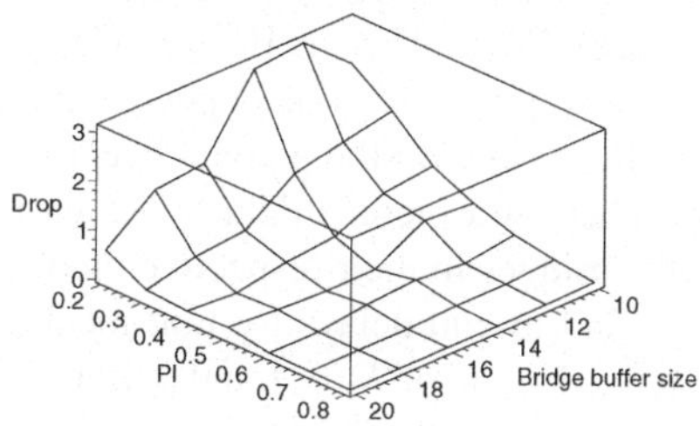

(f) Bridge from P2 to P1, $K = 2$.

FIGURE 14.6
Bridge buffer blocking probability with fixed $M_b = 12$.

to 0.8. We have considered both bridge buffers in the bridge B12 from Fig. 14.3 as representatives of bridges in an exterior and interior piconet, respectively.

The diagrams in Fig. 14.6 show the blocking probability for different values of K, with fixed $M_b = 12$. As could be expected, the blocking probabilities are lower in the piconet with lower load; in this case, this is the piconet P1 where the buffer from P1 to P2 'resides.' At the same time, the increase in K means that the bridge spends more time in the piconet (in fact, K piconet cycles as well as $K+1$ exchanges with the master). This increases the number of packets sent to the bridge and, consequently, leads to increased probability that the bridge buffer will be full and unable to accept more packets. As before, this effect is more pronounced in the piconet with the longer cycle time – piconet P2, in this case, where the buffer from P2 to P1 'resides.'

The diagrams in Fig. 14.7 show the dependency of blocking probability for different values of Mb, with fixed $K = 1$. As could be expected, the blocking probability decreases with M_b. Namely, larger number of packets per exchange means that the bridge queue will empty faster, and the probability that both the bridge queue and the outgoing queue at the master will empty increases. This event, as the reader may recall, will cause the bridge to terminate its residence and switch to another piconet.

It may be noted that the bridge blocking probability is lower in the bridge that 'resides' in the interior piconet P2, despite its higher total load and longer piconet cycle time. However, there are four bridges in that piconet, as opposed to only two in the exterior piconet P1, and the traffic per bridge is lower. Consequently, the blocking probability will be lower, too.

End-to-end packet delays

The next set of simulation results illustrate intra-piconet and inter-piconet end-to-end delays. The slave buffer size was fixed to $K_u = 30$ while the traffic locality P_l was varied in the range $P_l = 0.4 \,..\, 0.9$. The packet burst arrival rate λ_{iu} was varied in the range 0.001 .. 0.006 bursts per Bluetooth time slot. The polling parameters were fixed at $M_s = 5$ for ordinary slaves, and at $M_b = 15$ for the bridges.

End-to-end delays for local (intra-piconet) traffic are shown in Fig. 14.8 for values of $K = 0$, 1, and 2. The delays were averaged over all exterior and interior piconets, respectively. The interior piconet can be seen to exhibit much higher intra-piconet delays, which is to be expected since the traffic load of those piconets is higher, as is the case with the piconet cycle time. Although higher values of K correspond to higher intra-piconet delays (which is due to the increased residence times spent by the bridges in their respective piconets), the difference is not significant.

End-to-end delays for non-local (inter-piconet) traffic are shown in Fig. 14.9, for a fixed value of $K = 1$ and variable value of M_s. In cases where multiple paths from the source to the destination exist, delays were averaged over all the paths where both the source and destination are in exterior and interior piconets; the resulting delays are shown in the left and right column diagrams, respectively.

As before, the slave buffer size was fixed at $K_u = 30$, the master buffer was fixed at $K_d = 100$, and the bridge buffer had the length of $K_b = 40$ baseband packets. The

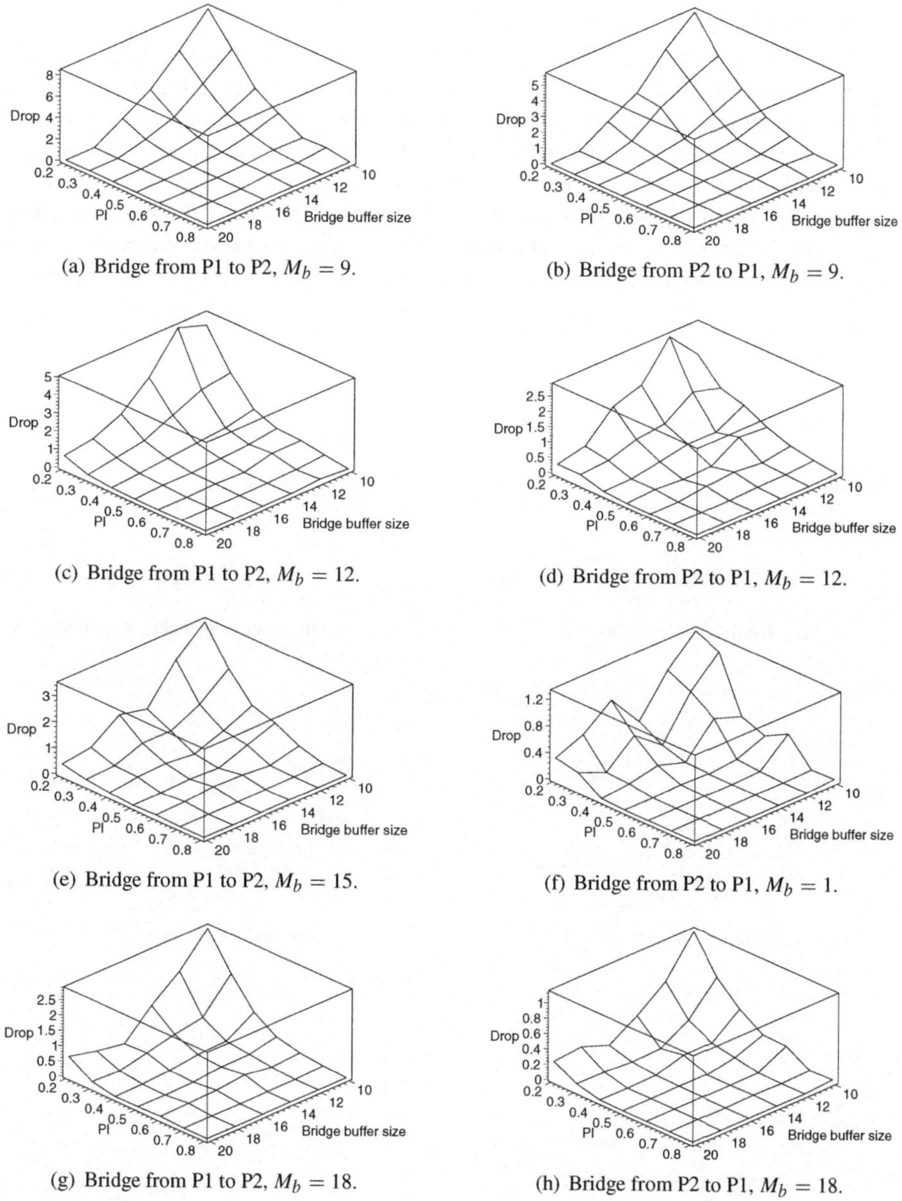

(a) Bridge from P1 to P2, $M_b = 9$.

(b) Bridge from P2 to P1, $M_b = 9$.

(c) Bridge from P1 to P2, $M_b = 12$.

(d) Bridge from P2 to P1, $M_b = 12$.

(e) Bridge from P1 to P2, $M_b = 15$.

(f) Bridge from P2 to P1, $M_b = 1$.

(g) Bridge from P1 to P2, $M_b = 18$.

(h) Bridge from P2 to P1, $M_b = 18$.

FIGURE 14.7
Bridge buffer blocking probability at fixed $K = 1$.

traffic locality P_l was varied in the range 0.4 . . 0.9, and the packet burst arrival rate λ_{iu} was varied in the range 0.001..0.006 bursts per Bluetooth time slot. The polling parameter for the bridges was $M_b = 15$.

As in the case of local delays, exterior piconet exhibit much lower delays than the interior ones, due to their smaller aggregate traffic load. However, the increase of the polling parameter for the slaves leads to increase in the delays; this change is much more pronounced in the exterior piconets. This effect can be explained as follows. For small values of M_s, the piconet cycle is short and the number of packets to be transmitted when the bridge is present is also small. Therefore, the bridge is likely to be able to exchange all the data in the first exchange and leave the piconet, even though $K = 1$ and it may stay for two exchanges if necessary. When M_s increases,

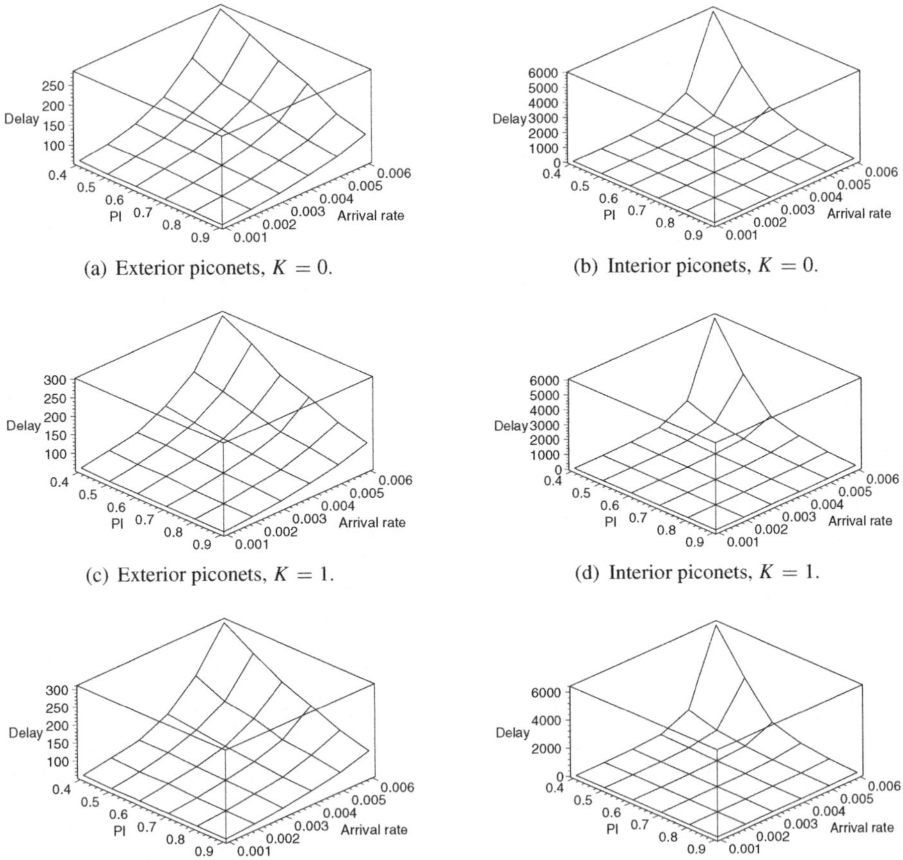

(a) Exterior piconets, $K = 0$.

(b) Interior piconets, $K = 0$.

(c) Exterior piconets, $K = 1$.

(d) Interior piconets, $K = 1$.

(e) Exterior piconets, $K = 2$.

(f) Interior piconets, $K = 2$.

FIGURE 14.8
End-to-end delay for local traffic with fixed $M_s = 5$.

more packets are picked up by the master for both local and non-local destinations, and the piconet cycle lengthens. The bridge will have more packets to exchange, and it may not succeed in doing so in only two exchanges; as a consequence, the queueing delays in the bridge will be long and the overall end-to-end packet delays will also increase.

Throughput

Our final set of measurements shows the throughput through the piconets. The slave polling parameter value was varied in the range of $M_s = 3 \,..\, 15$, while the corresponding bridge polling parameter was kept constant at $M_b = 15$. The packet burst

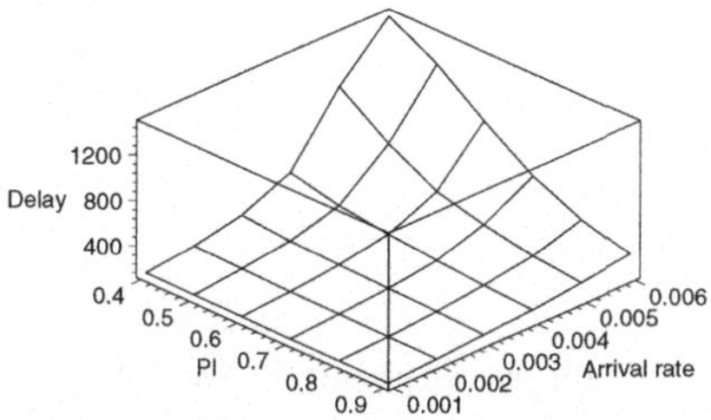

(a) Exterior piconets, $M_s = 2$.

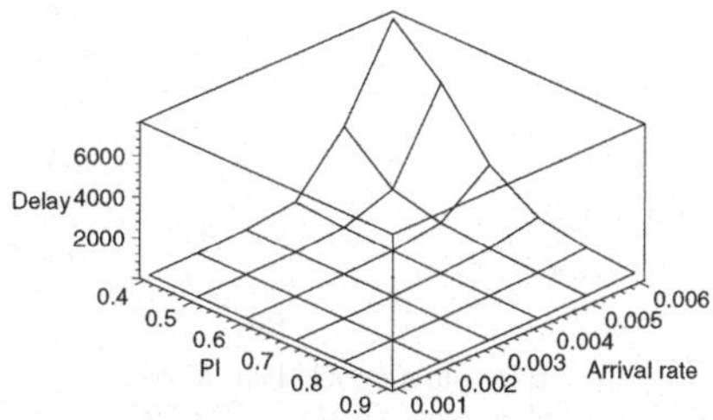

(b) Interior piconets, $M_s = 2$.

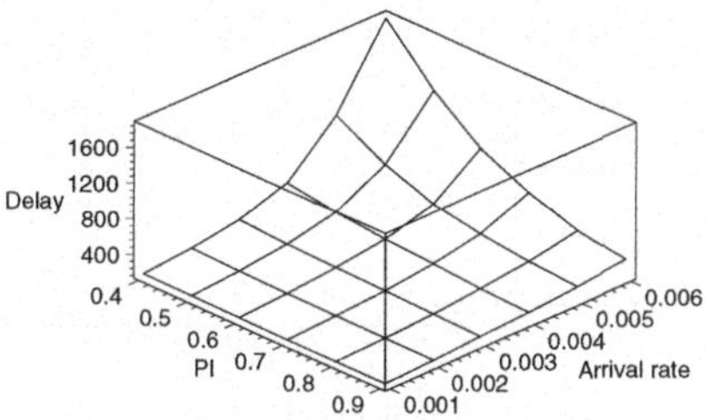

(c) Exterior piconets, $M_s = 5$.

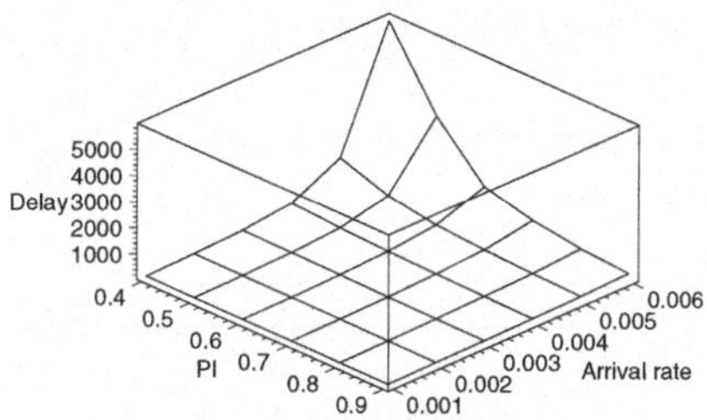

(d) Interior piconets, $M_s = 5$.

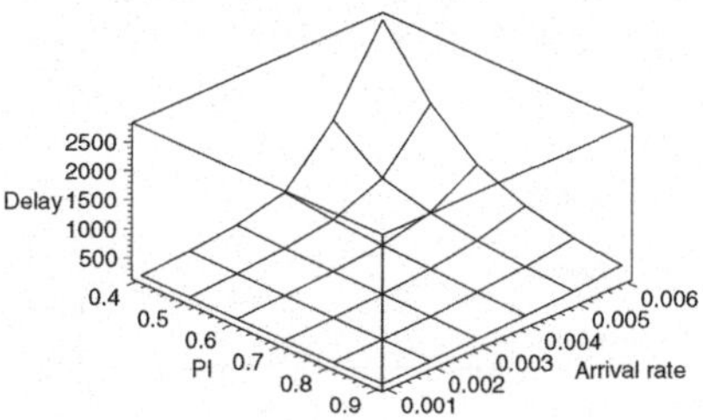

(e) Exterior piconets, $M_s = 8$.

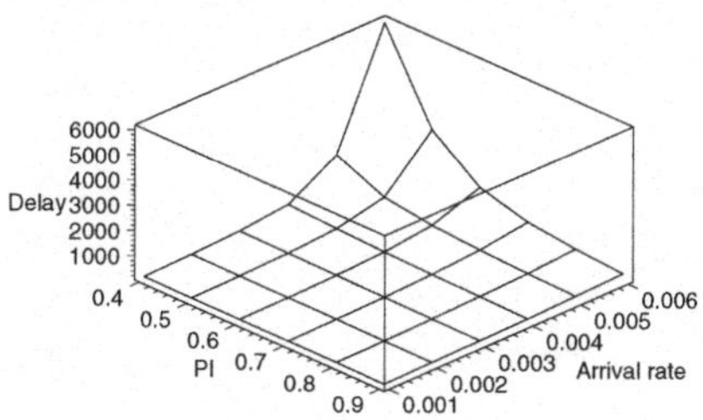

(f) Interior piconets, $M_s = 8$.

FIGURE 14.9

End-to-end delay for non-local traffic with fixed $K = 1$.

arrival rate per slave was varied in the range of $\lambda_{iu} = 0.001 \,..\, 0.01$ packet burst arrivals per Bluetooth time slot, and traffic locality was fixed at $P_l = 0.6$. The bridge residence parameter values were $K = 0, 1$, and 2. Buffer sizes were kept at values $K_u = 30$, $K_d = 100$, $K_b = 40$.

As can be seen from Fig. 14.10, the throughput for interior piconets is higher when compared to the exterior ones, which is again due to their higher traffic load. Note that exterior piconets carry their own traffic only, whilst the interior ones do perform the routing as well. For example, the interior piconet P2 has three slaves and approximately the same self-generated traffic load as piconets P1 and P3 – but

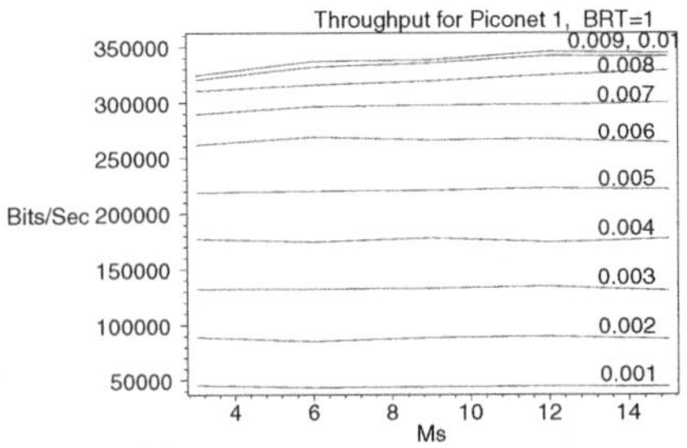

(a) External piconet (P1), $K = 0$.

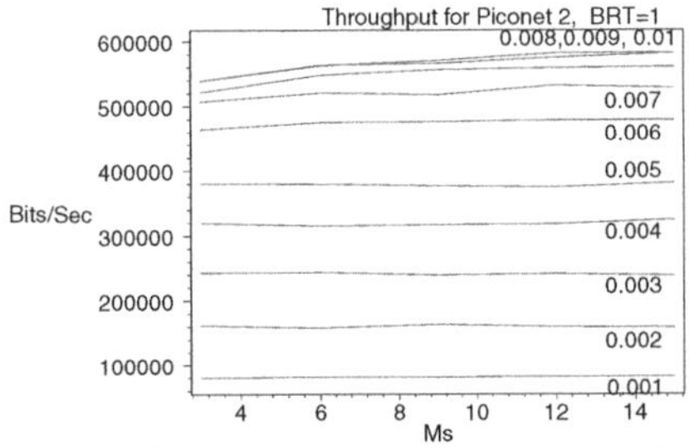

(b) Internal piconet (P1), $K = 0$.

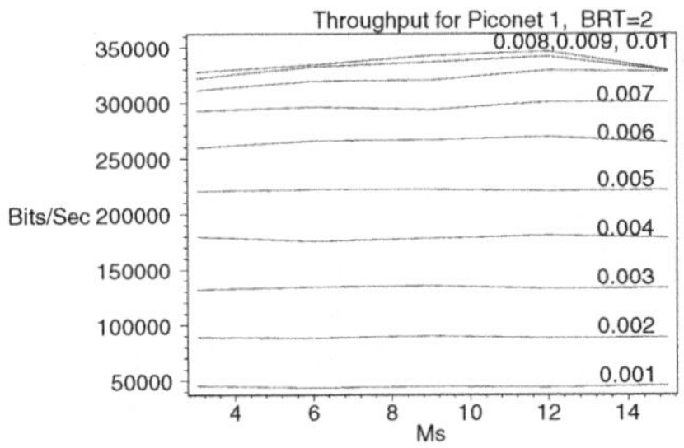

(c) External piconet (P1), $K = 1$.

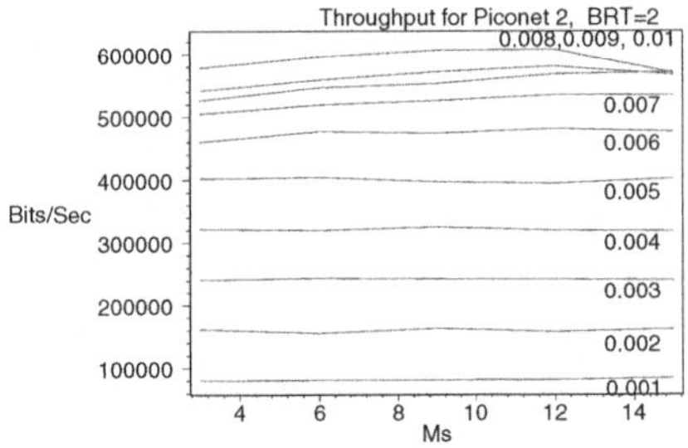

(d) Internal piconet (P1), $K = 1$.

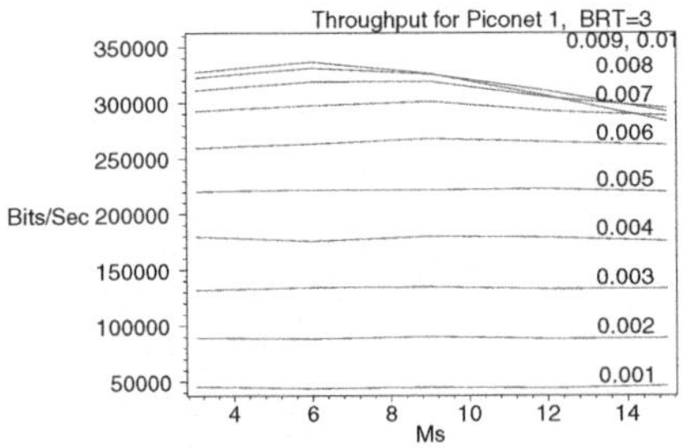

(e) External piconet (P1), $K = 2$.

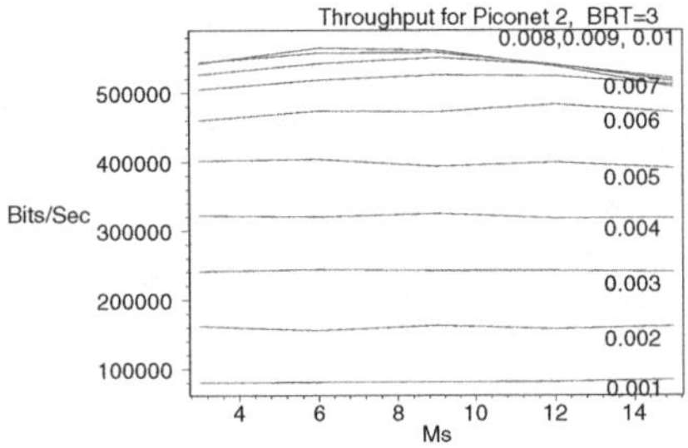

(f) Internal piconet (P1), $K = 2$.

FIGURE 14.10
Throughput in the example scatternet.

it also carries the traffic between piconets P1 and P3 as well as half of the traffic between piconets P1 and P4.

As can be seen from the diagrams, the effects of packet burst arrival rate, the slave polling parameter M_s, and the bridge residence time (through the value of K) on the throughput are interdependent. When M_s is small, the number of packets accumulated for the bridge in one piconet cycle is small, and they are likely to be delivered to the bridge within a single exchange. Therefore, the bridge cycle will be short, and the value of K does not matter much.

At higher values of M_s and higher packet burst arrival rates, there will be many packets to exchange and the bridge will be forced to stay longer. This will increase the duration of the bridge cycle, but also the losses at the slave and bridge buffers. In fact, the decrease in throughput at high arrival rates and large M_s are caused by packet losses due to the blocking at the slaves and bridges. (The similar effect has been observed in single piconets, *cf.* Chapter 4.)

We note that the maximum throughput in interior piconets is around 600Kbps, which is about one-third less than the theoretical maximum with five slot packets. This difference seems reasonable in view of the losses due to packet waste when the bridge is absent from the piconet and when there are no packets to exchange with a slave or a bridge, and packet blocking due to buffers filled up to capacity.

A

Probability generating functions and Laplace transforms

In this section we will briefly introduce the definitions and notation related to probability generating functions and their Laplace-Stieltjes transforms. For a more detailed introduction, the reader should consult one of the numerous texts on probability and queueing theory [Grimmett and Stirzaker, 1992; Kleinrock, 1972; Takagi, 1991; Wilf, 1994].

Random variables can be classified into discrete and continuous types. Discrete random variables take values from a countable set, while continuous random variables take values from a continuous range.

The probability distribution of a discrete random variable is determined by the probabilities p_k that the variable will take value k. Probabilities p_k are sometimes called mass probabilities.

The probability distribution function (PDF) of a random variable C is the function $C(x)$, defined as probability that the value of the random variable is less than some number x:

$$C(x) \stackrel{\text{def}}{=} P[C \leq x] \tag{A.1}$$

The PDF of a discrete variable is a jump function.

The probability density function (pdf) for a continuous random variable C is defined as

$$c(x) \stackrel{\text{def}}{=} \frac{dC(x)}{dx} \tag{A.2}$$

We also note that

$$C(x) = \int_{-\infty}^{x} c(y)dy \tag{A.3}$$

and

$$\int_{-\infty}^{\infty} c(x)dx = 1 \tag{A.4}$$

The probability generating function (PGF) of a discrete random variable G_p is defined as

$$G_p(z) \stackrel{\text{def}}{=} E[z^C] = \sum_{k=0}^{\infty} p_k z^k \tag{A.5}$$

i.e., it is a z-transform of the sequence of mass probabilities p_k. The PGF has the property that $G_p(1) = 1$. First and second moments of the probability distribution are given with

$$\overline{G_p} = G'_p(1) \tag{A.6}$$

$$\overline{G_p^2} = G''_p(1) + G'_p(1) \tag{A.7}$$

The Laplace-Stieltjes transform (LST) of the probability density function (pdf) of the random variable C is defined as

$$\begin{aligned} C^*(s) &\stackrel{\text{def}}{=} E[e^{-sC}] \\ &= \int_0^\infty e^{-sx} c(x)dx \end{aligned} \tag{A.8}$$

where s is a complex variable.

Moments of the random variable C can be further obtained as:

$$\begin{aligned} \overline{C^i} &= \int_0^\infty x^i c(x)dx \\ &= (-1)^i \left. \frac{d^i C^*(s)}{ds^i} \right|_{s=0}, \qquad \text{for } i = 1, 2, \ldots \end{aligned} \tag{A.9}$$

References

Abhyankar, S., Toshiwal, R., de Morais Cordeiro, C. and Agrawal, D. [2003], On the application of traffic engineering over Bluetooth ad hoc networks, in 'Proceedings of the 6th international workshop on Modeling analysis and simulation of wireless and mobile systems', ACM Press, pp. 116–123.

Ait Yaiz, R. and Heijenk, G. [2001], Polling best effort traffic in Bluetooth, in 'Proceedings 4th International Symposium on Wireless Personal Multimedia Communications (WPMC'01)', Aalborg, Denmark.

Ait Yaiz, R. and Heijenk, G. [2003], Providing delay guarantees in Bluetooth, in 'Proceedings 23rd International Conference on Distributed Computing Systems Workshops, 2003', Providence, RI, pp. 722–728.

Ajmone Marsan, M., Chiasserini, C. F., Nucci, A., Carello, G. and De Giovanni, L. [2002], Optimizing the topology of Bluetooth wireless personal area networks, in 'Proceedings Twenty-First Annual Joint Conference of the IEEE Computer and Communications Societies IEEE INFOCOM 2002', Vol. 2, New York, pp. 572–579.

Albrecht, M., Frank, M., Martini, P., Schetelig, M., Vilavaara, A. and Wenzel, A. [1999], IP services over Bluetooth: leading the way to a new mobility, in 'Proceedings of the 23rd Annual Conference on Local Computer Networks LCN '99', Lowell, MA, pp. 2–11.

Allman, M., Paxson, V. and Stevens, W. [1999], TCP congestion control, draft Internet standard RFC 2581, IETF.

Alzoubi, K. M., Wan, P.-J. and Frieder, O. [2002], Message-optimal connected-dominating-set construction for routing in mobile ad hoc networks, in 'Proceedings of the Third ACM Intl Symp. Mobile Ad Hoc Networking and Computing (MobiHoc02)', Lausanne, Switzerland, pp. 157–164.

Augel, M. and Knoor, R. [2004], Bluetooth scatternet formation, state of the art and a new approach, in 'Proceedings of the International Conference on Architecture of Computing Systems', Augsburg, Germany, pp. 260–272.

Avancha, S., Korolev, V., Joshi, A. and Finin, T. [2001], Transport protocols in wireless networks, in 'Proceedings Tenth International Conference on Computer Communications and Networks', Scottsdale, AZ, pp. 310–317.

Baatz, S., Bieschke, S., Frank, M., Martini, P., Scholz, C. and Kühl, C. [2002], Building efficient Bluetooth scatternet topologies from 1-factors, in 'Proc. IASTED Wireless and Optical Communications', Banff, Canada.

Baatz, S., Frank, M., Göpffarth, R., Kassatkine, D., Martini, P., Schetelig, M. and Vilavaara, A. [2000], Handoff support for mobility with IP over Bluetooth, in 'Proceedings 25th Annual IEEE Conference on Local Computer Networks. LCN 2000', Tampa, FL.

Baatz, S., Frank, M., Kühl, C., Martini, P. and Scholz, C. [2001], Adaptive scatternet support for Bluetooth using sniff mode, in 'Proceedings of the 26th Annual Conference on Local Computer Networks LCN 2001', Tampa, FL.

Baatz, S., Frank, M., Kühl, C., Martini, P. and Scholz, C. [2002], Bluetooth scatternets: An enhanced adaptive scheduling scheme, in 'Proceedings Twenty-First Annual Joint Conference of the IEEE Computer and Communications Societies IEEE INFOCOM 2002', New York, pp. 782–790.

Bahl, P., Adya, A., Padhye, J. and Walman, A. [2004], 'Reconsidering wireless systems with multiple radios', *SIGCOMM Comput. Commun. Rev.* **34**(5), 39–46.

Bak, J. and Newman, D. J. [1982], *Complex Analysis*, Springer-Verlag, New York, NY.

Bakker, D. M. and McMichael Gilster, D. [2002], *Bluetooth End-To-End*, Hungry Minds, Inc., New York, NY.

Balai, K., Kapoor, S., Nanavati, A. and Ramachandran, L. [2001], Scatternet formation algorithms in the Bluetooth network, in 'Manuscript'.

Balatti, E., Marzegalli, L. and Vitello, M. [2001], Increasing TCP/IP performance over home wireless networks (Bluetooth), in 'Proc. OPNETWORK conference'.

Barrière, L., Fraigniaud, P., Narayanan, L. and Opatrny, J. [2003], Dynamic construction of Bluetooth scatternets of fixed degree and low diameter, in 'Proceedings of the fourteenth annual ACM-SIAM symposium on Discrete algorithms', Society for Industrial and Applied Mathematics, pp. 781–790.

Basagni, S. [1999], Distributed clustering for ad hoc networks, in 'Proc. Int. Symp. Parallel Architectures, Algorithms and Networks ISPAN', Fremantle, Australia, pp. 310–315.

Basagni, S., Bruno, R., Mambrini, G. and Petrioli, C. [2004], 'Comparative performance evaluation of scatternet formation protocols for networks of Bluetooth devices', *ACM/Baltzer Wireless Networks (WINET)* **10**(2), 197–213.

Basagni, S., Bruno, R. and Petrioli, C. [2002*a*], Device discovery in Bluetooth networks: a scatternet perspective, in 'Proc. IFIP-TC6 Networking Conf., Networking 2002, LNCS 2345', Italy, pp. 1087–1092.

Basagni, S., Bruno, R. and Petrioli, C. [2002*b*], Performance evaluation of a new scatternet formation protocol for multi-hop Bluetooth networks, in 'Proc. IEEE Wireless Personal Multimedia Comm. WPMC', pp. 208–212.

Basagni, S., Bruno, R. and Petrioli, C. [2003*a*], Bluetooth scatternet formation in Bluetooth networks, in S. Basagni, M. Conti, S. Giordano and I. Stojmenovic, Eds, 'Ad Hoc Networking', IEEE Press, New York, NY.

Basagni, S., Bruno, R. and Petrioli, C. [2003*b*], A performance comparison of scatternet formation protocols for networks of Bluetooth devices, in 'Proc. IEEE PerCom', pp. 341–350.

Basagni, S., Bruno, R. and Petrioli, C. [2004], Scatternet formation in Bluetooth networks, in S. Basagni, M. Conti, S. Giordano and I. Stojmenovic, Eds, 'Mobile Ad Hoc Networking', John Wiley & Sons.

Basagni, S. and Petrioli, C. [2002], A scatternet formation protocol for ad hoc networks of Bluetooth devices, in 'Proc. IEEE Vehicular Technology Conf.', pp. 341–350.

Bertsekas, D. P. and Gallager, R. [1991], *Data Networks*, 2nd ed, Prentice-Hall, Englewood Cliffs, NJ.

Bhagwat, P., Korpeoglu, I., Bisdikian, C., Naghshineh, M. and Tripathi, S. K. [1999], BlueSky: a cordless networking solution for palmtop computers, in 'Proceedings of the fifth annual ACM/IEEE international conference on Mobile computing and networking', Seattle, WA, pp. 69–76.

Bhagwat, P. and Rao, S. P. [2001], 'On the characterization of Bluetooth scatternet topologies', *available from http://www.winlab.rutgers.edu/~pravin* .

Bhagwat, P. and Segall, A. [1999], A routing vector method (RVM) for routing in Bluetooth scatternets, in 'Proceedings Sixth IEEE International Workshop on Mobile Multimedia Communications (MOMUC'99)', San Diego, CA, pp. 375–379.

Bhatnagar, V. and Kesidis, G. [2002], Bluetooth scatternet formation using proximity information of an election protocol, in 'Joint 2nd IEEE Int. Conf. on Networking and IEEe Int. Conf. Wireless LANs and Home Networks,', Atlanta, GA.

Bluehoc [2003], *The Bluehoc Open-Source Bluetooth Simulator, version 3.0*, Software and documentation available from http://www-124.ibm.com/developerworks/opensource/bluehoc/.

Bluetooth SIG [2001*a*], Bluetooth Network Encapsulation Protocol (BNEP) Specification, Technical report, Revision 0.95a.

Bluetooth SIG [2001*b*], *Specification of the Bluetooth System*, Version 1.1.

Bluetooth SIG [2003*a*], *Specification of the Bluetooth System*, Version 1.2.

Bluetooth SIG [2003*b*], *Specification of the Bluetooth System – Architecture & Terminology Overview*, Vol. 1, Version 1.2.

Bluetooth SIG [2003*c*], *Specification of the Bluetooth System – Core System Package [Controller volume]*, Vol. 2, Version 1.2.

Bluetooth SIG [2003*d*], *Specification of the Bluetooth System – Core System Package [Host volume]*, Vol. 3, Version 1.2.

Bluetooth SIG [2004], *Core Specification of the Bluetooth System*, Version 2.0 + EDR.

Bose, P., Morin, P., Stojmenovic, I. and Urrutia, J. [2001], 'Routing with guaranteed delivery in ad hoc wireless networks', *ACM/Kluwer Wireless Networks*, **7**(6), 609–616.

Braden, R. T. [1989], Requirements for Internet hosts – communication layers, Internet standard RFC 1122, IETF.

Bruno, R., Conti, M. and Gregori, E. [2001], Wireless access to internet via Bluetooth: performance evaluation of the EDC scheduling algorithm, in 'Proceedings of the first workshop on Wireless Mobile Internet', Rome, Italy, pp. 43–49.

Bruno, R., Conti, M. and Gregori, E. [2002], 'Bluetooth: architecture, protocols, and scheduling algorithms', *Cluster Computing* **5**(2), 117–131.

Cai, L., Shen, X. and Mark, J. W. [2003], Delay analysis for AIMD flows in wireless/IP networks, in 'Proceedings Globecom'03', Vol. 6, San Francisco, CA, pp. 3221–3225.

Capone, A., Kapoor, R. and Gerla, M. [2001], Efficient polling schemes for Bluetooth picocells, in 'Proceedings of IEEE International Conference on Communications ICC 2001', Vol. 7, Helsinki, Finland, pp. 1990–1994.

Cassioli, D., Detti, A., Loreti, P., Mazzenga, F. and Vatalaro, F. [2002], The Bluetooth technology: state of the art and networking aspects, in 'Proceedings Networking 2002 – Second International IFIP-TC6 Networking Conference', Vol. 2345 of *Lecture Notes in Computer Science*, Springer-Verlag, Pisa, Italy, pp. 479–490.

Chan, K. L., Mišić, V. B. and Mišić, J. [2004], Efficient polling schemes for bluetooth picocells revisited, in 'Proceedings XXXVIIth Annual Hawaii International Conference on System Sciences HICSS-37 (CD-ROM), Minitrack on Wireless Personal Area Networks', Big Island, Hawaii.

Chan, W.-C., Chen, J.-L., Lin, P.-T. and Yen, K.-C. [2004], 'Quality-of-service in IP services over Bluetooth ad-hoc networks', *ACM/Kluwer Journal on Special Topics in Mobile Networking and Applications (MONET)* **8**(6), 699–709.

Chawla, S., Saran, H. and Singh, M. [2001], QoS based scheduling for incorporating variable rate coded voice in Bluetooth, in 'Proceedings of IEEE International Conference on Communications ICC 2001', Vol. 4, Helsinki, Finland, pp. 1232–1237.

Chen, L.-J., Kapoor, R., Sanadidi, M. and Gerla, M. [2004], Enhancing Bluetooth TCP throughput via link layer packet adaptation, in 'Proceedings of IEEE International Conference on Communications ICC 2004', Vol. 7, pp. 4012–4016.

Chiang, C.-C. and Gerla, M. [1997], Routing and multicast in multihop, mobile wireless networks, in 'Proceedings 1997 IEEE 6th International Conference on Universal Personal Communications', Vol. 2, San Diego, CA, pp. 546–551.

Chiasserini, C. F., Ajmone Marsan, M., Baralis, E. and Garza, P. [2003], Towards feasible topology formation algorithms for Bluetooth-based WPANs, in 'Proceedings XXXVIth Annual Hawaii International Conference on System Sciences (CD/ROM)', Computer Society Press, Big Island, HI.

Choi, C. and Choi, H. [2002], Dsr based bluetooth scatternet, in 'Proceedings ITC-CSCC'.

Cooper, R. B. [1990], *Introduction to queing theory*, 2nd ed, North-Holland, New York, NY.

Crovella, M. E. and Bestavros, A. [1997], 'Self-similarity in world wide web traffic: evidence and possible causes', *ACM/IEEE Transactions on Networking* **5**(6), 835–846.

Călinescu, G., Mandoiu, I., Peng-Wan, J. and Zelikovsky, A. [2001], Selecting forwarding neighbors in wireless ad hoc networks, in 'Proc. 5th International Workshop on Discrete Algorithms and Methods for Mobile Computing and Communications', Rome, Italy, pp. 34–43.

Cuomo, F., Melodia, T. and Akyildiz, I. [2004], 'Distributed self-healing and variable topology optimization algorithms for qos provisioning in scatternets', *IEEE Journal on Special Areas in Communications – Wireless Series* **22**(7), 1220–1236.

Daptardar, A. [2004], Meshes and cubes: Distributed scatternet formations for Bluetooth personal area networks, in 'Master of Science thesis', Washington State Univ., School of EECS.

Das, A., Ghose, A., Razdan, A., Saran, H. and Shorey, R. [2001], Enhancing performance of asynchronous data traffic over the Bluetooth wireless ad-hoc network, in 'Proceedings Twentieth Annual Joint Conference of the IEEE Computer and Communications Societies IEEE INFOCOM 2001', Vol. 1, Anchorage, AK, pp. 591–600.

de Morais Cordeiro, C., Abhyankar, S., Toshiwal, R. and Agrawal, D. P. [2004], 'BlueStar: enabling efficient integration between Bluetooth WPANs and IEEE

802.11 WLANs', *ACM/Kluwer Journal on Special Topics in Mobile Networking and Applications (MONET)* **9**(4), 409–422.

de Morais Cordeiro, C., Sadok, D. and Agrawal, D. P. [2001], Piconet interference modeling and performance evaluation of Bluetooth MAC protocol, in 'Proceedings Global Telecommunications Conference 2001 GLOBECOM '01', Vol. 5, San Antonio, TX, pp. 2870–2874.

Delurgio, S. [1998], *Forecasting – Principles, and Application*, McGraw-Hill/Irwin, New York.

Dong, Y. and Wu, J. [2003], Three bluetree formations for constructing efficient scatternets in Bluetooth, in 'Proc. of the 7th Joint Conference on Information Sciences', pp. 385–388.

Famolari, D. and Anjum, F. [2002], Improving simultaneous voice and data performance in Bluetooth systems, in 'Proc. Global Telecommunications Conference GlobeCom'02', Vol. 2, Taipei, Taiwan, pp. 1810–1814.

Fantacci, R. and Zoppi, L. [2000], 'Performance evaluation of polling systems for wireless local communication networks', *IEEE Transactions on Vehicular Technology* **49**(6), 2148–2157.

Ferraguto, F., Mambrini, G., Panconesi, A. and Petrioli, C. [to appear], 'Blue pleiades, a new solution for device discovery and scatternet formation in multihop Bluetooth networks', *ACM/Baltzer Wireless Networks (WINET)* .

Foo, C. and Chua, K. [2002], Bluerings Bluetooth scatternets with ring structures, in 'IASTED Wireless and optical Communications'.

Francia, G. A., Kilaru, A., Phuong, L. and Vashi, M. [2004], An empirical study of bluetooth performance, in 'MSCCC '04: Proceedings of the 2nd annual conference on Mid-south college computing', Little Rock, AK, pp. 81–93.

Gabriel, K. R. and Sokal, R. R. [1969], 'A new statistical approach to geographic variation analysis', *Systematic Zoology* **18**, 259–278.

Garg, S., Kalia, M. and Shorey, R. [2000], MAC scheduling policies for power optimization in Bluetooth: a master driven TDD wireless system, in 'Proceedings VTC2000-Spring IEEE 51st Vehicular Technology Conference', Vol. 1, Tokyo, Japan, pp. 196–200.

Ghosh, J., Kumar, V., Wang, X. and Qiao, C. [2003], BTspin – single phase distributed Bluetooth scatternet formation, tech. report TR 2004-06, Dept. Comp. Sci. & Eng., Univ. of Buffalo, Buffalo, NY.

Golmie, N. [2004], 'Bluetooth dynamic scheduling and interference mitigation', *ACM/Kluwer Journal on Special Topics in Mobile Networking and Applications (MONET)* **9**(1), 21–31.

Golmie, N., Chevrollier, N. and ElBakkouri, I. [2001], Interference aware Bluetooth packet scheduling, in 'Proc. Global Telecommunications Conference GlobeCom'01', Vol. 5, San Antonio, TX, pp. 2857–2863.

Golmie, N., Van Dyck, R. E. and Soltanian, A. [2001], Interference of Bluetooth and IEEE 802.11: simulation modeling and performance evaluation, in 'Proceedings 4th ACM international workshop on Modeling, analysis and simulation of wireless and mobile systems', Rome, Italy, pp. 11–18.

Gonzales-Valenzuela, S., Vuong, S. and Leung, V. [2004], Bluescouts: A scatternet formation protocol based on mobile agents, in '4th Workshop on Applications and Services in Wireless Networks', Boston, MA, pp. 293–303.

Grimmett, G. R. and Stirzaker, D. R. [1992], *Probability and Random Processes*, 2nd ed, Oxford University Press, Oxford, UK.

Guerin, R., Kim, E. and Sarkar, S. [2002], Bluetooth technology key challenges and initial research, in 'Proc. SCS Comm. Networks and Distributed Systems Modeling and Simulation CNDS', pp. 157–163.

Guerin, R., Sarkar, S. and Vergetis, E. [2002], Forming connected topologies in Bluetooth ad hoc networks.

Hahne, E. L. [1991], 'Round-roin scheduling for max-min fairness in data networks', *IEEE Journal on Special Areas in Communications – Wireless Series* **9**(7), 1024–1039.

Hightower, J. and Borriello, G. [2001], 'Location systems for ubiquitous computing', *IEEE Computer* **34**(8), 57–66.

Hodge, L. and Whitaker, R. [2004*a*], What are characteristics of optimal Bluetooth scatternets, in 'Mobiquitous', Boston, MA.

Hodge, L. and Whitaker, R. [2004*b*], What are the characteristics of optimal Bluetooth scatternets?, in 'Proc. First Annual International Conference on Mobile and Ubiquitous Systems: Networking and Services MOBIQUITOUS 2004', pp. 124–125.

Huang, L., Chen, H., Sivakumar, T., Kashima, T. and Sezaki, K. [2004], Impact of topology on Bluetooth, in 'International Conference of Embedded and Ubiquiotus Computing (EUC2004)', Aizu, Japan.

Huang, T., Yang, S., Bai, S. and Huang, C. [2003], Hierarchically grown bluetrees an effective topology for Bluetooth scatternets,, in 'Int. Symp. Par. & Distr. Proc. & Applications ISPA', AIZU-Wakamatsu, Japan.

IEEE [2001], IEEE standard for local and metropolitan area networks, IEEE Std 802-2001, Piscataway, NY.

IEEE [2002], Wireless PAN medium access control MAC and physical layer PHY specification, IEEE standard 802.15, New York, NY.

Jacobson, V. [1988], 'Congestion avoidance and control', *ACM Computer Communication Review* **18**(4), 314–329.

Jacobson, V. [1990*a*], Berkeley TCP evolution from 4.3-tahoe to 4.3-reno, in 'Proceedings of the Eighteenth Internet Engineering Task Force', Vancouver, BC.

Jacobson, V. [1990*b*], 'Modified TCP congestion avoidance algorithm', note posted on end2end-interest mailing list, available at ftp://ftp.isi.edu/end2end/end2end-interest-1990.mail.

Jayanna, D. and Zàruba, G. V. [2005], A dynamic and distributed scatternet formation protocol for real-life Bluetooth scatternets, in 'Proc. of the 38th Annual Hawaii International Conference on System Sciences HICSS'05 (CD-ROM)', Big Island, HI.

Johansson, N., Alriksson, F. and Jönsson, U. [2001], JUMP mode – a dynamic window-based scheduling framework for Bluetooth scatternets, in 'Proceedings 2001 ACM International Symposium on Mobile ad hoc networking & computing', Long Beach, CA, pp. 204–211.

Johansson, N., Kihl, M. and Körner, U. [2000], TCP/IP over the Bluetooth wireless ad-hoc network, in 'NETWORKING 2000 Broadband Communications, High Performance Networking, and Performance of Communication Networks', Springer-Verlag, pp. 799–810.

Johansson, N., Körner, U. and Johansson, P. [1999], Performance evaluation of scheduling algorithms for Bluetooth, in 'Proceedings of BC'99 IFIP TC 6 Fifth International Conference on Broadband Communications', Hong Kong, pp. 139–150.

Johansson, N., Körner, U. and Tassiulas, L. [2001], A distributed scheduling algorithm for a Bluetooth scatternet, in 'Proceedings of the International Teletraffic Congress – ITC-17', Salvador de Bahia, Brazil, pp. 61–72.

Johansson, P., Kapoor, R., Kazantzidis, M. and Gerla, M. [2002], Rendezvous scheduling in Bluetooth scatternets, in 'Proceedings of IEEE International Conference on Communications ICC 2002', New York, pp. 318–324.

Joung, Y.-J. and Huang, G.-D. [2004], A simple and fast algorithm for Bluetooth network formation, in 'manuscript'.

Kail, E., Németh, G. and Turányi, Z. R. [2001], Throughput of ideality routed wireless ad hoc networks, in 'Proceedings 2001 ACM International Symposium on Mobile ad hoc networking & computing', Long Beach, CA, pp. 271–274.

Kalia, M., Bansal, D. and Shorey, R. [1999], MAC scheduling and SAR policies for Bluetooth: A master driven TDD pico-cellular wireless system, in 'Proceedings Sixth IEEE International Workshop on Mobile Multimedia Communications (MOMUC'99)', San Diego, CA, pp. 384–388.

Kalia, M., Bansal, D. and Shorey, R. [2000], Data scheduling and SAR for Bluetooth MAC, in 'Proceedings VTC2000-Spring IEEE 51st Vehicular Technology Conference', Vol. 2, Tokyo, Japan, pp. 716–720.

Kalia, M., Garg, S. and Shorey, R. [2000], Scatternet structure and inter-piconet communication in the Bluetooth system, in 'IEEE National Conference on Communications', New Delhi, India.

Kapoor, R. and Gerla, M. [2003], A zone routing protocol for Bluetooth scatternets, in 'Proceedings Wireless Communications and Networking Conference WCNC 2003', Vol. 3, New Orleans, LA, pp. 1459–1464.

Kapoor, R., Jyh-Ling, C., Lee, Y.-Z. and Gerla, M. [2002], Bluetooth: carrying voice over ACL links, in 'Proc. 4th International Workshop on Mobile and Wireless Communications Network', Los Angeles, CA, pp. 379–383.

Kapoor, R., Sanadidi, M. and Gerla, M. [2003], An analysis of Bluetooth scatternet topologies, in 'Proceedings of IEEE International Conference on Communications ICC 2003', Vol. 1, Anchorage, AK, pp. 266–270.

Kawamoto, Y., Wong, V. and Leung, V. C. M. [2003], A two-phase scatternet formation protocol for Bluetooth wireless personal area networks, in 'Proceedings Wireless Communications and Networking Conference WCNC 2003', Vol. 3, New Orleans, LA, pp. 1453–1458.

Kazantzidis, M. and Gerla, M. [2002], On the impact of inter-piconet scheduling in Bluetooth scatternets, in 'The 3rd International Conference on Internet Computing IC 2002', Las Vegas, NV.

Kazantzidis, M., Zanella, A. and Gerla, M. [2002], End-to-end adaptive multimedia over Bluetooth scatternets, in 'Eurocomm 2002', Brussels, Belgium.

Kim, J., Lim, Y., Kim, Y. and Ma, J. S. [2001], An adaptive segmentation scheme for the Bluetooth-based wireless channel, in 'Proceedings Tenth International Conference on Computer Communications and Networks', Scottsdale, AZ, pp. 440–445.

Kleinrock, L. J. [1972], *Queuing Systems*, Vol. I: Theory, John Wiley & Sons, New York.

Kumar, A., Ramachandran, L. and Shorey, R. [2001], 'Performance of network formation and scheduling algorithms in the Bluetooth wireless ad-hoc network', *Journal of High Speed Networks* **10**, 59–76.

Kurose, J. F. and Ross, K. W. [2005], *Computer Networking: A Top-Down Approach Featuring The Internet*, 3rd ed, Addison-Wesley Longman, Boston, MA.

Lai, W. K. and Tan, D. H. [2003*a*], 'A novel scatternet scheme with IPv6 compatibility', *ACM/Kluwer Journal on Special Topics in Mobile Networking and Applications (MONET)* **8**(6), 675–685.

Lai, W. K. and Tan, D. H. [2003*b*], 'A novel scatternet scheme with IPv6 compatibility', *ACM/Kluwer Journal on Special Topics in Mobile Networking and Applications (MONET)* **8**(6), 675–685.

Lapeyrie, J.-B. and Turletti, T. [2002], Adding QoS support for Bluetooth piconet, Rapport de recherche n° 4514, INRIA, Sophia-Antipolis, France.

Lapeyrie, J.-B. and Turletti, T. [2003], FPQ: a fair and efficient polling algorithm with QoS support for Bluetooth piconet, in 'Proceedings Twenty-Second Annual Joint Conference of the IEEE Computer and Communications Societies IEEE INFOCOM 2003', Vol. 2, New York, NY, pp. 1322–1332.

Law, C., Mehta, A. K. and Siu, K.-Y. [2001], Performance of a new Bluetooth scatternet formation protocol, in 'Proceedings 2001 ACM International Symposium on Mobile ad hoc networking & computing', Long Beach, CA, pp. 183–192.

Law, C. and Siu, K.-Y. [2001], A Bluetooth scatternet formation algorithm, in 'Proceedings Global Telecommunications Conference 2001 GLOBECOM '01', Vol. 5, San Antonio, TX, pp. 2864–2869.

Lee, T.-J., Jang, K., Kang, H. and Park, J. [2001], Model and performance evaluation of a piconet for point-to-multipoint communications in Bluetooth, in 'Proceedings IEEE VTS 53rd Vehicular Technology Conference, Spring 2001', Vol. 2, Rhodes, Greece, pp. 1144–1148.

Lee, Y.-Z., Kapoor, R. and Gerla, M. [2002], An efficient and fair polling scheme for Bluetooth, in 'Proceedings MILCOM 2002', Vol. 2, pp. 1062–1068.

Leung, K. K. and Eisenberg, M. [1989], A single-server queue with vacations and gated time-limited service, in 'Proceedings of the Eighth Annual Joint Conference of the IEEE Computer and Communications Societies INFOCOM '89', Vol. 3, Ottawa, ON, pp. 897–906.

Leung, K. K. and Eisenberg, M. [1990*a*], 'A single-server queue with vacations and gated time-limited service', *IEEE Transactions on Communications* **38**(9), 1454–1462.

Leung, K. K. and Eisenberg, M. [1990*b*], A single-server queue with vacations and non-gated time-limited service, in 'Ninth Annual Joint Conference of the IEEE Computer and Communication Societies INFOCOM '90', San Francisco, CA, pp. 277–283.

Levy, H., Sidi, M. and Boxma, O. J. [1990], 'Dominance relations in polling systems', *Queueing Systems Theory and Applications* **6**(2), 155–171.

Li, N., Hou, J. and Sha, L. [2003], Design and analysis of an mst-based topology control algorithm, in 'Proc. INFOCOM', San Francisco.

Li, X., Stojmenovic, I. and Wang, Y. [2004], 'Partial Delaunay triangulation and degree limited localized Bluetooth scatternet formation', *IEEE Transactions on Parallel and Disributed Systems* **15**(4), 350–361.

Lin, C. R. and Gerla, M. [1997], 'Adaptive clustering for mobile wireless networks', **15**(7), 1265–1275.

Lin, C. R. and Wu, J.-W. [2002], Enhancing the channel utilization of asynchronous data traffic over the Bluetooth wireless ad-hoc network, in 'Proc. Global Telecommunications Conference GlobeCom'02', Vol. 1, Taipei, Taiwan, pp. 212–216.

Lin, T., Tseng, Y. and Chang, K. [2003], 'A new bluering scatternet topology for Bluetooth with its formation, routing, and maintenance protocols', **3**, 517–537.

Liu, Y., Lee, M. and Saadawi, T. [2003], 'Adaptive clustering for mobile wireless networks', **21**(2), 229–239.

Liu, Z., Nain, P. and Towsley, D. [1992], 'On optimal polling policies', *Queueing Systems Theory and Applications* **11**(1–2), 59–83.

Ludwig, R., Rathonyi, B., Konrad, A., Oden, K. and Joseph, A. [1999], Multi-layer tracing of TCP over a reliable wireless link, in 'Proceedings of the 1999 ACM SIGMETRICS', pp. 144–154.

Malpani, N., Welch, J. L. and Vaidya, N. [2000], Leader election algorithms for mobile ad hoc networks, in 'Proc. 4th International Workshop on Discrete Algorithms and Methods for Mobile Computing and Communications', Boston, MA, pp. 96–103.

Manfield, D. R. [1985], 'Analysis of a priority polling system for two-way traffic', *IEEE Transactions on Communications* **33**(9), 1001–1006.

Mazzenga, F., Cassioli, D., Loreti, P. and Vatalaro, F. [2002], Evaluation of packet loss probability in Bluetooth networks, in 'Proceedings 2002 IEEE International Conference on Communications ICC 2002', Vol. 1, New York, NY, pp. 313–317.

McDermott-Wells, P. [2004], 'Bluetooth scatternet models', *IEEE Potentials* **23**(5), 36–39.

Mehta, V. and El Zarki, M. [2004], 'A Bluetooth based sensor network for civil infrastructure health monitoring', *ACM/Baltzer Wireless Networks (WINET)* **10**(4), 401–412.

Melodia, T. and Cuomo, F. [2004*a*], 'Ad hoc networking with Bluetooth: key metrics and distributed protocols for scatternet formation', **2**(2), 189–202.

Melodia, T. and Cuomo, F. [2004*b*], 'Locally optimal scatternet topologies for Bluetooth ad hoc networks', pp. 116–129.

Miklós, G., Rácz, A., Valkó, A. and Johansson, P. [2000], Performance aspects of Bluetooth scatternet formation, in 'Proceedings 2000 ACM International Symposium on Mobile ad hoc networking & computing', pp. 193–203.

Miller, B. A. and Chatschik, B. [2000], *Bluetooth Revealed: The Insider's Guide to an Open Specification for Global Wireless Communications*, Prentice-Hall, Upper Saddle River, NJ.

Miorandi, D. A. and Zanella, A. [2002], On the optimal topology of Bluetooth piconets: roles swapping algorithms, in 'Proc. Mediterranean Conf. on Ad Hoc Networks', Sardinia, Italy.

Miorandi, D., Caimi, C. and Zanella, A. [2003], Performance characterization of a Bluetooth piconet with multi-slot packets, in 'Proceedings WiOpt'03', Sophia-Antipolis, France.

Miorandi, D., Trainito, A. and Zanella, A. [2003], On efficient topologies for Bluetooth scatternets, in '8th IFIP TC6 PWC, LNCS 2775', pp. 726–740.

Miorandi, D., Zanella, A. and Pierobon, G. [2004], 'Performance evaluation of Bluetooth polling schemes: an analytical approach', *ACM/Kluwer Journal on Special Topics in Mobile Networking and Applications (MONET)* **9**(1), 63–72.

Mišić, J., Chan, K. L. and Mišić, V. B. [2004], 'Admission control in Bluetooth piconets', *IEEE Transactions on Vehicular Technology* **53**(3), 890–911.

Mišić, J. and Mišić, V. B. [2003*a*], 'Bridges of Bluetooth county: topologies, scheduling, and performance', *IEEE Journal on Special Areas in Communications – Wireless Series* **21**(2), 240–258.

Mišić, J. and Mišić, V. B. [2003*b*], 'Modeling Bluetooth piconet performance', *IEEE Communication Letters* **7**(1), 18–20.

Mišić, J., Mišić, V. B. and Chan, K. L. [2005*a*], 'Talk and let talk: performance of Bluetooth piconets with synchronous traffic', *Elsevier Ad hoc networks* **3**(4), 451–477.

Mišić, J., Mišić, V. B. and Ko, E. W. S. [2004], 'Fixed cycles and adaptive bandwidth allocation can coexist in Bluetooth', *Canadian Journal of Electrical and Computer Engineering* **29**(1/2), 135–147.

Mišić, V. B. and Mišić, J. [2003*c*], 'Adaptive inter-piconet scheduling in small scatternets', *ACM MC^2R – Mobile Computing and Communications Review* **7**(2), 45–58.

Mišić, V. B., Mišić, J. and Chan, K. L. [2004], 'Performance of adaptive inter-piconet scheduling in a scatternet with a slave/slave bridge', *Wiley Wireless Communications and Mobile Computing Journal* **4**(1), 85–98.

Mišić, V. B., Mišić, J. and Chan, K. L. [2005*b*], 'Walk-in scheduling in Bluetooth scatternets', *Cluster Computing* **8**(2/3), 197–210.

Mogul, J. C. and Deering, S. E. [1990], Path MTU discovery, draft Internet standard RFC 1191, IETF.

Muller, N. J. [2001], *Bluetooth Demystified*, McGraw-Hill, Boston, MA.

ns2 [2003], *The Network Simulator ns-2*, Software and documentation available from http://www.isi.edu/nsnam/ns/.

Pabuwal, N., Jain, N. and Jain, B. [2003], An architectural framework to deploy scatternet based applications over Bluetooth, in 'IEEE Transactions on Communications'.

Padhye, J., Firoiu, V., Towsley, D. F. and Kurose, J. F. [2000], 'Modeling TCP Reno performance: A simple model and its empirical validation', *ACM/IEEE Transactions on Networking* **8**(2), 133–145.

Pagani, E., Rossi, G. and Tebaldi, S. [2004], An on-demand Bluetooth scatternet formation algorithm, in 'WONS, LNCS 2928', Madona di Campiglio, pp. 3336–3340.

Pamuk, K. and Karasan, E. [2003], Sf-devil: Distributed Bluetooth scatternet formation algorithm based on device and link characteristics, in 'IEEE Int. Symp. Computers and Comm. ISCC', Bodrum, Turkey, pp. 646–651.

Partridge, C. [1994], *Gigabit Networking*, Addison-Wesley, Boston, MA.

Pasolini, G. [2003], Bluetooth piconets coexistence: analytical investigation on the optimal operating conditions, in 'Proceedings of IEEE International Conference on Communications ICC 2003', Vol. 1, Anchorage, AK, pp. 198–202.

Pasolini, G. [2004], 'Analytical investigation on the coexistence of Bluetooth piconets', *IEEE Communication Letters* **8**(3), 144–146.

Paxson, V. and Floyd, S. [1995], 'Wide area traffic: the failure of Poisson modeling', *ACM/IEEE Transactions on Networking* **3**(3), 226–244.

Perkins, C. E., Ed. [2001], *Ad Hoc Networking*, Addison-Wesley, Boston, MA.

Persson, K., Manivannan, D. and Singhal, M. [2004], 'Bluetooth scatternets: criteria, models and classification', *Ad Hoc Networks* .

Peterson, B. S., Baldwin, R. O. and Raines, R. A. [2004], 'Inquiry packet interference in Bluetooth scatternets', *ACM MC^2R – Mobile Computing and Communications Review* **8**(2), 66–75.

Petrioli, C. and Basagni, S. [2002], Degree-constrained multi-hop scatternet formation for Bluetooth networks, in 'Proc. Global Telecommunications Conference GlobeCom'02', Taipei, Taiwan.

Petrioli, C., Basagni, S. and Chlamtac, I. [2003], 'Configuring bluestars: Multi-hop scatternet formation for Bluetooth networks', *IEEE Trans. Computers* **52**(6), 779–790.

Petrioli, C., Basagni, S. and Chlamtac, I. [2004], 'Bluemesh: degree-constrained multi-hop scatternet formation for Bluetooth networks', *ACM/Kluwer Journal on Special Topics in Mobile Networking and Applications (MONET)* **9**(1), 33–47.

Prabhu, B. J. and Chockalingam, A. [2002], A routing protocol and energy efficient techniques in Bluetooth scatternets, in 'Proceedings of IEEE International Conference on Communications ICC 2002', Vol. 5, New York, pp. 3336–3340.

Rácz, A., Miklós, G., Kubinszky, F. and Valkó, A. [2001], A pseudo random coordinated scheduling algorithm for Bluetooth scatternets, in 'Proceedings 2001 ACM International Symposium on Mobile ad hoc networking & computing', Long Beach, CA, pp. 193–203.

Ramachandran, L., Kapoor, M., Sarkar, A. and Aggarwal, A. [2000], Clustering algorithms for wireless ad hoc networks, in 'Proc. 4th International Workshop on Discrete Algorithms and Methods for Mobile Computing and Communications', Boston, MA, pp. 54–63.

Raman, B., Bhagwat, P. and Seshan, S. [2001], Arguments for cross-layer optimizations in Bluetooth scatternets, in 'Proceedings 2001 Symposium on Applications and the Internet SAINT 2001', San Diego, CA, pp. 176–184.

RSoft Design, Inc. [2003], *Artifex v.4.4.2*, San Jose, CA.

Salonidis, T., Bhagwat, P. and Tassiulas, L. [2001], Proximity awareness and fast connection establishment in Bluetooth, Tech. report TR-2001-10, University of Maryland, College Park, MD.

Salonidis, T., Bhagwat, P., Tassiulas, L. and LaMaire, R. [2001], Distributed topology construction of Bluetooth personal area networks, in 'Proceedings Twentieth Annual Joint Conference of the IEEE Computer and Communications Societies IEEE INFOCOM 2001', Vol. 3, Anchorage, AK, pp. 1577–1586.

Sangvornverphan, V. and Erke, T. [2001], Traffic scheduling in Bluetooth network, in 'Proceedings Ninth IEEE International Conference on Networks ICON 2001', Bangkok, Thailand, pp. 355–359.

Seth, A. and Kashyap, A. [2002], Capacity of Bluetooth scatternets, Tech. report, Indian Institute of Technology, Kanpur, India.

Shorey, R. and Miller, B. A. [2000], The Bluetooth technology: merits and limitations, in 'Proc. of IEEE International Conference on Personal Wireless Communications', Hyderabad, India, pp. 80–84.

Shreedhar, M. and Varghese, G. [1996], 'Efficient fair queueing using deficit round-robin', *ACM/IEEE Transactions on Networking* **4**(3), 375–385.

Siegemund, F. [2002], Kontextbasierte Bluetooth-scatternetz-formierung in ubiquitaren systemen, in 'First German Workshop on Mobile Ad hoc Networks'.

Siegemund, F. and Rohs, M. [2003], 'Rendezvous layer protocols for Bluetooth-enabled smart devices', *Personal Ubiquitous Computing* **7**(2), 91–101.

Sikdar, B., Kalyanaraman, S. and Vastola, K. S. [2003], 'Analytical models for the latency and steady-state throughput of TCP Tahoe, Reno, and SACK', *ACM/IEEE Transactions on Networking* **11**(6), 959–971.

Sivakumar, T., Chen, H. and Huang, L. [2004], Adaptive network formation protocol for Bluetooth scatternet, in 'The First IEEE and IFIP Int. Conf. on Wireless and Optical Communications Networks (WOCN 2004)', Muscat, Oman.

Snow, C. and Primak, S. [2002], Performance evaluation of TCP/IP in Bluetooth based systems, in 'IEEE 55th Vehicular Technology Conference VTC Spring 2002', Vol. 1, Birmingham, AL, pp. 429–433.

Son, L. T., Schiøler, H. and Madsen, O. B. [2001], Predictive scheduling approach in inter-piconet communications, in 'Proc. of the 4th international symposium on Wireless personal multimedia communications', Aalborg, Denmark.

Song, W., Li, X.-Y., Wang, Y. and Wang, W. [2005], 'dBBlue: Low diameter and self-routing Bluetooth scatternet', *Journal on Parallel and Distributed Computing* **65**(2), 178–190.

Song, W.-Z., Li, X.-Y., Wang, Y. and Wang, W. [2003], dBBlue: low diameter and self-routing Bluetooth scatternet, in 'Proceedings of the 2003 joint workshop on Foundations of mobile computing', ACM Press, pp. 22–31.

Sreenivas, H. and Ali, H. [2004], An evolutionary Bluetooth scatternet formation protocol, in 'Proceedings of the 37th Hawaii Int. Conf.on System Sciences', Hawaii.

Stojmenovic, I. [2002], Dominating set based Bluetooth scatternet formation with localized maintenance, in 'CD Proc. IEEE Int. Parallel and Distributed Processing Symposium and Workshops', Fort Lauderdale.

Stojmenovic, I. [2004*a*], Degree limited Bluetooth scatternet formation based on maximal independent sets, in preparation.

Stojmenovic, I. [2004*b*], Routing in Bluetooth with geometric structures enhanced with long links, in preparation.

Stojmenovic, I. and Wu, J. [2004], Broadcasting and activity scheduling in ad hoc networks, in S. Basagni, M. Conti, S. Giordano and I. Stojmenovic, Eds, 'Mobile Ad Hoc Networking', IEEE/Wiley, pp. 205–229.

Stojmenovic, I. and Xu, L. [2001], 'Power aware localized routing in wireless networks', *IEEE Transactions on Parallel and Distributed Systems* **12**(11), 1122–1133.

Sun, M.-T., Chang, C.-K. and Lai, T.-H. [2002], A self-routing topology for Bluetooth scatternets, in 'Proc. Int. Symp. on Parallel Architectures, Algorithms and Networks I-SPAN', Manila, Philippines, pp. 13–18.

Takács, L. [1962 (reprinted 1982)], *Introduction to the Theory of Queues*, Oxford University Press, New York, NY.

Takagi, H. [1986], *Analysis of Polling Systems*, The MIT Press, Cambridge, MA.

Takagi, H. [1987], 'Analysis and application of a multiqueue cyclic service system with feedback', *IEEE Transactions on Communications* **35**(2), 248–250.

Takagi, H. [1988], 'Queuing analysis of polling models', *ACM Computing Surveys* **20**(1), 5–28.

Takagi, H. [1991], *Queueing Analysis*, Vol. 1: Vacation and Priority Systems, North-Holland, Amsterdam, The Netherlands.

Takagi, H. [1993*a*], *Queueing Analysis*, Vol. 2: Finite Systems, North-Holland, Amsterdam, The Netherlands.

Takagi, H. [1993*b*], *Queueing Analysis*, Vol. 3: Discrete-Time Systems, North-Holland, Amsterdam, The Netherlands.

Tan, G. and Guttag, J. [2002], A locally coordinated scatternet scheduling algorithm, in 'Proceedings of the 26th Annual Conference on Local Computer Networks LCN 2002', Tampa, FL, pp. 293–303.

Tan, G., Miu, A., Guttag, J. and Balakrishnan, H. [2001], Forming scatternets from Bluetooth personal area networks, Technical Report MIT-LCS-TR-826, MIT, Cambridge, MA.

Toussaint, G. [1980], 'The relative neighborhood graph of a finite planar set', *Pattern Recognition* **12**(4), 261–268.

Vergetis, E., Guerin, R., Sarkar, S. and Rank, J. [2005*a*], 'Can Bluetooth succeed as a large-scale ad hoc networking technology?', *IEEE Journal on Special Areas in Communications – Wireless Series* **23**(3), 644–656.

Vergetis, E., Guerin, R., Sarkar, S. and Rank, J. [2005*b*], 'Can Bluetooth succeed as a large-scale ad hoc networking technology?', *IEEE Journal on Special Areas in Communications – Wireless Series* **23**(3), 644–656.

Wang, Y., Stojmenovic, I. and Li, X.-Y. [2004], Bluetooth scatternet formation for single-hop ad hoc networks based on virtual positions, in 'IEEE Symposium on Computers and Communications', Alexandria, Egypt, pp. 7–14.

Wang, Z., Thomas, R. J. and Haas, Z. [2002], Bluenet – a new scatternet formation scheme, in 'Proceedings of the 35th Annual Hawaii International Conference on System Sciences', Big Island, HI.

Whitaker, R., Hodge, L. and Chlamtac, I. [2005], 'Bluetooth scatternet formation: a survey', *Ad Hoc Networks* **3**(4), 403–450.

Whittaker, E. T. and Watson, G. N. [1952], *A Course of Modern Analysis*, Cambridge University Press, Cambridge, UK.

Wilf, H. S. [1994], *Generatingfunctionology*, 2nd ed, Academic Press.

Wu, J. and Li, H. [1999], On calculating connected dominating set for efficient routing in ad hoc wireless networks, in 'Proceedings 3rd International Workshop on Discrete Algorithms and Methods for Mobile Computing and Communications', Seattle, WA, pp. 7–14.

Wu, Y. and Todd, T. [2004], Link sharing in high capacity Bluetooth voice access networks, in 'Proc. Wireless Communications and Networking Conference', Vol. 1, Atlanta, GA, pp. 21–25.

Wu, Y., Todd, T. D. and Shirani, S. [2003], 'SCO link sharing in Bluetooth voice access networks', *Journal on Parallel and Distributed Computing* **63**(1), 45–57.

Xue, J. and Todd, T. [2001], Basestation collaboration in Bluetooth voice networks, in 'Proc. 26th Annual IEEE Conference on Local Computer Networks LCN 2001', pp. 533–538.

Yao, A. [1982], 'On constructing minimum spanning trees in k-dimensional spaces and related problems', *SIAM J. Computing* **11**, 721–736.

Yun, J., Kim, J., Kim, Y. and Ma, J. [2002], A three-phase ad hoc network formation protocol for Bluetooth systems, in 'Proc. 5th Int. Symp. Wireless Personal Multimedia Communications WPMC', Hawaii.

Zàruba, G. V., Basagni, S. and Chlamtac, I. [2001], Bluetrees – scatternet formation to enable Bluetooth-based ad hoc networks, in 'Proceedings of IEEE International Conference on Communications ICC 2001', Vol. 1, Helsinki, Finland, pp. 273–277.

Zhang, H. and Hou, J. C. [2002], A scheduling algorithm for transporting variable rate coded voice in Bluetooth networks, in 'Proceedings of the 5th ACM international workshop on Wireless mobile multimedia', ACM Press, pp. 25–32.

Zhang, H., Hou, J. C. and Sha, L. [2003], A Bluetooth loop scatternet formation algorithm, in 'Proceedings of IEEE International Conference on Communications ICC 2003', Anchorage, AK, pp. 1174–1180.

Zhang, W. and Cao, G. [2002], A flexible scatternet-wide scheduling algorithm for Bluetooth networks, in 'Proc. 21st IEEE International Performance, Computing, and Communications Conference IPCCC 2002', Phoenix, AZ.

Zhang, W., Zhu, H. and Cao, G. [2001], On improving the performance of Bluetooth networks through dynamic role management, Tech. report CSE-01-018, Pennsylvania State University, University Park, PA.

Zhang, W., Zhu, H. and Cao, G. [2002], Improving Bluetooth network performance through a time-slot leasing approach, in 'Proceedings of IEEE Wireless Communications and Networks Conference WCNC 2002', Vol. 2, Orlando, FL, pp. 592–596.

Zhen, B., Park, J. and Kim, Y. [2003], Scatternet formation of Bluetooth ad hoc networks, in '36th Hawaii Int. Conference on System Sciences', pp. 312–319.

Zürbes, S. [2000], Considerations on link and system throughput of Bluetooth networks, in 'Proceedings of the 11th IEEE International Symposium on Personal, Indoor and Mobile Radio Communications PIMRC 2000', Vol. 2, London, UK, pp. 1315–1319.

Index

For Product Safety Concerns and Information please contact our EU representative GPSR@taylorandfrancis.com Taylor & Francis Verlag GmbH, Kaufingerstraße 24, 80331 München, Germany

Printed and bound by CPI Group (UK) Ltd, Croydon, CR0 4YY
14/09/2026
02216223-0020